*Science,
Race, and
Religion
in the
American
South*

LESTER D. STEPHENS

Science, Race, and Religion in the American South

JOHN BACHMAN AND THE

CHARLESTON CIRCLE

OF NATURALISTS, 1815–1895

The University of North Carolina Press

Chapel Hill & London

This book was set in Monotype Bulmer

by Keystone Typesetting, Inc.

Book design by April Leidig-Higgins

The paper in this book meets the guidelines for permanence

and durability of the Committee on Production Guidelines

for Book Longevity of the Council on Library Resources.

Frontispiece: "Charleston, 1851: View from the Ashley River

at White Point Gardens," by William Hill. Photograph

courtesy of the South Carolina Historical Society.

Library of Congress Cataloging-in-Publication Data

Stephens, Lester D.

Science, race, and religion in the American South: John

Bachman and the Charleston circle of naturalists, 1815–1895

/ by Lester D. Stephens

p. cm. Includes bibliographical references and index.

ISBN 0-8078-2518-2 (cloth: alk. paper)

1. Naturalists—South Carolina—Charleston Biography.

2. Natural history—South Carolina—Charleston—History—

19th century. 3. Bachman, John, 1790–1874. I. Title.

QH26.S735 2000   508'.092'2757915—dc21   99-27008 CIP

04 03 02 01 00   5 4 3 2 1

CONTENTS . . . .

ILLUSTRATIONS ·

By the 1840s the city of Charleston, South Carolina, was gaining notice as an important center of natural history research in the Old South, and by the early 1850s no other city in the region matched it in the number and quality of scientific contributions. In addition, by midcentury no southern city could boast of a natural history museum that paralleled the one in Charleston, nor did any equal the productivity of its scientific society. The only regional rival to Charleston was the much larger city of New Orleans, which nourished the interests of an able group of naturalists and its active scientific society and encouraged the development of natural history collections. While their efforts are notable, however, the naturalists of New Orleans produced fewer scientific works than their Charleston counterparts, and they enjoyed less success than their sister-city compatriots in promoting a strong museum and in advancing the status of their scientific organization. That Charleston had attained status as the center of natural history research in the South is also evident from the volume and nature of the Charleston scientists' correspondence with leading American and European naturalists and from the expressions of interest shown by the great Swiss naturalist Louis Agassiz, who visited the city in 1847, soon after his arrival in the United States, and again every year between 1849 and 1853. Agassiz's repeated praise of the Charleston naturalists and the Charleston Museum and his assistance in getting the American Association for the Advancement of Science to hold its third meeting in Charleston, in 1850, indicate the national standing of the city in natural history research.

At that time, only three other cities in the United States—Philadelphia, Boston, and New York—exceeded Charleston in natural history studies. The standing of Charleston in the field is all the more notable when one compares its size to that of the northeastern cities, each of which was many times larger in population. Moreover, each of the northeastern cities could claim continuously thriving scientific organizations, and their museums of natural history, while not as old as the original museum in Charleston, had received better financial support. Still, the quality of natural history research in Charleston was fully equal to that done in the northeastern cities.

Interest in the physical sciences was generally lower in Charleston, however, and there were far fewer contributions to physics, astronomy, and chemistry in that city during the antebellum period. The reasons for this phenomenon are not fully clear, but almost certainly they were related to the increasing tendency of the physical sciences toward experimentation and the application of mathematics and away from largely descriptive studies. Thus, with few exceptions, the physical sciences became more and more the province of university-trained men of science, while research in natural history, still mainly descriptive by midcentury, remained accessible to planters, physicians, businessmen, and clergymen, who largely constituted the intellectual corps of the Old South. Since Charleston enjoyed the status of "Queen City" of the South, it was only logical that it would lead in the development of natural history, which had been cultivated there in the eighteenth century, and where, in 1773, one of the first museums of natural history in North America had been established. Perhaps, too, the region's abundance of scientifically undescribed species—living and fossil—helps to account for the dominance of interest in natural history, an interest that continued to follow the traditional notion of showing that God is manifest in nature.

In any case, Charleston was the center of natural history research in the Old South, and it played a truly significant role in advancing knowledge in that field. Yet the interests and activities of the major Charleston naturalists have received too little attention. The work of those men constitutes an important, but much neglected, part of the history of the South. My purpose in writing this book is to examine as fully as possible the interests and activities of the major Charleston naturalists in order to fill this gap in southern history and in the history of science. My study focuses on John Bachman, Edmund Ravenel, John Edwards Holbrook, Lewis Reeve Gibbes, Francis Simmons Holmes, and John McCrady, but I also explore the role of religion and culture in shaping their views toward natural history. Constituting an informal but nonetheless recognizable and real circle, these six men were bound together in spirit by their devotion to southern culture, their commitment to advancing science in their city and region, and their interest in natural history. All of them were dedicated to the study of animals, but two of them, Bachman and Gibbes, were also interested in botany. They were not as active in the latter as was their predecessor Stephen Elliott, who, though an excellent botanist, did his major work almost two decades before the circle of Charleston naturalists developed into a nationally recognized group and made Charleston an important center of scientific study. Only one of the major Charleston naturalists, Gibbes, paid particular attention to chemistry and physics.

Central to my account is the role of John Bachman. As the unofficial but generally acknowledged leader of the circle, this Lutheran clergyman turned in midlife to serious work in mammalogy and established himself as an internationally known authority on North American mammals. In a sense, the circle was formalized when the Elliott Society of Natural History was founded in 1853, and all six of the major naturalists joined it, with Bachman serving as the first president. Holbrook, however, was never very active in the organization, and Holmes, its founder, eventually withdrew as a result of a dispute with his fellow society members. The circle never lost its integrity, however, for all of its members continued to interact with the others in various ways, especially through the Charleston Museum, the College of Charleston, the Medical College, and the activities of Louis Agassiz. Although several minor naturalists stood on the periphery of the circle, they cannot be counted as members of the inner circle. The men in the inner circle were the most productive in natural history research, received the greatest acclaim for their scientific work, and served as the region's leaders in science. They were the ones who brought Charleston to its impressive status as a scientific center.

In the first three chapters of this account, I discuss Bachman's developing interests in mammalogy and his contributions to that field. The fourth chapter covers the scientific activity of Edmund Ravenel, who gained recognition for his work in conchology and paleontology. In the fifth chapter, I treat the interests and work of John Edwards Holbrook, whose pioneering volumes on the reptiles and amphibians of North America and studies in ichthyology earned for him a lasting place in the development of systematic zoology in those fields. The following chapter is devoted to the most versatile of the Charleston scientists, Lewis Gibbes. Recognized for his studies in botany, physics, and chemistry, Gibbes gained special notice for his descriptions of Crustacea. Francis Holmes is the subject of the seventh chapter. His studies of fossil invertebrates from the Charleston area, his long service as the curator of the Charleston Museum, and his role in founding the Elliott Society of Natural History made him a central figure in the advancement of science prior to the American Civil War. John Mc-Crady, the youngest, and perhaps most promising, member of the Charleston circle, is the subject of the next chapter. A discussion of his work in hydrozoan zoology, his association with the great naturalist and Charleston supporter Louis Agassiz, and his ideas on race and southern culture offers some indications of the place and promise of Charleston in American science while highlighting factors that hampered the circle in reaching its full potential. John Bachman reenters the picture in the ninth and tenth chapters as the leading advocate of the unity of all human races. His scientifically based argument that all humans are of one species brought him into controversy with Agassiz and others who advocated polygenesis, that is, the separate origin and distinctly specific nature of blacks. When the South decided to secede from the Union in 1861, however, Bachman, like the other members of the Charleston circle, placed regional patriotism above science and union and joined ardently in the Confederate cause. The views and activities of the circle members during the war that ensued and the devastating impact of that conflict on science in the South are the subject of the eleventh chapter. The final chapter returns to McCrady and the last links of the circle. McCrady's resistance to "Yankee" ways and to the theory of evolution form the heart of this chapter and provides insights, or so I hope, into the cultural predilections that played a part in delaying the recovery of science in the South.

In 1936, Thomas Cary Johnson Jr. published *Scientific Interests in the Old South*, which includes a chapter on the Charleston naturalists. His treatment served as a useful corrective to a view common among historians that scientific activity in the Old South was rare. It offered little critical analysis, however, and it barely sketched the specific work of the Charleston naturalists. My study aims at a deeper understanding of the interests and activities of each member of the circle and at viewing each one in comparison with his counterparts elsewhere in the United States. From the standpoint of a present-day zoologist, one can praise individual members of the Charleston circle for specific contributions to science, but to see them only in terms of individual achievements is to isolate them from their time and culture. The same would be the case if one tried to isolate John Bachman from the context in which he proclaimed that all human races belong to a common species. My study differs from the general approach to southern history by examining the specific scientific work of the Charleston naturalists. I make no claim to being a professional naturalist or zoologist, but my familiarity with natural history subjects, especially zoology and paleontology, and assistance of scientists in those fields have made it possible for me to discuss and evaluate the specific studies of each of the Charleston naturalists. The project I have undertaken seems to work best by employing the method of collective biography, which allows me to examine the scientific activities of each naturalist and to avoid the more generalized treatment offered by Johnson. At the same time, it places the debate over human species within the context in which it developed.

That context is crucial to understanding the essence of the debate. It was indeed a debate based upon the understanding and misunderstanding of scientific principles, not primarily on religious beliefs, as William Stanton contends in his influential book, *The Leopard's Spots: Scientific Attitudes Toward Race in America, 1815–59* (1960). Erring in his argument that southerners rejected polygenism as scientific justification for slavery and remained faithful to the biblical account of the curse of Ham, Stanton depicts Bachman as "half theologian, half scientist," confused by conflicting elements in his personality. In fact, Bachman was faithful to his theology, and he was a first-rate scientist. Stanton fails to understand that Bachman, like good mammalogists of his time and of the present, used diagnostic characters to classify mammalian species, including humans. I endeavor to show that Bachman relied upon the best scientific knowledge available and that, despite his firm belief in the biblical account of creation, he was unwavering in his use of science to argue for the unity of all human races as a common species. At the same time, contrary to Stanton's argument, I offer abundant evidence that most of the southerners who embraced the polygenist, or pluralist, view did so because they considered it to be a scientific, not a religious, argument. My account also corrects Stanton's factual errors regarding the Charleston naturalists.

During the American Civil War, John Bachman and John McCrady made the fateful choice of sending the bulk of their personal papers and manuscripts to Columbia, South Carolina, for safekeeping, but the items were destroyed by the fire that consumed a sizable segment of that city in February 1865. Yet a surprisingly large number of Bachman's prewar papers have survived, and an impressive collection of McCrady's postwar letters and manuscripts are extant. The abundant papers of Lewis Gibbes were placed in the Manuscript Division of the Library of Congress many years ago, but relatively few personal papers are included in that collection. Francis Holmes likely lost many of his papers when his office burned to the ground soon after the war. According to his descendant the late A. Baron Holmes III, others of his papers were probably destroyed as a result of a hurricane that flooded the Charleston carriage house in which they were stored. Baron Holmes related to me that he distinctly recalled wagonloads of soggy papers being hauled from the carriage house to a dumping ground. Still, more of the papers of Holmes are extant than is the case for those of Edmund Ravenel and John Edwards Holbrook. Quite naturally, my treatment of the life and work of each of the six major naturalists varies somewhat in relation to the availability of the primary sources, but I do not believe that more of their papers would alter my view of Bachman as the central figure in the Charleston circle of naturalists.

ACKNOWLEDGMENTS . . . .

It was my good fortune to have free access to the papers and diaries of John McCrady in possession of the McCrady family of Sewanee, Tennessee, and I wish to express my warmest thanks to Edith McCrady and her sons Edward McCrady and Waring McCrady for their generosity in allowing me to use them. The late A. Baron Holmes III was equally generous toward me, and I am grateful that I had the opportunity to become his friend.

A grant from the University of Georgia Humanities Center and a University of Georgia Faculty Fellowship for Study in a Second Discipline greatly aided me in preparing this book. The former assisted me in bridging the disciplines of history and science, and the latter made it possible for me to devote an entire year to the study of marine invertebrates. In connection with the fellowship, I am grateful to the zoologists William Fitt and James Porter, of the University of Georgia Institute of Ecology, for teaching me much about marine invertebrates—in the classroom, in the laboratory, and in the field. I owe a special debt to the late Joshua Laerm, who died as I was completing this manuscript. As director of the University of Georgia Museum of Natural History, he gave me a second home in the Museum, taught me much about mammals, and offered suggestions for improving my manuscript. Above all, he shared an interest in my work, and he provided inspiration. Too soon did a good man leave us. Thanks too must go to Timothy S. McCay, for assisting me in studying mammals, and to others in the Museum—Amy Edwards, Liz McGhee, and Elizabeth Reitz—for treating me as one of their own.

Enormously helpful to me was William D. Anderson Jr. of the Grice Marine Biological Laboratory, College of Charleston, who read the manuscript carefully and made many helpful suggestions. Albert E. Sanders, of the Charleston Museum, deserves special thanks for sharing his knowledge of the Charleston naturalists, aiding and encouraging my work, and offering indispensable criticisms of my manuscript. Ronald Vasile, biographer of William Stimpson, provided additional help. My colleague Thomas G. Dyer read the manuscript and made many useful suggestions for improving it. Ronald Numbers, of the University of Wisconsin, Madison, also offered valuable comments on ways to improve the manuscript, and I am greatly indebted to him for his help. To Dale R. Calder, I express genuine appreciation for directing my studies in the invertebrates laboratory of the Royal Ontario Museum, coauthoring articles with me, criticizing my manuscript, and befriending me. O. F. Schuette, emeritus professor of physics at the University of South Carolina, offered encouragement and kindly shared with me some Bachman materials he located in Berlin. Kraig Adler of Cornell University was very helpful to me on matters pertaining to Holbrook, and Raymond B. Manning of the National Museum of Natural History offered useful comments on my chapter on Lewis Gibbes. Any errors that may remain after the efforts of all of these generous scholars must be laid solely at my doorstep.

The former chairman of my department, David Roberts, and the current chairman, James Cobb, provided support and encouragement, for which I express my deep appreciation. My former graduate student Naomi Kyriacopoulos made some helpful comments on an early draft of the manuscript, and my colleague Edward Larson strongly encouraged my work, for which I am thankful. Bonnie Cary did a superb job of typing the manuscript, and I thank her. My University of North Carolina Press editor Kathy Malin did a splendid job of preparing the manuscript for publication, and I thank her for her diligence—and for her good cheer. To my dear daughters Karen and Janet, my pride and joy, I express gratitude for love and support over the years. Marie C. Ellis freely and frequently assisted me in locating sources, in offering wise advice, in criticizing my manuscript, in proofreading, and in supporting my efforts. The University of Georgia was fortunate to have her as a member of its Libraries faculty until her retirement after three decades of service, but I am even more fortunate to claim her as my beloved wife.

Readers unfamiliar with the scientific nomenclature appearing in this work should find the following explanations helpful.

Taxonomy is the theory and practice of classifying organisms. A taxon (plural, taxa) is a group of organisms that is distinct enough to warrant a distinguishing name and to be placed in a definite taxonomic category (e.g., in a species, genus, family, or order). The scientific name of a species (and not of any other taxon) consists of two words (a binomen)—the first word being the generic name, and the second the specific name (e.g., *Sylvilagus aquaticus*).

The person who first provides a scientific name for a taxon is the author of, or the authority for, that name. When the authority's name is cited, it follows the name of the taxon (e.g., *Rypticus maculatus* Holbrook, 1855). Citation of the date of publication of a scientific name is optional, but, if cited, it is placed after the name of the author.

As a result of additional study, a species may be referred to, or placed in, a genus different from that to which the original author assigned it. For example, the species *Serranus nigritus* Holbrook has been placed in the genus *Epinephelus* and is correctly cited as *Epinephelus nigritus* (Holbrook). To point out subsequent generic assignments, the following form is usually used herein: *Scalops breweri* (= *Parascalops breweri*). In standard form, the name today would be *Parascalops breweri* (Bachman, 1842).

A synonym is one of two or more names given to the same taxon. The oldest name for that taxon is ordinarily the valid name (i.e., the correct scientific name) of the taxon. A synonymy is a list of the synonyms applied to a single taxon.

Types or type specimens are biological objects that serve as the bases for the names of taxa. A type or type specimen as used herein refers to the actual material studied in describing and naming a new species.

The scientific name of a species, or the binomen, normally precedes the common name, but, for the benefit of the general reader and in order to provide variety, the common name sometimes appears first in this work.

The term Recent (initial letter capitalized) indicates species of animals living at present or still existing within the Holocene epoch. Prior geological epochs referred to include, in reverse chronological order, the Pleistocene, Pliocene (spelled "Pleiocene" in several major titles of the mid-nineteenth century), Miocene, Oligocene, and Eocene. In nineteenth-century usage, however, the distinctions were far less precise than they later became.

Science,
Race, and
Religion
in the
American
South

In a Singular Place

S ituated on a narrow finger of land separating the Ashley and Cooper Rivers just before they merge into a
natural harbor, Charleston, South Carolina, was already an old city when the young Lutheran clergyman John Bachman
(pronounced BACK-mun) stepped off the stagecoach there on January
10, 1815. Before the turn of the century, Charleston had been the fourth
largest city in the young republic, but, by the time Bachman arrived, it
had fallen far behind its northeastern sisters, Boston, New York, and
Philadelphia—not only in population and commerce but also in the
development of intellectual life. Still it was a great city, and the twenty-
five-year-old Bachman was happy to be there.[1]

Ailing from tuberculosis, Bachman had left his home in Dutchess
County, New York, exactly four weeks and one day earlier, riding day
and night in the coach. Tiring to even the most vigorous of passengers,
the grueling journey was especially hard on the young minister, whose
body was weakened by severe spasms of coughing and bleeding from
his diseased lungs. Indeed, it was the consumptive condition in his
chest that had compelled him to leave his native state. Upon the advice
of physicians, he was seeking a warmer climate, the only remedy for

tubercular patients known to contemporary medical doctors. Bachman had also selected that particular southern city because it was home to St. John's, the largest of the relatively few Lutheran churches in the region.[2]

No doubt Bachman had begun to feel better even as the coach entered the lowlands of South Carolina, for he could readily see that the region was rich in flora and fauna. In fact, he was not only a student of the Bible but also a student of, as he called it, "the Book of Nature." The notion of "two books," clearly espoused by Sir Francis Bacon in 1605 and reiterated by Galileo a decade later, had long served as an effective compromise between the creation account in the book of Genesis and studies by geologists, and it remained essentially unquestioned until around 1830. A Baconian from the beginning of his interest in nature to the end of his life, Bachman firmly believed that by opening the second book through careful observation one gained a better understanding of the Holy Book. To the devout minister, God was the author of both books, designer of the rules of conduct in one and of the plants and animals in the other. In his view, the Great Architect had not only created every species but also placed them in their appointed geographic regions. Some flourish in more than one region, but others do not range beyond a limited area. Thus, as he gazed from the coach, Bachman would have noticed familiar groves of pines, but he could also have seen unfamiliar live oak trees, their grand limbs draped with gray beards of Spanish moss. Botany was a favorite subject of interest to him, and he could observe a wealth of plants in this new country.[3]

The vertebrate animals were no less fascinating to the dedicated naturalist. After all, he had begun to study birds and mammals for hours upon end during his boyhood days, and whenever he could get into a town, he had searched in bookstores for copies of works in natural history. Long before he entered South Carolina, he was already an expert on birds. Thus, as he entered the state in early January 1815, he would have observed the familiar male cardinal, its flashing plumage of summer now a duller red, and, near a stream, he may have spied the male painted bunting, which he later described as a "livery of bright purplish lilac, vermillion and glossy green." Coloration in birds and mammals had long interested Bachman, and he intended at some point to study the phenomenon in detail. As the coach swayed and bounced along the rutted road toward the city of his destination, the dedicated man of God and enthusiastic observer of Nature also watched for mammals, a third major area of interest to him. At dawn he had likely observed numerous white-tailed deer browsing on woody twigs, and as the sun rose higher, he could hardly have missed seeing gray squirrels scampering up the massive trunks of live oaks. As dusk began to settle upon the land, he had opportunities to glance upon eastern cottontail rabbits hopping about the fringe of thickets. So intrigued by nature was John Bachman that he began to collect and study specimens soon after his arrival in Charleston.[4]

During the first weeks of his ministry in Charleston, Bachman traveled up and down the straight, sandy streets and along the few paved avenues of Charleston, and he learned much about the culture of the city. On the street called Vendue Range he could witness an auction of slaves. It is doubtful that he would have lingered, however, for he condoned slavery. In fact, his own father had once owned slaves in New York and taught him that the Sacred Scriptures sanctioned the practice. To John Bachman, it was the duty of the bondsman to follow the admonition of the Apostle Paul to be content with his station in life. Surely, the disobedient slave must be punished. Thus, Bachman would have held no reservations about the South Carolina law that permitted Charleston authorities to apply nineteen lashes to the back of a black lawbreaker. Nor is it likely that he later opposed the punishment of a convicted black by placing him on a treadmill, which, said a contemporary writer, would break his "idle habits" and teach him morality.[5]

Bachman would soon discover that punishment of blacks could be especially severe in Charleston. White Charlestonians had long been uneasy over the great number of blacks in their city. Ever since the mid-1720s, at least one-half of the city's population had consisted of black inhabitants, and now, in 1815, of the nearly 25,000 people residing in the city, almost 56 percent were black. Although the majority of those African Americans were slaves, approximately 1,400 of them were free. An appreciable number were mulattoes, for, while interracial mating was generally disfavored by most Charlestonians, the practice was neither illegal nor harshly condemned. Certainly, the young Lutheran minister noted the number of mulattoes in the city, and he was no doubt already interested in the biological success of such racial crossings, which, for him, confirmed that God had created but a single species of humans, though they were by that time separated into varieties. Everywhere Bachman traveled in the city during those early weeks in January 1815, he saw hundreds of blacks—peddling goods, driving coaches, cleaning streets, or serving as skilled artisans. By the time the clock struck 9:00 in the evening, however, only whites remained on the streets. The reason for the rapid change, Bachman learned, was a law forbidding blacks to be on the streets after that hour. Charleston vigorously enforced the long-standing statute, employing a police force comparably larger than that of any of its sister cities in the Northeast in order to do so. Even when a building caught fire in Charleston, a fairly frequent occurrence, armed militia accompanied the fire engines for fear that the event might be the beginning of an insurrection. Perhaps Bachman pondered the enormous expense of night patrols and the harsh punishment of black lawbreakers and wondered whether Charlestonians should not give more attention to the moral and spiritual instruction of blacks. As the pastor of St. John's, he quickly made it part of his mission to bring blacks into the fold, and by his second year in Charleston he was encouraging their simultaneous presence with whites in the sanctuary.[6]

As he traversed the streets of Charleston early in 1815, Bachman would become aware of two vices that he especially deplored: dueling and drinking. The practice of dueling to settle questions of honor among Charleston's aristocratic gentlemen was already in disfavor, but, as Bachman noticed, recently enacted laws providing punishment for offenders had not completely suppressed the strong southern penchant for pistols when a man's honor was at stake. The abuse of spirituous liquors distressed him also, and eventually it would compel him to support a local temperance movement. Among Charleston's elite, however, liquor was not a pleasure to be abandoned merely because it met with disapproval by a minister of the Gospel. As Bachman learned, Charleston's aristocrats relished many other pleasures, especially balls, dancing, horse racing, and theatrical presentations. He did not view those activities as necessarily sinful, but he found it more interesting to study the flora and fauna of the region. All around Charleston were marvels of nature. Bachman could lift his eyes to the city's skyline and see one of those marvels. There, perched on the gables of roofs or circling above the city for carrion, were scores of turkey buzzards and black vultures. Along with virtually every foreign traveler, Bachman had immediately noticed that vultures abounded in

Charleston, where they enjoyed the full protection of law because of their role in disposing of a portion of the city's offal. To some, the vultures may have been unsightly and repugnant, but to the naturalist Bachman, there were mysteries to be solved about their morphology and habits. Did only a few Charlestonians care about such matters of the mind? Bachman was uncertain, but he would soon come to believe that many people could be guided to such interests.[7]

He already knew that interest in scientific matters had a long tradition in Charleston. The great English naturalist Mark Catesby had resided with honor in Charleston from 1722 to 1725, and thirty years later the Scottish physician-naturalist Alexander Garden had received acclaim from the townspeople during the three decades he had studied the region's plants and animals. During his stay in Charleston, from 1796 to 1798, the French naturalist Louis Augustin Guillaume Bosc had collected specimens and described several new species, both marine and terrestrial, and in 1802 the Charlestonian John Drayton had described a number of vertebrate fossils from the region. Meanwhile, in 1748, a group of citizens had formed the Charles Town Library Society, which served not only as a repository of learning but also as the locus of meetings for the study of literary, philosophical, and scientific subjects. Twenty-five years later, several enterprising members of the Library Society, taking note of the abundance of flora, fauna, and fossils in the area, had decided to establish collections of specimens for their own edification and for the enlightenment of the public. Thus, in 1773, the Library Society had created a museum, which eventually came to be called The Charleston Museum. Focusing upon collections of natural objects and "curiosities," the institution waxed and waned over the years. When Bachman arrived in Charleston in 1815, it was resurging under the aegis of the Literary and Philosophical Society, founded only two years earlier by several of the city's professionals, businessmen, and literati. Foremost among that group was the banker and botanist Stephen Elliott, who was then preparing a study that would be published in thirteen parts between 1816 and 1824 and bound as the first and second volumes of *A Sketch of the Botany of South Carolina and Georgia* in 1821 and 1824, respectively. Also interested in conchology, ichthyology, and other areas of natural history, Elliott served as the first president of the Literary and Philosophical Society. Indeed, he became the first highly accomplished naturalist in Charleston after Independence, and Bachman no doubt sought him out soon after his arrival in Charleston. The newcomer relished the thought of participating in the affairs of the Literary and Philosophical Society.[8]

As Bachman became familiar with Charleston, he noticed not only that many of the residents kept flower gardens but also that they built their houses with piazzas, or verandas, which served as the place for the family to retreat on the hot and humid days of spring and summer. Many of the silent and shuttered homes of the well-to-do were unoccupied, however, for their owners and families were away in the country. Upon their return they could revel in the activities of the Jockey Club and other social organizations. Of more concern to Bachman, however, were Charleston's provisions for paupers and orphans. The city operated a poor house that was caring for more than 100 paupers in 1815, and it supported an asylum for orphans that housed as many as 200 children at a time. For the more fortunate, the South Carolina Society provided the costs of educating the minor children of a deceased member. Charlestonians did not neglect the New Testament admonition to feed the hungry and clothe the poor.[9]

Charleston was also a city of churches—nearly two dozen in all. Most prominent among the houses of worship was St. Michael's Episcopal Church, the tip of its grand, white steeple standing more than 180 feet from the ground. Not far away stood another elegant edifice, St. Philip's Episcopal Church. To the congregations of those two Anglican cathedrals belonged a large number of the city's elite and their families—some rich, some of modest means, but none poor. Likewise drawing many of Charleston's prominent citizens were churches for Congregationalists, Unitarians, Presbyterians, Quakers, French Huguenots, and, of course, Bachman's own Lutheran congregation. The poor could mingle with those of modest income in churches for Baptists and Methodists, and Catholics and Jews could worship in the edifices they had erected. Bachman would soon discover, however, that, despite the number of churches in Charleston, attendance was relatively low. Nevertheless, the sheer number of religious sanctuaries doubtless impressed him, as likely did the great variety of denominations. He disagreed with the doctrines of other Protestant groups, and he believed that the descendants of Abraham had failed to recognize Jesus as the Messiah, but he had no desire to quarrel with any of them. As he passed the city's Roman Catholic Church, however, he likely experienced a surge of intolerance, for he considered its leaders as potential assailants of Luther's doctrines, though he possessed no ill will toward its communicants and would go his own way as long as the Catholic clergy kept silent about the founder of Protestantism. His duty as a man of God was to minister to the needs of his congregation and to strengthen Lutheranism in Charleston, in South Carolina, and in the South as a whole.[10]

During his first weeks in Charleston, Bachman would have had occasion to walk along the row of wharves on the east side of the city. There he likely witnessed little activity, for most of the year's harvest of cotton and rice had already been shipped out. Nevertheless, he could gain some measure of appreciation for the city as the commercial center of the South. Even before he left New York, he had probably heard that Charleston was a thriving city, but he was likely unaware that it was rapidly falling behind other major cities in America. The slow growth of Charleston probably meant little to Bachman, for neither he nor his contemporaries fully understood the correlation between the size of an urban area and the strength of its scientific activity. As far as he was concerned, Charleston's population was entirely sufficient to support science. To him, science meant primarily natural history. He understood the importance of natural philosophy (the physical sciences), but, like most of his peers in Charleston—indeed in most of the South—he considered natural history to be the queen of sciences. In his view, the Creator had populated the earth with living plants and animals, and to study His material manifestations was to gain deeper appreciation of His omnipotence.[11]

From the southernmost point of the peninsula on which Charleston was located, John Bachman could cast his eyes directly southward across the harbor and see the towering trees and grassy marshes of James Island. In wild places still remaining there, he could perhaps find species yet undescribed by naturalists. As his eyes swept northward up the island, he could see the vast marshes west of the Ashley River. Not long before his arrival he could have crossed over by a 2,200-foot bridge, but a powerful storm had since wrecked nearly all of the enormous wooden structure. Still, Bachman could cross over by boat and find wonders of nature to observe. By turning his head back to the southeast, he could make out the fringe of Sullivan's Island. A popular retreat during the summer season for Charlestonians seeking pleasure in surf and sand, it too was rich in flora and fauna. As Bachman had already learned when he first passed through the "neck" of the Charleston peninsula, the fields and forests north of the city contained a treasure trove of species that could be collected, examined, and described. To serve science and God simultaneously would require rigor, discipline, and wise use of his time. Bachman envisioned such an opportunity, and the challenge beckoned more strongly as winter faded and spring burst forth with the renewal of life.[12]

The Charleston Library Society had been in existence for over half a century, of course, but it contained fewer than 9,000 volumes on all subjects. In all likelihood it would have been somewhat larger had not a fire in 1778 destroyed all but a handful of the volumes, but it would not have matched the holdings of the libraries in Philadelphia, where Bachman had spent some time as a student. Even though Charleston had established the Library Society in 1748, and even though the city could boast of a number of intellectuals, its leaders had not placed as much emphasis on books, journals, and magazines as had those in Boston, Philadelphia, and New York. In fact, journals published in the major city of the South tended to enjoy a relatively short life. As a writer noted a decade after Bachman's arrival, "the love of literary fame has not yet aroused the energies of our citizens." Some personal libraries were impressive, but the institutional ones left much to be desired. Literate Charlestonians did read local and other newspapers, but the Charleston newspapers rarely devoted much space to matters of science. Bachman could hope, however, that the fledgling College of Charleston, which had opened in 1790, would elevate interest in the mind and build a sizeable library. That point could not be reached soon, for the College was essentially no more than a preparatory school in 1815. Later, the naturalists in Charleston often complained of the want of current books and journals, leading them in several cases to describe species that had already been characterized and named by naturalists elsewhere in the nation or in Europe. Such deficiencies prompted the well-to-do to send their sons to colleges in the North or in Europe. Because he knew the literature of botany, ornithology, and mammalogy, however, and since he suspected that scores of species in the South had never been described, Bachman likely gave little thought to those matters in 1815.[13]

Charleston, the New York native could see, was a city of contradictions, but the contrasts in its character were generally no greater than those of other large cities in the nation. As in Boston, New York, and Philadelphia, virtue and vice lived side by side in Charleston, as did charity and crime, enlightenment and ignorance, tolerance and intolerance, and science and myth. For Bachman, the first character of each couplet would make the city a desirable place to plant his roots, while the second would offer an inviting challenge to his mission in life. In 1815, however, he was unable to fathom the ways in which Charleston differed drastically from its northeastern counterparts. The city was, like the region in which it lay, a peculiar place. Above all, more than half of its inhabitants lived in either a state of enslavement or of limited freedom, and the minds of most of its free citizens were closed completely to criticism of the system. Moreover, in the view of both its aristocratic leaders and its ordinary citizens, Charleston was the Queen City of the South, the heart of a unique culture—indeed, of a special civilization. While generally receptive to a variety of ethnic groups, a host of religious denominations, and open discussion of virtually any topic, save slavery and southern ways, most well-to-do Charlestonians viewed their city as superior to any elsewhere in the nation or the world. Their unity lay in the special culture they shared. Except for his stand against a later movement to represent blacks as a separate creation by God, Bachman would soon come to share that spirit.[14]

Contrary to older historical interpretations, however, the institution of slavery did not deter scientific inquiry and activity. In fact, Charleston produced a group of naturalists equal in ability and accomplishments to any elsewhere in the nation. The number it nurtured was relatively small, however, because its population was insufficient to produce a sizeable cadre of men who held scientific inquiry as a great mission of life. Inadequate libraries, insufficient outlets for publishing articles in the region, limited financial support, and relative isolation from the hub of intellectual activity made it no easier for the naturalists of Charleston. In Bachman's eyes, however, good science could be done by men of ability and dedication. Unobserved and undescribed species were everywhere in the area, and both a society for the discussion and criticism of scientific papers and a museum for holding and displaying specimens of natural history existed in Charleston. Moreover, the opportunity for a fruitful ministry and for strengthening Lutheranism in the South promised fulfillment of his mission. In addition, he soon found the local climate conducive to his health. Despite occasional longings to see his parents and the land where he had begun to study natural history and had received the call to serve his Heavenly Father, John Bachman quickly came to believe that he could find no better place than Charleston for his health and for his work, and when he returned to New York during the following summer to visit his ailing father, he wrote to a friend, "Charleston I consider as my home." Indeed, it was, and for almost six decades the transplanted New Yorker labored to advance science and minister to the spirit in the singular city called Charleston.[15]

Exalting Two Books

D uring his boyhood, even the coldest days of winter
had not kept John Bachman confined to the indoors
of his family's farm home, located near the village of
Schaghticoke in eastern New York, a short distance from the Hoosic
River and within walking distance of the great Hudson River. A year
after he was born, in a "humble stone house" in the town of Rhinebeck, New York, his father Jacob, grandson of an immigrant from
Switzerland, and his mother Eva Shop, whose ancestors were natives
of Würtemburg, had moved up the Hudson. There "Johny," as his
parents called him, freely explored the surrounding hills, valleys, rivers, and woods. By the time he had celebrated his twelfth birthday
on February 4, 1802, Johny had a room cluttered with plant and animal specimens, and he kept a number of live birds and mammals in
captivity.[1]

After a snowfall, he took special delight in tracking animals across
open fields and into the woodlands. Among the mammals that intrigued him especially were the ermine, the snowshoe hare, the shrews
and moles, and squirrels and other rodents. Garbed in its winter-white
coat, the ermine, *Mustela erminea*, blended almost imperceptibly into

the snow-covered ground, but it could not escape the sharp eye of young Johny, for he could spot the black-tipped tail of the small carnivore. Soon he would know that it was the same species he had captured during the spring, when the upper part of its long body bore a coat of dark brown. Standing in silence, he could watch it pounce upon its prey, perhaps a small meadow vole, *Microtus pennsylvanicus*, or a tiny shorttail shrew, *Blarina brevicauda*, foraging desperately to satisfy its voracious appetite. Before long, as he moved into the woods, he might catch a glimpse of another mammal whose pelage changed from brown to white as winter approached, the snowshoe hare, *Lepus americanus*. Rabbits and hares also fascinated Johny, and he was especially intrigued over *Lepus americanus*, for its molt seemed to differ from that of *Mustela erminea*, both of which he had held in confinement and about whose molts he made notes in a little journal.[2]

The young student of nature would often walk deeper into the woods, where he might hear the familiar barking of the gray squirrel, *Sciurus carolinensis*, perched on a barren limb, flicking its tail and scolding the intruder. Johny would already have known the squirrel was near, for he could readily spy open patches in the snow where the frisky sciurid rodent had tunneled its way down to a nut stored long before the first frost. Equally interesting to the boy were the birds of the region. He could see a small flock of birds or hear them calling as they flew toward a grove of conifers and know instantly that they were red crossbills, *Loxia curvirostra*, which inhabit the area throughout the year. He would not, however, expect to see the Baltimore oriole, *Icterus galbula*, for he knew that the species migrated from the region during the fall. By the time he was thirteen, Johny already had a number of notes in his journal about the seasonal departure and arrival of the region's avifauna.[3]

As young John Bachman savored those scenes in 1803, he knew that he would soon be enrolled in a school in Philadelphia, a long distance from his beloved land. His father and mother, who indulged his interest in nature and taught him that animals were God's creations, had often permitted one of their slaves to assist him in trapping mammals. As devout Lutherans, they had reared their son to follow the teachings of the Bible and the writings of the great Reformer. Johny had already learned much about religion, languages, and nature from the local Lutheran pastor, Anton Braun, but now he needed more formal schooling, and his parents chose Philadelphia for accomplishing that goal. The choice proved profitable for John Bachman in several ways.[4]

By 1804, Johny was in Philadelphia. There he visited the famous Bartram Garden, met the elderly William Bartram, and developed his interest in botany. He also met and became a friend of the struggling but superb ornithologist Alexander Wilson and, through conversations and collecting trips with him, learned even more about birds. Within two years after entering a college in Philadelphia, however, Johny was coughing frequently, and the bloody sputum indicated that he was yet another victim of the disease that seemingly consumed its victims. Unable to find relief, even at the hands of a leading Philadelphia physician, Johny had to quit his studies and head for home, either in late 1806 or early 1807.[5]

Fortunately, the disease soon entered a state of temporary arrest, and within a year or so, Bachman was again studying nature. But he was also reading the Bible and the works of Luther, especially Luther's thoughts on the Book of Galatians. Around 1809, Bachman decided that he would become a Lutheran minister, and once again he became a student of Braun. For a brief time, he also studied with the Lutheran theologian Frederick Henry Quitman, who lived in the vicinity. By 1810, Bachman was back in Pennsylvania, serving as a teacher of Latin, French, and German in the Elwood School in Milestown, and a year later he was in Philadelphia, studying theology and teaching at the school sponsored by the city's St. John's Lutheran Church. During the spring of 1813, Braun died, and Bachman was called to take his place as the minister of three small congregations in the region of his home.[6]

By 1814, however, tuberculosis was again plaguing Bachman, and he traveled to the West Indies to seek relief in the warm and sunny climate of a Caribbean isle. Soon after his departure, his former theology professor Philip Frederick Mayer had recommended him as the pastor of St. John's Lutheran Church in Charleston, South Carolina. Upon learning of the recommendation when he returned to New York in the fall, Bachman began to weigh his prospects, and he and his congregations agreed upon a nine-month leave of absence, during which time Bachman would make a trial of the situation in the South. He departed for Charleston in mid-December 1814.[7]

During the four years immediately prior to Bachman's arrival in Charleston in January 1815, St. John's had been without a pastor and had depended upon other Protestant ministers in the city to conduct services. Eager for one of their own denomination to fill the vacant pulpit, the congregation warmly welcomed the twenty-five-year-old New Yorker. They soon knew he was likely to be with them for a long time, for in January 1816 he married Harriet Martin, the granddaughter of a former pastor of the church. During the first two decades after he set foot in Charleston, Bachman devoted the bulk of his energy to his ministerial duties, preaching three sermons on each Sabbath, one in German for the congregants who continued to identify themselves as German Lutherans. Within three years after arriving, Bachman had persuaded his charges to admit blacks to fellowship in St. John's, established a benevolent fund to aid the poor, organized an academy for girls, and superintended the construction of a new edifice for his congregation. Between 1824 and 1837, he baptized more than 500 whites and 300 blacks and brought the number of communicants, white and black, from 62 to 425. Meanwhile, the busy pastor exerted extraordinary efforts to strengthen Lutheranism in South Carolina and elsewhere in the South, leading a movement to establish a synod and a theological seminary in the state. He succeeded in the former in 1824 and served as its president for the first ten years and for three years later on. By 1831, Bachman's dream of a Lutheran theological seminary in South Carolina had also been realized. It was located in the small town of Lexington, only a short distance west of the state capital, Columbia. The seminary board, of which Bachman was a member, appointed twenty-three-year-old John G. Schwartz as the first professor of theology. A member of the congregation of St. John's, Schwartz had been tutored by Bachman and become like a son to the ardent promoter of Lutheranism. Unfortunately, Schwartz was struck by an illness in mid-August 1831 and died on the twenty-sixth day of that month. Though deeply saddened over the loss of his young charge, Bachman viewed it as the will of God, and continued without relent to carry forward with his main mission in life. In the meantime, he was encouraging promising young black men to minister to the free and enslaved African Americans in the South.[8]

A growing family of eight living children by 1830 also demanded much of Bachman's time. Soon he had to move his burgeoning brood, along with his wife's sister and mother, into more spacious quarters, and he purchased a fifteen-room house to accommodate them. Some of the burdens of tending to affairs associated with a large home, along with cultivating a spacious garden and minding a menagerie of birds and beasts, were alleviated by the help of four slaves, the property of his wife and part of the settlement of her father Jacob Martin in 1810. While recuperating from an illness in Philadelphia, Martin, a successful businessman, had become involved in an extramarital affair with a woman named Elizabeth Pennington. Secretly, Martin arranged for his lover to come to Charleston and to live in "an obscure part of the city." His frequent visits to the abode of his paramour eventually caused the secret to be known, but, unwilling to end his relationship with Pennington, he returned with her to Philadelphia. Jacob Martin's nephew Martin Strobel placed chief blame upon Pennington, calling her "one of the most dissolute, and prostitute wretches that any community was ever cursed with." Bachman concurred in the view of his relative-in-law, and he wanted nothing to do with Jacob Martin, who, in his eyes had severely injured his family and sinned against God.[9]

After her marriage to John Bachman in January 1816, however, Harriet Martin no longer brooded over the father who had abandoned her and her three sisters, for she found strength in a kind and caring husband. Moreover, within eleven months after her marriage she had borne a lovely daughter, whom she named Maria Rebecca. Less than a year later, Mary Eliza was born, followed in the next year by Jane, and in January 1820, by Cordelia, who died about six months afterwards. Then, early in 1822, came the first son, named John. He too died halfway through the first year of life. In 1823, Harriet Bachman gave birth to her sixth child, Harriet Eva. During the same year, Bachman's father died of a stroke. Bachman had gone back to New York earlier when his father was ill, but he did not try to settle his father's affairs until much later. Meanwhile, his only sister, Eva Dale, who resided in Troy, New York, looked after their mother.[10]

By 1826, two more children had made their entrance into the Bachman home, but Henry, a son born in 1824, lived only a few months. Julia, born in 1826, was acknowledged by all to be the most beautiful of the Bachman daughters. Within a year after the birth of Julia, Harriet Bachman, suffering from frequent illnesses, fatigued by the duties of tending to five children, burdened over the loss of three others, and pregnant again, was in great need of assistance. Thus, she asked her younger sister, Maria Martin, to live with and assist the family. Soon afterwards, John Bachman set off for New York to settle his father's affairs and to travel for respite. With sister-in-law Maria, he departed on June 27, 1827, to visit the "deserted home" of his youth and to settle his late father's estate. From there, in mid-July, he and Maria headed northward, traveling via the Erie Canal, visiting Niagara Falls, sailing on Lake Erie, and ending up in Montreal on August 1st. Bachman carefully observed the flora and fauna along the way. Word from home informed him that Harriet had borne twin daughters. Upon returning to New York, Bachman and Martin visited with his mother's brother and with his sister Eva Dale, but on the tenth day of August, he became seriously ill, suffering high fever, severe headaches, and, soon thereafter, total blindness. Maria Martin accompanied him to the city of New York, but a leading physician there was unable to aid the gravely ill traveler. At one point Bachman was certain that he would die, but by late August the symptoms had subsided. His strength returned only gradually, however, and, upon the urging of Harriet, Maria Martin, and his congregation, he decided to remain in the North until the "fever-season" had passed in Charleston. Bachman and Martin did not return to Charleston until November 1827.[11]

Tragedy struck the Bachmans once again on June 20, 1828, when one of the twins died, and, again on July 14, when the other one succumbed to an illness. Bachman viewed the losses as God's will, just as certainly as were the respective births of Lynch, Samuel Wilson, William Kunhardt, and Catherine during the next five years. By 1833, the Bachman home consisted of the parents, nine of their offspring, Maria Martin, John's mother, who had come there after the death of his sister Eva Dale late in 1832, and the mother of Harriet and Maria.[12]

That John Bachman could find time to pursue his interests in natural history is little short of wonder. But find time he did, for his interest in plants and animals came only slightly behind his interest in his family and his commitment to God. Indeed, in his judgment, to study nature was to study God's handiwork. Much of his early scientific work in Charleston was devoted to experimentation in hybridization. In fact, Bachman conducted a number of successful experiments in crossing species, especially ducks, one of his favorite animals, and he read widely upon the subject of hybridization among domestic species. But he continued to observe and collect specimens of plants and animals, giving most of his attention to birds and mammals, especially the insectivores, rodents, rabbits, and bats. He was particularly intrigued by the Virginia opossum, *Didelphis virginiana*, the only marsupial native to North America. In addition, he collected insects.

That John Bachman was an able naturalist who knew more about birds and mammals than any of his contemporaries in the South around 1830 was generally known by the learned citizens of Charleston, but his reputation as an authority on those subjects had hardly spread any further.[13]

Then, on October 23, 1831, fortuity played a role in the life of John Bachman and ultimately set him on the path of publishing important works in natural history. On that day John James Audubon arrived in Charleston and happened to be introduced to Bachman in the streets of the city. Bachman was already aware of the success of the gifted artist and author of *The Birds of America*, of which the first volume and parts of the second volume had already been published. He insisted that Audubon and his two assistants must lodge and board at his home. As Audubon told his wife Lucy in a letter written later that day, "we removed to his house in a crack—found a room already arranged for Henry [Ward] to skin our Birds, another for me & [George] Lehman to Draw, and a third for thy Husband to rest his bones in on an excellent bed." It was the beginning of a friendship that, though sorely tested in later years, would endure until the death of Audubon.[14]

But the ornithologist Audubon found more than friendship in his association with Bachman, for he quickly discovered that the Lutheran clergyman was intimately familiar with the region's avifauna, including their morphology, calls, behavior, migratory routes, breeding habits, and nesting sites. Indeed, Bachman could supply information on the passerines, or perching birds, the vultures and hawks, the owls, the doves and pigeons, the shorebirds, the gulls, and the herons around Charleston, and he could tell him a thing or two about the woodpeckers as well. Audubon was quick not only to take advantage of his new friend's knowledge but also to enlist him in collecting avifaunal specimens and in taking him to places for collecting and observing birds and their nests. For Bachman, the opportunity to discuss birds with a knowledgeable ornithologist and to witness his exceptional skill in drawing and coloring them opened a new door—and doubtless evoked memories of the field trips he had taken with Alexander Wilson a quarter of a century earlier. Certainly, he had conversed often with the Charleston naturalists Edmund Ravenel, John Edwards Holbrook, and Stephen Elliott, and he was interested in their fields of study. Until Audubon's arrival, however, he was in want of an equal in ornithology. Moreover, Audubon knew quite a bit about mammals, though, as it turned out, far less than he claimed. For Audubon, the union opened the door to valuable knowledge, and for Bachman, the association provided an incentive to publish works in natural history.[15]

For several years, Bachman had been building a herbarium and working up a list of the phanerogams, or flowering plants, in the area of Charleston, but he had refrained from publishing anything in botany while Stephen Elliott was alive. After the death of Elliott in 1830, he decided to produce a list of some of the local flora. Further encouragement came from young Lewis Reeve Gibbes, whom Bachman had met when Gibbes was a student at the state's medical college in Charleston in 1830–1831 and who had since moved to Columbia to serve as a tutor and instructor in mathematics at South Carolina College. At work on a catalogue of the phanerogamous plants around Columbia, Gibbes was already the best botanist in the state after Elliott, and he and Bachman agreed that Elliott's two-volume *Sketch of the Botany of South Carolina and Georgia* stood in need of revision, but neither of them chose to undertake the task. In 1834, Bachman published his *Catalogue, Phanerogamous Plants and Ferns, Native or Naturalized, Found Growing in the Vicinity of Charleston, South Carolina*, which included approximately 1,100 species, arranged alphabetically. Although the list constituted only a minor contribution to the study of botany, it reflects the range of Bachman's knowledge of natural history. By 1837, Elliott's daughter had placed her father's herbarium in the care of Bachman, who kept it in his possession for many years and allowed others, including the noted New York botanist John Torrey, to make use of it.[16]

Far more important was Bachman's work in ornithology, though his contributions to that field were later partially obscured by Audubon's fame. Certainly, Bachman was in part responsible for the concealment because of his eagerness to assist Audubon and his willingness to let his friend receive the credit. In some measure, the relative obscurity of Bachman's contributions emanated from the strong desire of the author of *The Birds of America* and *Ornithological Biography* for acclaim. Audubon explicitly credited Bachman with the species collected by the clergyman and even named three species after him. Moreover, he paid ample tribute to "my worthy and generous friend John Bachman" for providing information about some of the birds included in his volumes, referring to him at least eighty times in the 1840 edition of *The Birds of America*. Yet, his references to Bachman failed to convey fully the magnitude of his friend's assistance.[17]

During the early years of their friendship, Bachman verged upon the fulsome in his praise of Audubon, wielding the pen in his defense like a sword in the hands of a crusader, and he solicited subscriptions to Audubon's work with the fervor of a missionary. Without Audubon, Bachman likely would not have emerged from his clerical cocoon. Indeed, his metamorphosis into a major figure in natural history owes much to his association with Audubon. Despite his truly superb talent for painting birds and despite his admirable knowledge as an ornithologist, Audubon was often careless, and some of his depictions and claims created controversy. In other cases, however, attacks upon the credibility of Audubon stemmed from dislike of his audacity and from envy of his success, none more so than those directed by the Philadelphia naturalist George Ord, who viewed Audubon as a rival to his late friend Alexander Wilson, whose unfinished work Ord had himself completed in 1813. Ironically, Bachman, who had known Wilson well and accompanied him on collecting trips, now found himself in the position of defending the man who was rapidly replacing Wilson as the leading ornithologist and bird artist in America. But John Bachman never retreated when he believed he was right nor spared the feelings of friend or foe when he thought he had detected errors in the work of either.[18]

By the time he entered Charleston, Audubon had already been under attack for claiming that vultures locate their food solely by sight, not by smell. Relentless in his criticism of Audubon, Ord had played a significant role in denying him membership in the Academy of Natural Sciences of Philadelphia (ANSP), and he renewed his assault upon the artist's reputation after the friendship developed between Audubon and Bachman. Convinced that he and Audubon could determine the sense vultures use for location of food, Bachman proposed a series of careful experiments, which, with Audubon, he conducted in late December 1833. Like other contemporary naturalists, Bachman subscribed to the Baconian philosophy of inductive inquiry that avoided hypothesizing. Unlike his fellow Charleston naturalists, however, he readily used the experimental method whenever appropriate. In this case, he placed a pile of offal in an open area and erected a covered frame above the heap so that vultures could smell but not see it. Of an estimated 100 vultures in the vicinity, including both the turkey buzzard, *Cathartes aura*, and the black vulture, *Coragyps atratus*, none had approached the stinking mess by noon, whereupon Bachman then removed some of the offal and set it thirty feet away from the covered pile. Eight vultures quickly alighted to seize the rotting meat. Two days later, he placed a small portion of the offal in the open but near the covered pile. Vultures quickly landed and consumed the visible portion but never discovered the large pile. Audubon later painted a sheep "cut open, with the entrails hanging out," and, soon after placing it in the open, a turkey buzzard landed to inspect it.

Although Bachman believed the evidence clearly showed that vultures detect food by sight alone, he desired more conclusive proof. Thus, a day later he placed over the reeking pile of offal a net cloth that hid it but allowed its odor to escape. He also set a portion of fresh meat nearby. Several vultures of each species alighted and devoured the fresh meat, but, although their beaks were within an inch of the covered pile, none discovered the offal. Bachman therefore concluded that vultures detect sources of food solely by sight. He sent a paper on the experiments to be read before the ANSP and thus hushed criticism of Audubon on one count. Many decades passed before better studies established that *Coragyps atratus* relies mainly upon vision and *Cathartes aura* primarily upon its sense of smell for locating food.[19] In any case, experimentation in natural history was uncommon at the time, and Bachman's study thus represented a rather remarkable approach to the acquisition of knowledge.

While Ord subdued his criticism of Audubon on the subject of vultures, he continued to attack Audubon for depicting a rattlesnake in a tree, saying that reptile does not have the ability to climb. Early in 1834, Bachman sent his "Remarks in Defence of the Author of 'The Birds of America'" to the Boston Society of Natural History, before which it was read and then published in the Society's *Journal*. He also sent it to England for publication in *Loudon's Magazine*, in an effort to squelch attacks on Audubon by some British naturalists. In the paper he again defended Audubon's claim about vultures, but he also assailed those who ridiculed Audubon for painting a rattlesnake in a tree. Reminding his readers that there are several species of rattlesnakes, he cited reliable witnesses to the climbing ability of one species of that reptile, but he could offer nothing more concrete. In mid-1835, Ord, still in pursuit of a way to damage Audubon's standing, sent a letter to the *Bucks County (Pennsylvania) Intelligencer*, attacking the character of the ornithologist-artist by referring to a charge by the English naturalist Charles Waterton that Audubon was guilty of exaggerations.

Bachman minced no words in his response, in which he flailed Waterton for his own gross exaggerations.[20]

More important, however, was the scientific assistance Bachman gave to Audubon, providing him with specimens of birds, furnishing him with valuable information on their habits, and criticizing his friend for committing careless errors. Motivated by this opportunity to share his knowledge of birds with the most widely recognized ornithologist in America at that time, Bachman promised in a letter on December 2, 1831, to "astonish" Audubon with new species. During the next three months Bachman sent many detailed letters to his friend about several local species. He also kept his promise to astonish Audubon, for he collected two new species of birds within the next two years. Audubon named one of them Swainson's warbler, *Sylvia swainsonii* (= *Limnothlypis swainsonii*), and the other, Bachman's warbler, *Sylvia bachmanii* (= *Vermivora bachmanii*). So rare are both species that the former was not rediscovered until fifty years later, and the latter (now an endangered species) not until 1901. In due course, Bachman also collected a specimen that he and Audubon believed to be a new species, and Audubon named it Bachman's finch, *Fringilla bachmanii*, but it was later identified as *Aimophila aestivalis*, though it became known commonly as Bachman's sparrow. In addition, Bachman provided Audubon with considerable information on other South Carolina avifauna. For example, he told him about the South Carolina nesting sites of the American bittern, *Botaurus lentiginosus*, a secretive species; the hooded merganser, *Lophodytes cucullatus*; the gray kingbird, *Tyrannus dominicensis*; the northern rough-winged swallow, *Stelgidopteryx serripennis*; and the black-crowned night heron, *Nycticorax nycticorax*. In addition, he furnished Audubon with records on his sightings in South Carolina of such uncommon species as the black-shouldered kite, *Elanus caeruleus*; the Eskimo curlew, *Numenius borealis* (at the time an occasional transient on the South Carolina coast but now verging on extinction); the stilt sandpiper, *Calidris himantopus*; and the avocet, *Recurvirostra americana*.[21]

Equally important, as a leading ornithologist has noted, were the "steadying qualities" that Bachman exerted upon Audubon in his later publications on birds. Indeed, he "pointed out some mistakes and erroneous conclusions" made by his often too hasty friend. In a letter on January 15, 1835, for example, Bachman told Audubon that he "must be cautious" in generalizing about European and American species of swifts and swallows, and he directed him to "examine the eggs and nests carefully." In a letter of September 17, 1836, Bachman told Audubon that he could "show what true greatness is by doing all in your power to correct every error" before issuing new editions of his works, but, as Bachman would learn when they commenced a joint work on mammals, Audubon often seemed indifferent to mistakes. In some instances, Audubon was not above trying to finesse a statement when he lacked facts. Such cover-ups did not escape Bachman's eye. In a letter of April 24, 1837, regarding Audubon's comments on some of the wading birds, Bachman told his colleague that he had "manage[d] the article cunningly but not ingeneously [sic]." Bachman urged him to quit searching for new species and to learn more about those already known.[22]

By mid-1832, Bachman was preparing a paper on bird migration to read before Charleston's Literary and Philosophical Society. He did not present the paper until March 15, 1833, however. Much later he sent the manuscript on to the editor of the *American Journal of Science*, who published it in the July 1836 issue. Noting a paucity of literature on the subject in the United States, Bachman observed that he had long studied the migration of birds in three regions of the country. Viewing birds as endowed by God with "instincts which cannot be equaled by the boasted reason of man," Bachman noted that certain species migrate "either to avoid the cold of winter, or to find more congenial, or more abundant[,] food." Some species, he noted, continue their journey by night. For example, he had observed that "the Great Whooping Crane [*Grus americana*] scarcely ever pauses in his migrations, to rest, in the Middle States." He knew, he added, for he had heard the "hoarse notes" of the crane while it was flying "over the highest mountains of the Allegheny." Citing numerous species and their patterns of migration, Bachman demonstrated an exemplary and long familiarity with the subject. He noted also that when he released several species he had held in captivity since they were young, they would immediately fly in the direction of their instinctive migration. His study, he added, was far from complete, but, in his words, it nonetheless showed "a wise arrangement in nature which governs instinct."[23]

Of equal interest to Bachman was the nature of molting in birds and mammals, and soon after meeting Audubon he renewed a study that he had begun as a boy. In a letter to Audubon on February 18, 1833, he

John Bachman. Reproduced from an oil painting on canvas by John Woodhouse Audubon, ca. 1837. Courtesy of The Charleston Museum, Charleston, S.C.

mentioned several species of molting birds he had been observing, including the exquisitely colored painted bunting, *Passerina ciris*, which, he said, "staggers me [for] it does not change like the rest." By May of 1837, Bachman had drafted a lengthy manuscript titled "Observations on the Changes of Colour in Birds and Quadrupeds," read it before the Literary and Philosophical Society, and sent it off to the American Philosophical Society. The Society referred the manuscript to a committee on publication, which, ironically, included George Ord, who had himself recently published an error-filled article on molting in birds. Ord rose to a professional level, however, and not only agreed with his colleagues that Bachman's paper should be published but also admitted his own errors. Telling Bachman that he was mortified by his own "silly paper," he said that he was "totally in error with respect to spring molt in some birds." He had, he confessed, relied upon people who did not really know the subject, and he told Bachman, "you listen to all, but you examine for yourself, and report the result[s] only of your own investigations."

Published in the *Transactions of the American Philosophical Society* in 1838, the monograph devoted twenty-eight pages to birds and fourteen to mammals. Citing numerous cases of changes in the plumage of specific birds, Bachman reiterated his commitment to the Baconian philosophy, which stressed the discovery of facts by rigorous observation and eschewed the development of hypotheses as guides to the study of scientific phenomena. Said Bachman, "in our investigations of nature, we are perhaps too prone to build our theories first, and afterwards seek for the facts which are to support them." Obviously, the Baconian approach discouraged pure speculation, but it was inadequate for determining why or how some phenomena work. In fact, Bachman had recognized the necessity of establishing experiments designed to ascertain how the vultures locate their food. In the case of bird molting, however, he correctly emphasized careful observation as necessary to the development of a generalization. Among the species he discussed was the now extinct Carolina parakeet, *Psitacus carolinensis* (= *Conuropsis carolinensis*). Because that species was already, as he noted, "rare in Carolina," Bachman observed some individuals held in captivity by a friend. Many of the other species he had observed in his own aviary or collected during molting seasons. In nearly every case, he provided dates and the number of individuals of each species studied. Despite the sizable number of species he observed, Bachman cautioned that others must also be studied.[24]

Given the pioneering studies by Wilson and the contemporary work of Audubon in ornithology, Bachman apparently realized that he could make a greater contribution in the field of his other passion in natural history, namely mammalogy. He was familiar with the works of European mammalogists and with the relatively few contributions of American naturalists to the study of the North American species of mammals. Of the latter group, perhaps the most accomplished of Bachman's contemporaries was Richard Harlan, a physician and naturalist in Philadelphia. The author of *Fauna Americana* (1825), *American Herpetology* (1827), and many articles, Harlan was an able student of mammals, but, in his eagerness to claim credit for describing new species, he sometimes resorted to cheating and deception. It is doubtful, however, that Bachman was aware of Harlan's flaws during the first four years of their acquaintance, though he knew that Harlan had not done a thorough job in his *Fauna Americana*. In any case, Bachman had begun to correspond with Harlan early in 1832, telling him about two leporids, the "Swamp Rabbit" and the "Marsh Hare," which he thought were new species, and providing information about the gestation of the white-tailed deer, *Cervus virginianus* (= *Odocoileus virginianus*). Harlan responded on February 6, 1832, expressing interest in Bachman's comments, especially about the rabbits. On March 28, Harlan notified the clergyman that the Academy of Natural Sciences of Philadelphia "honoured itself last night by enrolling the name of the Revd. J. Bachman among its correspondents." Bachman was no doubt pleased, for, aside from the older organizations that promoted all of the sciences—namely, the American Philosophical Society and the American Academy of Arts and Sciences—the ANSP, founded in 1812, was the most important society with which contemporary American naturalists could be associated. The Boston Society of Natural History, founded in 1830, was then only a fledgling organization.[25]

Within a few months after his first contact with Harlan, Bachman was seriously engaged in collecting more specimens of mammals and in learning as much as he could about them. On January 20, 1833, he informed Audubon that he had "already added *one* Hare to the Fauna of the United States [and] hope to add another very soon," and late in the following year he complained to Audubon that the latter had taken with him an important British work on mammals and that he could not get a copy of it in the United States. Perhaps Bachman and Audubon were already discussing a joint work on the mammals of North America, to parallel *The Birds of America*. Certainly, Audubon had suggested earlier that his sons might do such a volume, and he must have been aware that neither he nor they were prepared to write descriptions of all North American mammals. In any case, Bachman pushed forward with his own study of mammals, while continuing his work on birds. Audubon, on the other hand, thought only of the latter. On October 17, 1837, Bachman responded to Audubon's obviously vague reference to beginning the work soon: "About the Quadrupeds I can say nothing till you see fit to enlighten me in regard to your plans. . . . I am left . . . much in the dark about the matter . . . size of the work, probable price, the kind of engraving [and] the plan of publication." He noted also that he had "commenced preparing" a work on the quadrupeds several years earlier, but, since he had heard that someone was at work on such a project, he "gave the matter up."[26]

In the meantime, Bachman had completed a preliminary description of a rabbit and, on January 7, 1836, sent it to Harlan to read before a meeting of the ANSP. From Charles Pickering, a splendid naturalist and member of the ANSP, Bachman received a letter, dated January 19, 1836, indicating that Harlan had read the initial description before the Academy, but he added that the Academy desired a drawing and specimens of the species. Bachman arranged for someone to make the figure, though it was not a truly good representation. The final paper was read before the Academy on May 10, 1836. Titled "Description of a New Species of Hare found in South Carolina" and published in the *Journal of the* ANSP in 1837, the paper represented Bachman's second major scientific publication. It described a rabbit he first called the swamp hare and later, the marsh hare (now, marsh rabbit). Bachman named it *Lepus palustris*, but the species has since been referred to the genus *Sylvilagus* since it is in fact a rabbit and not a hare. In the paper, Bachman noted that he had discovered the species some fifteen years earlier (i.e., ca. 1821), and he quoted from his diary of April 10, 1832, that he had seen "a number of these animals wherever the marsh grass and reeds protruded above the water." Although he provided an excellent description of the pelage and morphology of the animal, including measurements, Bachman offered no information on the skull and the dental formula, an important omission he would later remedy.[27]

Soon after he sent that paper, Bachman completed three more papers and submitted them to the ANSP. The minutes of the Academy indicate that they were read before the organization, and, on March 28, 1837, the review committee reported in favor of publishing them. Two of them appeared in the Academy's *Journal* later in 1837, but the third manuscript fell victim to the duplicity of Harlan. Three years earlier, Harlan had told Bachman he was "among the most favored of his correspondents" and that Bachman's observations would be very useful to him in the revision of *Fauna Americana*. In this instance, it was indeed useful. Bachman's paper described a new murid rodent, the rice rat, for which he proposed the name *Arvicola oryzivora*. Bachman had sent a specimen to Harlan in May 1836, requesting that he compare it with a similar species. Harlan soon realized that the Academy already possessed a specimen of Bachman's new species, collected in New Jersey. Cleverly, he proceeded to describe the Academy specimen as *Mus palustris* (= *Oryzomys palustris*), adding that "a similar specimen was sent to me by Dr. Bachman," and thereby effectively placing Bachman in a secondary position. Bachman had in fact collected the first specimen of his *A. oryzivora* during the winter of 1816, but he waited until late in 1836 or early 1837 to submit his description. By the time the ANSP received Bachman's description, the one by Harlan was already in press for the *American Journal of Science*. Bachman learned from that experience, and later told another member of the ANSP, "I do not mean to wait so patiently or confide so implicitly" in the case of other descriptions.[28]

The other two papers published in the Academy's *Journal* were "Observations on the different species of Hares (genus Lepus) inhabiting the United States and Canada" and "Some Remarks on the Genus Sorex, with a monograph of the North American Species." In the former, Bachman offered descriptions of eight North American lagomorphs (rabbits, hares, and pikas), though, like his contemporaries, he placed all of them in the genus *Lepus*. The lengthy article provided excellent information on the known species, showing that Bachman was an authority on the order Lagomorpha. In every case, he offered body measurements, pelage color(s), habits, and geographical distribution (insofar as it was known), and the dental formula. The pika, or ochotonid, it turned out, had already been described, but the prodigious collector and able Philadelphia naturalist John K. Townsend had sent a new specimen from the Rocky Mountains to Bachman, an act that suggests that Bachman's status as a mammalogist was rising. Thought then to be a very small hare, the pika was called the "Little Chief Hare," *Lepus* (or *Lagomys*) *princeps* (= *Ochotona princeps*). In the article and in correspondence, Bachman expressed puzzlement over that lagomorph, but he nevertheless followed others who viewed it as a leporid. Bachman also described two new rabbits. One, based on specimens sent to him from Alabama and Mississippi, he described as *Lepus aquaticus* (= *Sylvilagus aquaticus*). The animal, which he referred to as the swamp hare but is known commonly today as the swamp rabbit, is a fascinating animal that swims with ease and eludes predators by submerging to its nose. The other, a specimen sent to him by Thomas Nuttall from the western United States, he described as "Nuttall's Little Hare," *Lepus nuttallii* (= *Sylvilagus nuttallii*), now commonly Nuttall's cottontail, an inhabitant of sagebrush, on which it also feeds. Thus, by 1837, with three new species of leporids to his credit and a lengthy and informed discussion of other known species of lagomorphs, the heretofore local naturalist had gained national recognition, and his reputation soon became known abroad.[29]

Bachman's standing was also enhanced by his superb paper on North American shrews, nervous little mammals that are in almost constant search of food because of their high metabolic rate. While he was refining the paper and having figures drawn, Bachman sent some specimens to the ANSP, but this time he avoided Harlan, sending them instead by a friend to another Academy member, Samuel George Morton. In a letter to Morton he said, in reference to Harlan, "he who takes my shoes will make no bones in taking my stockings also." Titled "Some Remarks on the Genus Sorex, with a Monograph of the North American Species," the paper described thirteen species of the tiny insectivores in the family Soricidae, commonly called shrews. Although current taxonomists consider only seven of them to be distinct species, that number includes two described by Bachman: *Sorex carolinensis* (= *Blarina carolinensis*) and *Sorex longirostris*. The former, now commonly called the southern short-tailed shrew, Bachman had first observed within a year or two after settling in Charleston. Later, he received a specimen from Dr. Alexander Hume, who had taken it in the swamps of South Carolina's Santee River. He collected a second specimen while visiting in Colleton County, South Carolina, when he noticed "a protuberance on the throat" of a dead hooded merganser

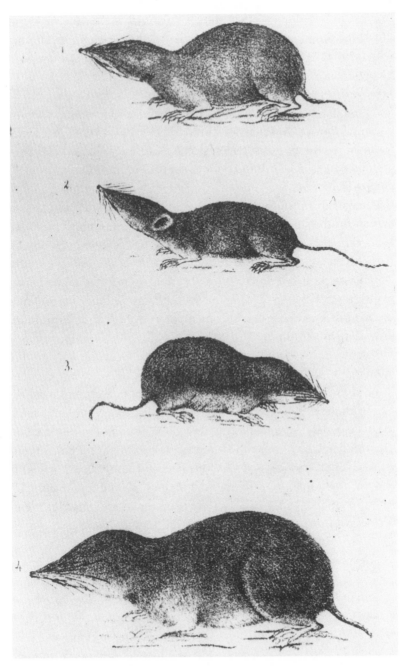

Four species of shrews figured for John Bachman, "Some Remarks on the Genus Sorex," *Journal of the Academy of Natural Sciences of Philadelphia* 7 (1837): plate 23. Bachman published original descriptions of *Sorex carolinensis* (= *Blarina carolinensis*) and *Sorex longirostris*.

brought in by a hunter. Upon slitting the throat of the bird, he found "this little shrew . . . neither much mutilated or destroyed." Much later, while visiting "the public house at the Talullah falls in the mountains of Georgia," he collected another specimen that he identified as *Sorex carolinensis* (= *Blarina carolinensis*), but it was almost certainly *B. brevicauda*, a closely allied species that inhabits that area. In his monograph on the North American shrews, Bachman provided the dental formula, dentition, measurements, color of pelage, habits, and geographic range (as known then) for each species. In addition, he made comparisons wherever appropriate. As recent authorities have noted, Bachman's treatment of the shrews was "thorough," including information on their tiny teeth. Although all members of the family have teeth with chestnut-colored tips, some can be identified only by the size of their unicuspids. It would not be long before the Charleston mammalogist offered descriptions of several species in another family of insectivores, the Talpidae, or moles, and thus solidified his standing as the most knowledgeable of contemporary American authorities on the order Insectivora.[30]

Meanwhile, in mid-1837, Bachman was diligently at work on his monograph on molting in birds and mammals, and, after it appeared in print in 1838, he gained additional recognition as a mammalogist. In that work, he demonstrated understanding of a very important principle of taxonomy: be careful about designating varieties as new species. "It is difficult," he wrote, "to expunge a species once admitted into books." Thoroughness and care characterized his work in taxonomy, and no doubt he was worrying a bit about a potential problem with Audubon in the joint work they were planning. Fearing that his friend was less committed to accuracy, he asked him in a letter on April 8, 1837, "Do you not see . . . that our mammalia requires [*sic*] as much examination as the birds?" But Bachman pushed on with his studies, collecting specimens as he found time in a truly busy schedule, while Audubon continued to search for the ever-elusive new species of bird.[31]

In May 1837, while attending a denominational meeting in Philadelphia, Bachman took advantage of the opportunity to visit with some of the city's naturalists. He also visited the headquarters of the American Philosophical Society, where, he proudly wrote home to his wife, he sat in Benjamin Franklin's chair several times and met "old General Clarke, the companion of Lewis." It was apparently then that the indefatigable collector John Townsend invited him to describe a number of the other specimens he had brought back from the Rocky Mountains. He soon became a "constant" correspondent of Townsend. His impression of several of the Philadelphia naturalists was favorable, but, he later told one of them, Edward Harris, "If I were a phrenologist, I should exceedingly like to examine [the] heads [of others], for I think there must be some queer bumps." He was particularly critical of Ord, whom he viewed as a "third rate ornithologist." Bachman said he had to show Ord that many "additions had been made in true species" since Wilson's day, to which Ord "exclaimed 'My God, where am I[?]' " Bachman responded, as he put it, "in as good natured a way as I knew how," by saying, "Mr. Ord[,] you are about 50 years behind the ornithological knowledge of the day." Yet, to Harris, he said that Audubon must make a number of corrections in the new edition of his *Birds*, noting that he had "written under each disputed species my observations."[32]

In addition to all his other activities, Bachman served as president of the Literary and Philosophical Society in 1836–1837. Determined to reinvigorate the organization, he arranged for a series of eighteen lectures, to run from May 17 through June 10, 1836. All of the lectures in that ambitious schedule dealt with science, the majority of them on topics in natural history. Bachman himself presented one of the lectures. In "The Morals of Entomology," he spoke of the wonders of the insect world as a manifestation of the Creator's work, citing the metamorphosis of the caterpillar into a butterfly as analogous to resurrection. Other lectures dealt with such topics as the nature of animal intelligence, conchology (presented by Edmund Ravenel, a longtime member of the Society), fossils, and, not unexpectedly, "on the negro race." Members received a rare treat in May 1837, when Bachman presented his paper on color changes in birds and mammals. Soon thereafter he presented an address on the status of knowledge in the various branches of natural history in South Carolina in which he noted that, while the quadrupeds were "hardly described," the reptiles and amphibians had been thoroughly covered in a pending publication "by a member of our Society," which was, of course, a reference to Holbrook. He commented that while the birds had been pretty well described by Audubon, much more needed to be done. In reference to mollusks, Bachman observed that a member of the society had built a huge collection of shells, but he added that studies should be made of "the living animal that inhabits the shell." He was, of course, referring to Ravenel, who, in fact, had been making drawings of the living animal but never attempted to publish any of them. In May 1837, shortly before his term as president was to end, Bachman presented another address on entomology. On that occasion, however, he emphasized how little attention had been paid to the study of insects in the South, and he called upon the members to add more specimens to "rich and accumulating collections in the Museum of this society."[33]

One of the most controversial lectures in the series arranged by Bachman was a presentation by James H. Smith, entitled "Geology and Revelation." Little is known about Smith, but he was obviously among that group of American and British intellectuals who advocated one of "the various long-day theories" to explain how geological evidence and the Genesis account of creation could be harmonized. According to the *Charleston Courier*, Smith displayed a diagram of the strata of the earth, of which the topmost "showed the deposit of *the last* or Noachic deluge." He called the lower layers "the results of other deluges," each of which contained fossils, indicating the presence of animals before the original pair of humans. Thus, Smith argued, the six days of creation in the Book of Genesis should be read as "six periods," each of considerable duration. The skeptical *Courier* writer said that Smith claimed his scheme would "make the glorious radiance of the Geologists blend sweetly and harmoniously with the twilight of the Scripture account of the creation." The notion, asserted the journalist, "is monstrous," and he criticized the geologists who subscribed to it. In his judgment, the Mosaic account was either "true or not" and should not be "frittered away by hard names and imputations of mistaken translations." Bachman did not enter into this controversy, but other comments indicate that he held mixed views on the subject of an ancient earth populated before humans by a vast number of animals. Without doubt, Bachman believed that Adam was the first human and that God had created him less than 6,000 years ago. Yet he found it difficult to reconcile the fossil record with such a young earth. Eventually, during a debate over the question of the non-Caucasian races as separate species of humans, he came down on the side of the biblical literalists. In the interest of science, however, he never attempted to stifle honest debate.[34]

By the fall of 1837, the greatly overworked Bachman was "suffer[ing] much from debility" and complained that "the least fatigue puts me in bed." His condition had not improved by March 1838, when he told Audubon of his "shattered and . . . declining health." Still, he was hard at work in describing the specimens collected by Townsend, and he was talking enthusiastically of the joint work with Audubon. By mid-May, however, he had asked the vestry of his church to grant him a leave of absence for six months so that he could travel to Europe for rest. The vestry readily consented and volunteered to pay him a quarter of his salary, expressing regret that funds were not available to pay his full salary. Before he left in June, however, Bachman had completed a paper describing the Townsend specimens, the ANSP minutes indicating that it was read before the Academy on August 7, 1838, while Bachman was in Europe. Published in 1839 in the Academy's journal, it was titled "Description of several New Species of American Quadrupeds." The descriptions were short, but Bachman noted that he planned to write more detailed ones and provide figures later. In fact, he had begun the complete descriptions, but, because he had not been able to consult recent scientific journals, he observed that "some of our enterprising naturalists abroad may have already published the species here described." One was a new species of mole, *Scalops townsendii* (= *Scapanus townsendii*). Two of the species were murid rodents, but only one of them, *Arvicola oregoni* (= *Microtus oregoni*), had not been previously described. The remaining six species were sciurid rodents, two of which were new species: *Spermophilus townsendii* and *Tamias minimus*. Two other species he described as new were later determined to be subspecies. Thus, by the time he was halfway through his tour of Europe, Bachman had added yet another important work to his list of contributions.[35]

Bachman kept a journal of his trip, but it was destroyed during the Civil War. Fortunately, a journal kept by his traveling companion, fourteen-year-old Christopher Happoldt, a member of St. John's, provides some useful information, though, of course, it reflects the interests of an adolescent and omits some things of importance. In any case, Bachman and young Happoldt embarked from Charleston on June 6, 1838, bound for Liverpool. They arrived on July 2, and before the day was over, Bachman had visited the city's zoo and botanical garden. During his six-week sojourn in Britain, he spent three weeks in Edinburgh and most of the remainder in London. He was particularly eager to meet the noted naturalists of those cities, and, in the latter, to examine specimens in the collections of the Zoological Society of London, where he found that, as Happoldt stated it, the curators "had miscalled several animals . . . [and] beged [*sic*] him to name them." Despite his illness, the enthusiastic mammalogist readily consented, taking great delight in the effort. Indeed, Bachman spent most of his time in London in identifying and labeling the specimens and in writing descriptions of them, mostly North American squirrels. In fact, he even delayed his departure for Hamburg so that he could complete the task. He was especially intrigued over some animals he had never seen before. In Happoldt's opinion, the health of his companion was "quite restored now, [though he has] a great deal of fatigue."[36]

Bachman was determined to visit the German states because of his maternal and language kinships. That trip, he believed, would be his only opportunity, for he was certain that he had not long to live. The vigor of his activity during the five weeks he traveled in the German and adjacent states belies his morbid view. From Hamburg, where he arrived on August 17, he traveled about Prussia, spending as much time in Berlin as he could, in order to meet naturalists and examine specimens in the museum. He was well received by the naturalists of that city, including the great naturalist Alexander von Humboldt, whom he had met in Philadelphia in 1804 and with whom he had exchanged correspondence since their meeting. Traveling on to the Czech city of Prague, through Bavaria to Munich, on to the Swiss city of Zurich, Bachman visited every museum of natural history he could manage along the way. Then he headed northwestward to the German city of Freiburg, arriving there on September 17. On the next day he attended an international meeting of naturalists, and a day later, by spontaneous invitation, he presented a brief paper on the state of natural history in the United States. Among the 600 naturalists in attendance at the meeting were many of the world's leading zoologists, including Louis Agassiz. Bachman left Freiburg on September 21, headed toward Heidelberg to see one of Luther's documents, and on to Frankfurt, Antwerp, and Brussels. On October 3, he arrived in Paris, where he stayed for ten days, spending much of his time at the famous Jardin des Plantes and in the Muséum d'Histoire Naturelle. On October 21, he and Happoldt reached London, and soon thereafter Bachman was again at the zoological museum. Bachman and his companion departed for home on November 18, and arrived in Charleston on December 28.[37]

Looking back upon his trip, Bachman was pleased, noting that he learned much in Europe and that he had been "cheerfully" received by naturalists. In fact, he had been especially favored in London. While he was there, G. R. Waterhouse, curator of the museum of the Zoological Society of London, named a new leporid after him, *Lepus bachmani* (= *Sylvilagus bachmani*), and members of the Zoological Society invited him to present a paper on the North American squirrels. The Society published his paper in its *Proceedings*, and several months later, Bachman expanded the paper and published it in the *Magazine of Natural History*, under the title "Monograph of the Genus *Sciurus*, with Descriptions of new Species and their Varieties, as existing in North America." Providing excellent, detailed descriptions of seventeen North American squirrels, Bachman noted the difficulties associated with determining whether some squirrels were actually species or merely varieties. He was aware, for example, that some specimens resembled the fox squirrel, *Sciurus niger*, in morphology and dentition but varied greatly in the color of pelage. "I am far from supposing," he said, "that I have noticed all the true species . . . nor can I say with positive certainty that I have in every case been able to draw the line of separation between varieties and true species . . . [even after examining] several hundred specimens." Bachman nevertheless classified seventeen as species, of which he thought six were new. Ultimately, however, none of the six proved to be distinct species, but five of them are currently considered valid subspecies. Thus, given the state of contemporary knowledge about the sciurid rodents, Bachman had again demonstrated that he stood high on the list of mammalogists, both in the United States and abroad. Within three more years, he would enhance that reputation with additional descriptions of new mammalian species.[38]

By the end of 1839, despite his continuing illness and fatigue, Bachman could take pleasure in reflecting upon the fame he had earned in the space of only a few years, and he could rejoice that each of his older daughters had married a promising young man, though he remained disconsolate for many weeks over their departure. The eldest Bachman daughter, Maria Rebecca, wedded the younger Audubon son, John Woodhouse, on June 21, 1837, and the second, Mary Eliza, married the older Audubon son, Victor Gifford, on December 4, 1839. Both daughters soon fell victim to tuberculosis, however. Maria died on December 18, 1840, after a lengthy period of intense suffering. Fate was no kinder to Mary Eliza, who succumbed to her debilitating illness on May 21, 1841. For John Bachman, however, the period of tribulation had just commenced. By 1839, he was aware that the "troublesome cough" afflicting his beloved wife indicated the likelihood of tuberculosis. Harriet Martin Bachman's condition was made worse by the tic douloureux, or trigeminal neuralgia, that brought fits of agonizing pain to her face. Reminiscent of Job, Bachman wrote on May 25, 1841, to his daughter Jane, who was visiting the Audubons in New York, where she had gone for treatment of a severe eye ailment: "The ways of God are dark and incomprehensible to us poor short sighted mortals, and it is our duty not to murmur, but to pray for submission." Harriet experienced periods of intense suffering over the next few years, spending much of her time in bed, and by April 1, 1845, Bachman said he was gravely concerned over her condition: "I cannot look forward to the event of losing my beloved wife without feelings bordering on despair. . . . My prospects are dark, very dark." A month later, Bachman noted that "the phlegm continues collecting on her chest . . . [and] she has also fevers and debilitating night sweats," signs, he knew from the cases of Maria Rebecca and Mary Eliza, that signaled the triumph of tuberculosis. On July 16, 1846, after several days of continuous fever and neuralgic pain in her face, Harriet Bachman died. John Bachman wanted "to be left alone for a few days," but he accepted the loss as the will of God, who "rules our destiny and orders our lot."[39]

His troubles were not over, however, for daughter Julia was showing signs of a "very distressing" cough soon after the death of her mother. Poor Bachman watched as the dreaded disease began to consume the body of his most beautiful daughter. Upon a physician's advice, he took her to Red Sulphur Springs, in Virginia, in July 1847. By then, Julia was so weak that she could sit up for only an hour each day. Bachman found solace in the surrounding mountains, but, upon returning from his walks, he saw the "pallid faces" of numerous other patients who were seeking a cure at the springs. "How terrible," he wrote, "is consumption—holding on with a deadly grasp, weakening the cords of life from day to day." Julia grew weaker as the weeks passed, and died on September 7, 1847. "Providence," lamented the grieving father, "calls us to suffer as well as to enjoy." He buried her in a grave two miles away, and, as he left, he mused, "I suppose I shall never see it again."[40]

Meanwhile, during his period of tribulation, Bachman had continued to describe new mammals and had begun the long-awaited project with Audubon. After the trip to Europe, where he apparently consulted some of the older works in natural history unavailable to him in Charleston, he drafted a paper in 1839 that corrected some of the errors of synonymy he had made in the lengthy article on leporids, elaborated upon previous descriptions of hares and rabbits, and described, as he believed, two new species of hares. In addition, he appended descriptions of three more of the sciurid rodents sent to him by Townsend. As usual, the article reflects Bachman's concern for accuracy and thoroughness. Grateful to Townsend for his recognition, the Charleston naturalist named yet another species after his faithful supporter: *Lepus townsendii*, Townsend's hare. He also described another of the Townsend specimens as Richardson's hare, after the renowned Scottish naturalist and Arctic explorer, John Richardson, who had befriended him during his sojourn in London. *Lepus richardsonii* was later determined, however, to be a subspecies of *L. californicus*. In his descriptions of three Townsend specimens, Bachman included two new species, but one of them, which he named the Oregon flying squirrel, *Pteromys oregonensis*, is actually a subspecies of the northern flying squirrel, *Glaucomys sabrinus*. The other, which he named *Geomys townsendii* (= *Thomomys townsendii*) was indeed a new species of the exclusively North America rodents known as pocket gophers, whose cheek pouches account for their common name.[41]

The illness and subsequent death of his eldest daughter and the illness of his second daughter had caused Bachman to delay a paper on the moles. He had already described one before he left for Europe, of course, but he did not submit a paper on the four North American talpid insectivores with which he was familiar until the fall of 1841. In this case, he chose to submit the paper to the Boston Society of Natural History, at the invitation of the young ornithologist Thomas Mayo Brewer, who sent him a specimen of an undescribed mole. Brewer collected many other mammals for Bachman between 1837 and 1841, and Bachman reciprocated by sending bird eggs to Brewer. Bachman's paper, containing short descriptions of four moles, was read before the Society on October 6, 1841, and published in its journal in 1842. In exemplary fashion, Bachman provided descriptions of four North American moles, two of which were new species. Of those small subterranean creatures with tiny eyes and large, clawed forefeet, he described Brewer's shrew mole, *Scalops brewerii*, now called *Parascalops brewerii*, the hairy-tailed mole. The other new species first came to the attention of Bachman at the Berlin Museum, whose director, M. H. Liechtenstein, had asked him to describe the little animal. Bachman named it *Scalops latimanus* (= *Scapanus latimanus*), or the broad-footed mole. Now he was clearly the authority on both families of North American insectivores.[42]

In the Shadow of Audubon

E ven though he harbored some misgivings, Bachman decided to collaborate with Audubon in doing a work on the mammals of North America. On July 5, 1839, he wrote to Audubon in Britain, telling him to talk with the Scottish naturalist John Richardson and to "take notes and write down everything." He also advised Audubon to go to the museum in London to examine specimens. "Study the skulls and teeth a little," he said, for "you take these things by intuition." In addition, he urged Audubon to read "pertinent books and articles." Their joint work, said Bachman, "must be original and creditable," not a mere compilation of the works of others. Furthermore, it must contain "no humbug." Bachman added that they could hardly call it a work on mammals if they excluded the cetaceans, or whales and dolphins, and the "antediluvian animals," or the fossil vertebrates, though he noted the latter would be "troublesome."[1]

That Audubon essentially ignored the advice soon became obvious, and on September 13, 1839, Bachman told him he was moving too quickly by issuing a prospectus of the work. Before proceeding to enlist subscribers, said Bachman, they needed to complete one vol-

John James Audubon. From an engraving by John Sartain after a painting by Frederick Cruikshank. Courtesy of the New-York Historical Society, New York, N.Y.

ume. The undertaking would not be easy, he warned, but Audubon could not comprehend the quality and scope of the work Bachman had in mind. Despite the informal agreement of long standing, Audubon seemed at one point to be thinking of doing the work alone. In a letter to Thomas Brewer on September 15, 1839, he wrote, "Now that I am about to commence the publication of the *Quadrupeds of North America*, I will expect your assistance in . . . [obtaining specimens]." Then he told Brewer that "John Bachman is about to give the whole of his collections and notes to me." Bachman was unaware of Audubon's notion of doing the work alone, and it is almost certain that he never promised to lend his specimens and notes to Audubon. After all, it was Bachman, not his friend, who had established himself as a leading authority on the mammals of North America. In the quite extensive and extant correspondence between the two men not even a hint appears that Bachman would lend his notes to Audubon. In any case, Audubon soon dropped the idea of doing the work alone. Perhaps Audubon did not relish the kind of exacting work that Bachman was sure to expect; perhaps he did think momentarily that he could do the work alone; or perhaps he was simply promoting himself to Brewer. No doubt he realized that Bachman was right when, in a letter to him on December 24, 1839, his friend bluntly said, "you cannot do without me in this business."[2]

Early in January 1840, Audubon wrote to Bachman, "I have thought very deeply on this [matter of the book] . . . and know . . . that such a publication will be fraught with difficulties innumerable, but I *trust* not insurmountable, provided We Join our Names together." Bachman replied on January 13, "I am not ashamed to let my name stand along with yours, . . . [and] I am also anxious to do something for the benefit of John and Victor. . . . The expense and the profit will be yours, or theirs, which is the same thing." It seemed to be understood that Audubon's name would be listed first on the publication, but Bachman was still worried that his friend did not understand fully how much work would be required. "Don't flatter yourself that this Book is child's play," he cautioned. Once more he reminded Audubon that "the skulls and teeth must be studied—and colour is as variable as the wind." Later, he told Audubon that he should record data and information he obtained and "try to put the habits [of mammals] under the right species." He was also concerned that Audubon intended to publish a number of plates in May. Meanwhile, aware of the amount of time required to prepare the text of *Quadrupeds*, Bachman decided to publish descriptions of some new species he had obtained in order to establish right of priority. In a spirit of generosity, however, he placed the name of Audubon as first author, although every description was his own. The paper, titled "Descriptions of New Species of Quadrupeds inhabiting North America," was read before the Academy of Natural Sciences of Philadelphia at its meeting on October 6, 1841, and published in its proceedings. Included in it were brief descriptions of two bats, a weasel, five mice, a marmot, and six squirrels. Two of the specific taxa were indeed new: the yellow-bellied marmot, *Arctomys flaviventer* (= *Marmota flaviventris*), and the eastern harvest mouse, *Mus humulis* (= *Reithrodonotomys humulis*). Bachman had originally described the latter in 1837 as *Mus humilis* [*sic*], in a paper sent to the ANSP, according to Academy minutes. He thought it had been published, but there is no record to confirm that it was. Three of the taxa of Audubon and Bachman—or, more accurately, of Bachman—are currently valid subspecies (a sciurid rodent and two murid rodents).[3]

At some point soon thereafter, the ANSP published in its journal an article by Audubon and Bachman under the same title, but it contained seven additional taxa and longer descriptions of each of the animals than those given in the Academy's proceedings. The descriptions appear to have been written solely by Bachman, though it is probable that Audubon supplied some of the specimens. In any case, two of the specific taxa are currently valid: Leib's bat, *Vespertilionis leibii* (= *Myotis leibii*), now commonly called the eastern small-footed myotis, and *Spermophilus annulatus*, now commonly called the ring-tailed ground squirrel. The bat was sent to Bachman by George C. Leib, who had collected it in Erie County, Michigan, and the ground squirrel, a western species, was sent by Spencer Baird, who had received it from another collector. In reference to the latter, the authors—or, more correctly, Bachman—added a statement that bordered on acceptance of the idea of transmutation of species, namely, "In every department of natural history a species is occasionally found which forms the connecting link between two genera." Later, Bachman would again verge upon acceptance of that notion but reject it when he realized the full import of the idea. Eventually, he expressed strong opposition to the Lamarckian theory of evolution.[4]

While Bachman was grieving over the loss of his daughters and worrying about the protracted illness of his wife during the years 1842–1845, he underwent further strain by working with John James Audubon and Victor Audubon on the North American mammals volume. Clearly, the Audubons viewed the book primarily as a work of art and a source of revenue, while Bachman envisioned it as a scientific treatise in which the descriptions advanced knowledge of mammals and the drawings gave form and substance to the information provided. The latter required extensive research in the pertinent, and often not readily accessible, literature, examination of many skins and skulls, and accurate, carefully crafted descriptions. Neither the elder Audubon nor son Victor, who was charged with labeling the plates and preparing copy for the printer, fully appreciated Bachman's intention. Indeed, in their view, the sooner the plates could be issued the better, for they would produce income. The numerous exchanges between Bachman and the Audubons between 1842 and 1845 clearly reveal the differences in their views. Certainly, John J. Audubon worked hard during that period, for he drew and painted many mammals, made long and tiring journeys to enlist subscriptions to the book, and went on collecting expeditions far from home. But he ignored most of Bachman's requests for books and articles unavailable in Charleston. He also ignored Bachman's plea for specimens of some animals and for more than one skin of every species. Moreover, during his travels, Audubon was constantly looking for new birds, and he paid little attention to the smaller mammals. In addition, he kept inadequate notes, often omitting the locality of the collected specimen or citing it so broadly that it was nearly useless. The importance of skulls and teeth of specimens he never understood. After all, it was the stuffed skin that was to be painted.[5]

As early as October 1842, Bachman was lamenting that, during a collecting expedition to eastern Canada, his partner had taken no specimens of pocket gophers or murid rodents, which were especially important because, said Bachman, "there exists great confusion and uncertainty about them." A few months later, in January 1843, Bachman noted the need for additional specimens, especially of the squirrels, in which intraspecific variation could be enormous. In addition, he reminded Audubon that most of all he needed books. "How can I get along" he asked, "without having the old Books where these species were first given[?]" He had long since sent Audubon a list of the books that he considered indispensable, but he had received none of them. Then, in June 1843, Victor asked Bachman to send a list of the mammals they were describing, to which the much-taxed naturalist replied, "If you had looked at your father's letters you would have seen all the species . . . named at least twice if not thrice." He would soon learn that Victor rarely or never looked at a letter a second time. Bachman prepared a new list, but, he noted, it "cost me no small degree of trouble." Before the summer had passed, Bachman was arising at 5:00 A.M. every day except Sunday in order to work on the descriptions.[6]

After Audubon returned from a collecting expedition along the Missouri River in early November of 1843, Bachman anticipated receipt of many mammal skins, but Audubon replied that "the variety of quadrupeds is small in the Country visited, and I fear that I have not more than 3 or 4 new ones." Happily, reported Audubon, "I have no less than 14 new Skins of Birds." Somehow, Bachman was able to subdue his disappointment, but he could not keep quiet when, in late December 1844, he heard from "a mutual friend" that Audubon "never referred to [Bachman's] name as co-author of the Quadrupeds." Perhaps the mutual friend was Brewer, who could have innocently mentioned what he had heard long ago in reference to Audubon's claim that he would be sole author, or perhaps Audubon simply puffed himself in talking about the work. In any case, Audubon adamantly denied the charge and declared that he always mentioned Bachman's role in the project when he was soliciting subscribers. His mind eased by that assurance and at last in receipt of some articles copied by Victor, Bachman pushed forward, and by November 1845 was arising daily at 4:00 A.M. to work on the manuscript. "Nothing but ill health or domestic affliction will keep me back," he wrote to Victor. But he informed Victor that he and his father must carry some of the load. Unless they did, he added, "I fear I may break down." The task had proven to be even more demanding than Bachman had anticipated. As he noted, "Writing descriptions is a slow and fatiguing process." Bachman said he was still waiting for Victor to send him specimens: "I cannot see a needle in a hay stack and cannot give a name to a *mouse* (and there are many of them) without knowing what it is." In an effort to slow Victor down and to avoid errors, as had been made in the first fifty plates, he complained, "I cannot work harder," and warned Victor, "*I will not publish [any of the text] a day before I am ready.*" The frustrated mammalogist urged Victor to write the information he copied from other works so that he could include it without rewriting it. Such expectation, as he would learn, exceeded Victor's ability.[7]

The saga continued, and on December 5, 1845, Bachman complained to Victor that he had omitted the measurements from some of Bachman's descriptions and failed to give the geographic range in others. "I am ashamed of the carelessness exhibited in these matters," sighed Bachman. Once again, he asked Victor to write synopses of information and descriptions so that they could be used without Bachman's having to revise them: "I scarcely have a page left in a whole of fifteen pages and thus the very correcting and arranging costs more trouble than the writing [of] the whole affair myself." Crying that his want of books and specimens was "somewhat like the Children of Israel . . . who had to make bricks without straw," Bachman urged Victor to be more diligent. The situation did not improve, and on December 24, 1845, the exasperated clergyman wrote to Edward Harris, complaining, "I find the Audubons are not aware of what is wanted in the publication of the Quadrupeds. All they care about is to get out . . . engravings. . . . On many of the Quadrupeds [Audubon] has not sent me one line and on others he has omitted even the geographical range." Bachman also noted the failure of the Audubons to meet most of his requests for pertinent books and articles. He added, "I am willing to write every description and every line of the book . . . without fee or reward," but was adamant that he could not do so unless he had specimens and reference sources. Furthermore, he declared, Audubon must not publish anything without his explicit approval. Bachman also insisted that Audubon must arrange for someone, perhaps his son John Woodhouse, to collect specimens in the West, especially the smaller rodents. Noting that he would soon complete the first volume of text, he vowed that he would not write a single line of the second unless the Audubons followed his stipulations. He had not, he said, reached this decision in haste but as a "result of four years [of] remonstrance, mortification, and disappointment." Bachman apparently hoped that Harris could persuade the Audubons to mend their slovenly ways.[8]

Whether or not Harris spoke to them about Bachman's concern or showed them the letter is uncertain, but Bachman received a letter from Victor Audubon, dated December 27, 1845, in which he expressed regret over "any mistakes that occurred or may occur, in our nomenclature." He downplayed the matter, however, by saying that others had made worse blunders. There followed a promise to make some corrections and to send some of the specimens Bachman had begged for. "I regret, my dear friend," added Victor, "your sensibility on these subjects . . . , [and] I assure you we are as much mortified as yourself." Seemingly oblivious to the real trouble he had caused Bachman and ignoring the repeated plea for books and articles, Victor reminded the exasperated naturalist of his "long silence" at one point, claiming that it had delayed the project. That Victor could so easily dismiss errors was especially troubling to a scientist so devoted to faithful inquiry. Like any other serious investigator, Bachman realized that he could make a slip, but he declared that such would not happen often if one pursued a subject with care and deliberation. In fact, the high esteem he enjoyed was largely a result of his commitment to accuracy and candor.[9]

Such qualities manifested themselves in other ways during the same period. Late in 1842, for example, the Literary and Philosophical Society (by then commonly called the Literary Club or the Conversation Club) invited its longtime associate and former president to present yet another address before the Society, on this occasion his views on a much-discussed topic in the state, namely, the potential value of a geological and agricultural survey of South Carolina. In his address, which gained considerable notice, Bachman carefully outlined the advantages to be gained by examining the physical resources of the state, but he also used the talk to promote understanding of the botanical and zoological treasures in South Carolina. While the northern states were advancing in agriculture, he noted in a patriotic appeal, "South Carolina has deteriorated." As state and local leaders were praising Bachman's defense of the value of a scientific inquiry, however, a charlatan was playing a hoax upon the citizens of Charleston. On January 16, 1843, the *Charleston Courier* announced that Alanson Taylor would soon exhibit an "anomaly in nature . . . the Fejee Mermaid." Taylor, an uncle of P. T. Barnum, was touring the South with his anomaly—and reaping profits for himself and his nephew. The *Courier* had predicted that the mermaid would "doubtless create much interest" among the city's scientists. Indeed, it did, with John Bachman taking the lead in denouncing Taylor and his specimen, which consisted of the head and torso of a small monkey carefully sewn to the tail of a fish and placed in a preservative in a glass container.[10]

Incensed over this corruption of natural history, Bachman, writing under the pseudonym "No Humbug," raged when the prominent but erratic Charleston attorney Richard Yeadon took the mermaid claim seriously and criticized Bachman—and science in general. A long, drawn-out controversy filled the pages of both Charleston newspapers and ended only when Taylor pocketed his profits and moved on to

Augusta, Georgia, to fatten his wallet further. Bachman fretted over the "egregious ignorance of the first principles of natural science," and three years later, as the mesmerist movement entered Charleston, he was still expressing concern over the "tendency [of our community] to swallow every humbug. . . . No one has done anything to stay the torrent . . . sweeping [us] back to the days of ignorance." Bachman could, however, find solace in the recognition he received from fellow scientists, including the renowned British geologist Sir Charles Lyell, who sought him out during visits in 1842 and 1845. Lyell was impressed with Bachman's knowledge of mammals and expressed a particular interest in the marsh rabbit, *Sylvilagus palustris*, first described by his host.[11]

Partially placated by Victor Audubon's conciliatory letter in late December 1845, Bachman entered the new year with guarded optimism that he could proceed with the scientific work he so dearly cherished. Arising before daylight on January 3, 1846, he began to pen a long letter to Victor, spelling out what his young collaborator must do in order to keep him in the project. "I have never doubted the dispositions of any of you," he began, "but you could not be made to comprehend that I could do nothing without books of reference to settle the species. You seemed to think that as we knew more than the old book makers it was not worth while to consult them." Bachman then proceeded to prescribe the conditions under which he would continue. First, having conceded that Victor was incapable of writing a suitable description, he instructed him not to try. Instead, Victor must send him the measurements, geographic range, place and date of collection, and any special information for each species. A week later, the irritated mammalogist continued his complaint: "It is [your] want of precision, [your] loose, rambling way of answering my questions that renders my situation in this matter so unpleasant." In the case of "the common grey squirrel," he asserted, "you certainly don't understand me." Bachman was concerned that Victor and his father did not recognize that a mature individual of that species possesses ear tufts during the winters and is thus not a separate species as the elder Audubon believed. Bachman requested specimens taken during the winter before they had become a year old. Also bothering him was the elder Audubon's decision to list "*Sciurus rubricaudatus*" as a species. Audubon "has a right to use his own discretion in spite of my criticism," said Bachman, "but in naming species with my name, my consent must be obtained." The name was nevertheless published, but it proved to be invalid. Then Bachman summed up his demands to Victor: "you will answer my letters with them before you"; "you will use precision, which is everything in natural history"; and, "as you have not studied the species, you will just attend to my advice." Once more he criticized the Audubons for rushing into the work. "You had better make good use of me," he warned, "for I am perfectly convinced that you cannot make a step by yourself."[12]

Apparently perturbed by a reply from Victor suggesting that the diligent mammalogist was excessively fatigued, Bachman asserted that he never shunned any work. The problem, he aptly opined, lay with the Audubons, who had never "read enough to know what belongs to such a work." Moreover, he expressed disappointment that they had freely substituted their misinformation for his knowledge. Bachman continued to write descriptions and to tell the Audubons when they were wrong. For example, when Victor argued about a certain squirrel, Bachman said, "before you worry me any more with your nonsense[,] will you do me the favour to look at my monograph of Sciurus published in 1838 . . . and then say, after looking at page 12th whether I am mistaken." By mid-February 1846, it was clear to Bachman that the Audubons had not reformed. Desperate for reliable reference sources, he reminded Victor that he had access to "only poor works" in Charleston. In addition, he scolded Victor for requesting names so that he could complete his drawings when he should be doing more to obtain specimens and necessary books. Again he urged the Audubons to "issue your numbers less frequently." Soon thereafter he wrote directly to the elder Audubon about his Missouri River journal, which his friend had withheld from him as long as he could: "it has been a very interesting one although there is less in it of the Quadrupeds than I had expected to find." Bachman criticized Audubon for leaving "a most interesting country unexplored" and for ignoring the smaller rodents, the latter of which was "a terrible error."[13]

To Bachman's surprise, Victor sent page proofs of the first part of the second volume in May 1846. Bachman had explicitly told Victor that he must send the galley proofs to him first. Aside from his dismay over seeing numerous typographical errors, he was upset that Victor had added to his descriptions "worn out old sayings—such as 'kind reader,' " which, Bachman moaned, are "rather below the dignity of such a work." The problems continued, and by July, Bachman's interest in the book had flagged. No doubt, his concern for his dying spouse accounted in part for the decline in his enthusiasm, but it was due much more to the continuing failure of the Audubons to comply with his requests. Victor expressed concern over the delay, and the discouraged naturalist replied, "My interest would be increased if I could see the work go on as I wish." Especially discouraging to Bachman was Victor's decision to alter Bachman's style and to insert his own observations. Bachman had shown the altered descriptions to fellow naturalists, and they had, he said, "unhesitatingly advised me not to write another line for the work." Then he asked Victor what he would think of him if he altered the drawings. Concluded Bachman, "I must either do the work and take the responsibility or you must do it. . . . Your answer will determine my future operations." Victor knew the project was in jeopardy and promised Bachman he would make no further changes. On August 15, 1846, the still seething naturalist reiterated his vow to quit the project if Victor altered anything else in his descriptions. Because of the final illness and subsequent death of his wife soon thereafter, Bachman did not resume his old pace in writing until November, when he was once more arising between 4:00 and 5:00 in the morning and working until afternoon and then, after discharging his pastoral duties, going back to the manuscript in the evenings.[14]

Although Victor did better, the problems continued, and Bachman resumed his verbal pummeling of the young man, which finally stirred up the elder Audubon, who criticized his old friend for not complaining directly to him. Audubon seemed to be unaware that he had not only turned most of the project over to Victor but also that he gave even less information to Bachman. "When I write to you," Bachman replied to his coauthor on December 23, 1846, "I write to my equal in some things, my superior in others." He added that he could speak freely and bluntly to Victor: "I can order him to copy books, to get specimens, to call on people . . . [though] not always . . . as quickly as I wish." He noted, moreover, that he could scold Victor because the younger man was "too respectful or too good natured to scold back." In fact, Bachman's harsh words never seemed to upset the pleasant and gentle Victor. Whether or not Bachman's reply satisfied the elder Audubon is uncertain. Perhaps it did, or perhaps he quickly forgot it, for he may have already been in the early stages of the disease that eventually robbed him of his memory. Apparently, Audubon had been

upset earlier in the year by Bachman's criticisms of some of his drawings. For example, Bachman noted that his friend had omitted the long
tail hairs of Richardson's squirrel (now the red squirrel, *Tamiasciurus
hudsonicus richardsoni*), to which the artist replied, "I can't help
copying nature." The problem, wrote Bachman in a letter to Victor,
was that Audubon was distorting nature. For example, the "tremen-
dous scrotum" that Audubon had drawn on one of the squirrels "was
not given to it by the creator . . . but was stuffed out of character" by
the collector. Bachman cited other cases in which Audubon had failed
"to copy nature accurately." His displeasure was ameliorated somewhat, however, by a favorable review of the first volume of the text of
The Viviparous Quadrupeds of North America in December 1846.[15]

The commentary, published in the *American Review*, was unsigned,
but its author displayed familiarity with Bachman's scientific contributions and understood the role Bachman had played in producing the
volume. While he praised Audubon for his illustrations of mammals,
he noted that the artist "could not have wisely undertaken such a task"
without the assistance of Bachman. Indeed, he asserted, "We may
say . . . that in the letter-press, for which Dr. Bachman is mainly
responsible, we find a greater precision of style than characterizes"
Audubon's *Ornithological Biography*. He concluded the lengthy review with a statement that the book would take a place "at the head of
Illustrative Mammalogy in the world." A writer for the *Literary World*
praised the volume and called Bachman "the most distinguished student of mammalogy in this country." Without Bachman, he added,
Audubon could never have succeeded, for his "technical knowledge of
mammals was insufficient for the task." The work was also lauded by
the *Southern Review*, but, in a letter to Bachman, Victor said, "I do not
think it gives *you* credit enough for the descriptive portion of the
work." Indeed, it did not, but, aside from Victor's private assertion,
the Audubons never went out of their way to tout the contribution of
the greatest mammalogist in America at the time.[16]

As the year of 1847 rolled in, Bachman could take pleasure in the
favorable reception of the first volume and in the excellent artistic work
of John Woodhouse Audubon, who had largely taken over from his
father in 1846. The continuing carelessness of Victor once more got
under the skin of Bachman, however, and on February 10, 1847, he
groused about "the helter skelter manner in which you read my let-

ters." He said, "I have written you page after page and letter after letter," giving names of species, but "scarcely is the work done before I receive another letter asking me once more for the identical names." Furthermore, although he had told Victor to omit the names of troublesome species until he could obtain better information, Bachman complained, "you repeat the inquiries and are ready to publish directly contrary to my express injunction." Victor's actions, moaned Bachman, were "almost enough to make a man strike his uncle." Nevertheless, the exasperated naturalist pushed on without pause until the fall, when he had to take his terminally ill daughter Julia to the health resort in Virginia. At least he had restrained Victor from getting the second volume into print, and on November 17, Victor asked, "Do you not think it would be best to *begin* to print the Second Volume *now*?" Before he received the letter, Bachman penned one to Victor, telling him "we are doing nothing—nothing to advance science in the great book" and that he must have specimens of some of the lesser-known mammals. He had hoped that John Woodhouse would be able to furnish some information, for the younger Audubon son had been abroad during the year to draw illustrations from specimens in the museums of London and Berlin. Alas, although he was talented with the brush, John Woodhouse knew even less about mammals than did his brother. Bachman had expressed fear that John was "intent only on making drawings without making or sending any notes in regard to the species." Indeed, although the younger Audubon brother did make some notes, he later told Bachman that the European scientific journals "are so voluminous, that for so poor a reader as I am . . . [the task] would have taken me months."[17]

Of course, Bachman had not expected John Woodhouse to copy everything, though likely he viewed this occasion as a unique opportunity to secure every bit of the much-needed information. The greater problem, however, was young Audubon's inability as a mammalogist. "You suppose me to have far more knowledge of the smaller animals than I can boast of myself," apologized the meek and sweet-spirited artist, adding, "I never thought I should have to become in any degree [a] naturalist as well as an artist." That he was indeed no naturalist was already clear to Bachman, for John Woodhouse had speculated that "the *Ixalus probator* . . . [is] a hybrid . . . [whose] father was a Fallow deer and its mother a female antelope." He was probably referring to the Rocky Mountain goat, which had been originally described as an antelope. In any case, his notion of such a hybrid no doubt pained Bachman.[18]

Exasperated though he was, Bachman never lost his affection for either of the widowers of his daughters, and he readily accepted the new wife of each. During all those years of strain, some of the Bachman clan visited with the Audubons, and daughter Jane stayed with them for many months while seeking medical help for her acute eye problem. Moreover, the former father-in-law shared his grief with them as though they were his own sons. In a letter addressed to both Victor and John Woodhouse, on October 28, 1847, shortly after the death of Julia, he lamented that the present illness of daughter Lynch worried him much, as he feared she might also be infected with tuberculosis: "I dread . . . the disease from which our family has suffered so much. . . . One by one my children have been attacked and swept to the grave." He was not unaware that the Audubon sons and their mother Lucy Audubon were then facing a loss of their own—not the death of the great Audubon but something closely akin to it: the deterioration of his mind. Bachman bade them to tell "my old Friend" that he wished God's blessing upon him.[19]

Meanwhile, Bachman struggled to carry out his personal and ministerial duties and to write descriptions for the second volume of *Quadrupeds*. He could not claim to know much about fossils, and it seems clear that by 1847 he no longer entertained the thought of including fossil mammals in the work. In fact, given the difficulties of the enterprise, he had conceded that he must also exclude the whales and dolphins and the manatee, and he was beginning to think it unlikely that he could write descriptions of the bats and the seals, though he knew quite a bit about the former.

Many matters delayed the *Quadrupeds* project, though some were of Bachman's own making. In fact, at least one of them was unnecessary and wasteful of his time, but to Bachman it represented another challenge to the ideal of fairness. In late 1847, Robert Gibbes publicly denounced the claim of the German traveler Albert Koch, who was then touring South Carolina with a special "exhibit"—the skeleton of an extremely long saurian fossil, which was in fact a composite of the bones of several animals collected from different sites. However, the case against Koch had not been fully proven in November 1847, when Gibbes attacked the pseudo-paleontologist in the pages of the *Charles-*

ton Mercury. Bachman felt he could not allow the charge to pass and wrote a letter to the *Mercury*, saying, "it is the duty of the friends of science at least to suspend their judgment before they pronounce the sentence of condemnation against Dr. Koch." In addition, he maintained that Gibbes had attacked Koch in part merely because he was a foreigner. Gibbes might have let the criticism pass had Bachman left the matter there, but the champion of "Justice," as Bachman signed his name, pointed out errors made but later corrected by Gibbes. Koch was entitled to the same opportunity, he said. The unrelenting mammalogist had not finished, however, and went on to note that Gibbes had attributed the find of a fossil whale skull near Charleston solely to Michael Tuomey, whereas Francis S. Holmes and Lewis R. Gibbes had made the discovery.[20]

Chagrined and incensed, Gibbes replied that he had not intentionally omitted the names of the real discoverers, and he denied that he had criticized Koch because he was a foreigner. Privately, he wrote to S. G. Morton, however, that "Dr. Bachman's German blood is up, because Koch is a German." Gibbes was especially upset that this controversy should occur on the eve of the visit of the great Louis Agassiz to South Carolina. On December 21, after several letters had appeared in the *Mercury*, Bachman suggested dropping the matter, for he had no desire to continue with it and because "there is just now a general jubilee among the naturalists of Charleston" with the imminent arrival of Agassiz. He called for "a harmony of feeling" and referred to Gibbes as "an associate and fellow-laborer." No doubt Bachman had overreacted, but he had made some valid criticisms.[21]

As with all of the Charleston naturalists, Bachman looked forward to the visit of Agassiz, for the presence of the distinguished naturalist in the city suggested that Charleston was equal to Boston and Philadelphia as a center of scientific inquiry. Moreover, it offered an opportunity for Bachman to have Agassiz for "a long conversation in my study." He even told Victor that he valued the opinion of Agassiz more than that of anyone else in America. The visit took place as expected, but Bachman came away from it with a lower opinion of Agassiz, for he had found that the renowned naturalist possessed relatively little knowledge of the mammals. Within two years, he would find himself standing alone among the Charleston scientists and medical doctors when he challenged Agassiz's views that human races constitute separate creations.[22]

The tribulations of 1847 had not weakened the resolve of John Bachman, and as he moved into the new year, he continued to push forward with the second volume of the *Quadrupeds*. Among the specimens sent to him by U.S. Army Lieutenant John J. Abert was a white hare that Bachman believed to be a new species. He would not attempt to describe it as new, however, for he did not have its skull and was aware that its pelage color might be seasonal. "It will be a hundred years," he opined, "before our Hares are settled and plenty of blunders we will yet make between old and young, winter and summer specimens." Meanwhile, Bachman had prepared a list of the mammals of Georgia, and, early in 1848, he completed a paper on the reproductive system of the Virginia opossum, *Didelphis virginiana*, a species that had intrigued him for over three decades. During the late winter and spring in 1846, 1847, and 1848, he obtained numerous females of the species for the purpose of observing the stages of embryological development, and when he finished a paper on the subject, he sent it to the ANSP. Published in the Academy's proceedings in April 1848, the paper was an excellent piece of scientific inquiry. Bachman sacrificed some of the animals in order to study embryos *in utero*. He also observed some of them giving birth and watched the embryos crawl up the mother's abdomen. Since he did not see the tiny, naked embryos enter the marsupium, however, he maintained that the mother must have shoved them into it and then helped each to fix on a teat. Bachman also removed some embryos from teats to see if they would reattach themselves, and he exchanged embryos between mothers to see if they would be accepted. He observed success in both instances. In addition, Bachman carefully examined the dissected females and found no placenta. As an authority on *Didelphis virginiana* noted a century later, Bachman "made astute observations on the opossum and . . . brought up to date the existing knowledge about that animal. . . . Seventy years passed before the problem of marsupial birth was again taken up."[23]

The trustees of the College of Charleston had been taking note of the scientific standing of Bachman, and in April 1848 they invited him to serve as professor of natural history. Granted approval by the vestry of his congregation, Bachman accepted. Along with all of his other duties, he discharged those of his post at the college for five years. From the standpoint of knowledge of his subject and enthusiasm for imparting it to others, Bachman was an excellent choice. From the standpoint of active participation in faculty affairs and of maintaining discipline, however, the choice left something to be desired. The faculty minutes show that the busy clergyman rarely attended faculty meetings, and the minutes and other accounts indicate that, while rigorously disciplined himself, the professor set no examples in the behavior of his charges. For instance, when the young men, by prior agreement, pushed together at once to force the last, hapless student off a school bench, Bachman seemed perplexed. Other cases of disorder increased his reputation among the faculty as an ineffective disciplinarian. In May 1850, a student stole his horse and buggy and drove it around the city, but Bachman was only momentarily incensed over the inconvenience. When the faculty sentenced the young man to two weeks of suspension, Bachman made "an earnest request that the punishment be mitigated or if possible entirely withheld." The faculty stuck to the original sentence, however.[24]

Meanwhile, in May 1848, before he began teaching at the college, Bachman had traveled to New York and there visited his old friend Audubon. He found Audubon's "noble mind all in ruins" and noted that "he is like a crabbed, restless, uncontrollable child." Those and other indications suggest that Audubon was suffering from what later came to be called Alzheimer's disease, but mild strokes or another form of mental impairment may have caused the changes in his mind and behavior. In any case, the visit was an agonizing experience for Bachman, and he wrote to his sister-in-law Maria Martin, "I turn away . . . with a feeling of indescribable sadness." Bachman continued the work he had begun with the great artist, but during the latter part of 1848, his eyes bothered him so much that he had to lay it aside for awhile. The nature of the ailment is unknown, but certainly a frightful incident that had occurred five months earlier could have been the cause. In December 1847, Bachman was sitting near the fireplace in his home when a young slave set a concoction of lard and gunpowder, intended for treating mange on Bachman's dogs, on the fire to melt the mixture. It exploded in Bachman's face, and only his spectacles saved his eyes from severe damage. In any case, by the fall of 1848 his eyes were "running water all the time," and for several weeks he had to dictate letters and descriptions to Maria. Victor expressed regret, but impatiently urged him to work "now!" Said Bachman in reply, "Your Father began the work before he or I was ready. . . . I have imperative duties. My life is worth something, to my children at least." He added that he could not risk permanent damage to his eyes "to oblige even you, for whom I will do more than for any other human being. . . . Whilst you are thinking of yourself, think also of me." His eyes were still bothering him by December, but he told Victor he would finish the work, as Maria had promised to help him. Then, he announced that "Maria has been weak enough to consent to take the old man . . . for better or worse." On December 28, 1848, the talented and self-effacing woman who had given herself to the Bachman family for two decades became a member of the family in name when she married the man whom she had diligently assisted in his rise to scientific fame.[25]

Maria Bachman continued to aid her new husband and to encourage him when he lost heart in his effort to complete the second volume of the *Quadrupeds.* As Bachman told Victor in June 1849, in reviewing their work, Maria "corrects, criticizes, abuses, and praises us by turns." But, fondly added the new husband, "she does wonders." He expressed concern, however, that the waters of Madison Springs, Georgia, to which he had taken Maria because her health seemed to be declining, were doing no good. Maria had often suffered from migraine headaches, and now she was experiencing chronic fatigue. Bachman himself seemed to be improving, however, and he informed Victor that he was working twelve hours each day on the descriptions. Once more Victor was drafting some parts for Bachman, and the caustic old critic admitted that "some of them are far better than they formerly appeared, [but] a few are lame enough." The latter he had rejected and "prepared new ones." Bachman lamented the decision to omit the bats, but, he told Victor, "I found you were determined on giving an imperfect work, [and] I could not help myself and was silent and sulky." Nevertheless, he expressed hope that he might at least offer descriptions of the bats in the third volume even if they were not figured. He was determined, moreover, to include "the rascally seals," for it would be "folly" to exclude them. "We might as well," he said, "omit the Deer or the Bear." Weary from his labors and aware of the difficulty in describing the numerous bats and of the paucity of information on the seals, Bachman eventually gave up on doing them, as he had earlier done on the cetaceans.[26]

Gaps in the correspondence between Bachman and Victor in 1850 and part of 1851 make it impossible to know the precise development of the project, but on August 25, 1851, Victor informed Bachman that the second volume had been published. Meanwhile, in March 1850, at the third meeting of the American Association for the Advancement of Science (AAAS), held in Charleston, Bachman, like his fellow naturalists, had played a prominent role. He presented a paper titled "On the American species of the Genus Putorius," or on the skunks in the United States. Involved in many other matters, including a controversy over the question of separate species of humans, Bachman never found time to prepare the paper for publication in the AAAS *Proceedings*. He had been interested in the skunks and other mustelids since he was a boy, but it is unknown, of course, whether he offered anything new. Of the three genera of skunks in the United States (*Mephitis*, *Spilogale*, and *Conepatus*) only one species, *Spilogale putorius*, bears the epithet in common use at the time. Bachman also believed he had discovered among specimens sent to him a new species of fox, which he described as *Vulpes utah*. He published the description in the ANSP *Proceedings*, with the name of the by-then mentally incompetent Audubon listed first as coauthor. The claim was countered by Baird in the very next issue of the ANSP *Proceedings*, for Baird had already described it as *Vulpes macrourus*, which, as it turned out, is a synonym for *V. vulpes*. In any case, his report indicates that Bachman was still keeping his name in secondary position.[27]

As he labored on with the third volume of the text of the *Quadrupeds*, Bachman grew more concerned over the declining health of Maria and over the fatigue he was himself suffering because of overwork. He decided, therefore, to find respite in travel. In mid-May 1851, he and Maria departed for New York, and after spending ten days at the Audubons' home, "Minnie's Land," they went by railway to Detroit, where, Maria wrote, they found little of interest "in the midst of abolition." Encountering many black servants in the busy city, Maria referred to them as "genteel *niggers*" who could never attain "higher stations than Barbers and waiters." In her view, the free blacks of Detroit had reached the highest level of which they were capable. Blinded by bias and angered by abolitionists, Maria Bachman was happy to return to her home.[28]

Back in Charleston in July, Bachman soon began to receive proofs of the first part of the third volume. He was unable to complete the work, however, until early in 1852, and then only after Victor came to Charleston to "hold the pen whilst I dictated with specimens and books before me." He informed Edward Harris on March 13, 1852, that "we worked hard, and now we are at the end of our labours." The volume appeared in 1853. In the meantime, Victor had been working on an octavo edition that combined plates and text, Volume I of which was published in 1849 and Volume II in 1851; the third appeared in 1854. No correspondence has been found to show why Bachman or Victor Audubon or both decided to drop the adjective "Viviparous" from the title, but the later volumes appeared as *The Quadrupeds of North America*. Its omission was meaningless in any case, for the only mammals to which it does not apply are two living species of monotremes in Australia and one in New Guinea. Soon after completing the final folio volume, Victor set out to seek subscribers to the new edition, and Bachman informed a friend in Savannah that "the figures were made by the Audubons and the descriptions and letter press were prepared by myself." He added, "I have no pecuniary interest in this work, as I have cheerfully given my labors . . . [with] the hope of having contributed something towards the advancement of . . . Natural History in our country."[29]

Prior to the publication of *The Quadrupeds of North America*, relatively little work had been done on the mammals of the United States, at least not as systematic treatments of all of the mammals in the country. The efforts of Richard Harlan, in his *Fauna Americana*, published in 1825, and John Godman, in his *American Natural History*, published in three volumes between 1826 and 1828, had fallen short. Harlan's volume was essentially a compilation that drew from the works of others, while Godman's volumes were more comprehensive, based on some original studies by the author himself, and nicely illustrated—but much in the tradition of the old-style, romantic naturalist. Thus, the work by Bachman represented a truly significant contribution to the field of North American mammalogy, though, as Bachman would realize, his work was overshadowed by the reputation of Audubon. Much later, in reference to biographies of Audubon, Bachman lamented that he had hardly been noticed as coauthor of *Quadrupeds*, even though he had "written every line." Champions of Audubon may view Bachman's lament as sour grapes and his claim as exaggerated. Quite the contrary is true. Bachman made *Quadrupeds* a reality and elevated it to high standing by his admirable labors. Moreover, he did in effect write every line of the descriptions, without which the work would have been only a compilation of illustrations, and he edited the extracts from Audubon's journals, making them suitable for the text. In addition, he forfeited claim to royalties and allowed Audubon's name to go first. His lament was justified, his claim warranted.[30]

By the time the work was completed, however, Bachman had been recognized many times over for his accomplishments. Pennsylvania College of Gettysburg (later Gettysburg College) had bestowed the honorary degree of Doctor of Divinity upon him in 1835. His daughter Catherine later claimed that he was awarded an honorary Ph.D. degree while he was in Berlin in 1838 (presumably by the University of Berlin), but no evidence has been found to support the claim. South Carolina College had invited him to become its president in November 1841, but he had declined because "the state of my health and the fatigues attendant on the duties would soon finish me." Nine years later, that institution conferred an honorary doctorate upon him for his "meritorious service to science." But events of the 1850s quickly turned Bachman's attention to other concerns. The growing issue of slavery and states rights, debates over whether human races were a single or separate species, and quarrels over the character of Martin Luther—all of these issues kept Bachman so deeply engaged that he had little time to devote to the study of mammals after completion of the *Quadrupeds* volumes.[31]

Treasures of Earth and Sea

As the morning sun broke over the distant horizon on a mild summer morning in 1820, five years after Bachman arrived in Charleston, a slender young man strolled along the shore of nearby Sullivan's Island, stooping frequently to pick up mollusk shells. During the previous year he had completed his medical studies at the University of Pennsylvania and returned to his home in Charleston. A descendant of one of the many Huguenots who had fled France upon the Revocation of the Edict of Nantes in 1685, he was Edmund Ravenel.[1] While he maintained a home and a medical practice in Charleston, Ravenel fell in love with the island and purchased a home there in 1823. He was attracted to the place for several reasons, chief of which were the mollusk shells lying on its shores and the warm waters that lapped upon its beaches and, so he believed, stimulated the health of anyone who bathed in them. In addition, the naturalist-at-heart was fascinated by the fishes that thrive in the waters surrounding the island. Above all, the traces of God in nature were clearly discernible there. Ravenel could think of no better place to serve as his second home, and so, again and again over the next forty-five years, he came to Sullivan's Island whenever he could.

The sixth of ten children born to Daniel Ravenel and Catherine Cordes Prioleau, Edmund had come into the world on December 8, 1797. His father owned a plantation called Wantoot, located several miles north of Charleston, but had moved to Charleston before the birth of his sixth child. Financially successful, Daniel Ravenel also owned the plantations known as Somerton and Hog Swamp as well as property in Charleston. When he died in 1807, he left to his heirs three plantations, sixteen slaves, a mansion at Wantoot, and £7,600. Thus, his nine surviving children were assured of the means of obtaining a good education. Edmund received the 1,295-acre Hog Swamp plantation and £1,000. Sons James and Edmund eventually decided to pursue careers in medicine. James, seven years the senior of Edmund, completed the M.D. degree in the medical department of the University of Pennsylvania, and Edmund followed in his steps. The other brothers—Daniel, John, Henry, and William—turned to business. Daniel became a very successful banker in Charleston and eventually one of that city's most prominent promoters of educational and cultural affairs. Later, however, Edmund would himself show that he not only had a good head for matters of business but also a talent for serving his fellow citizens. Certainly, he had enough money by 1820 to set up a medical practice in Charleston and to reside on Sullivan's Island during the summers.[2]

On April 16, 1823, Edmund Ravenel married Charlotte Matilda Ford, eldest daughter of Timothy Ford. A native of New Jersey and a graduate of Princeton College, Timothy Ford had studied law and been admitted to the bar in New York. He moved to Charleston in 1785, however, and became a successful attorney. At various times he had served as a member of the city council, the South Carolina legislature, the board of trustees of the College of Charleston, and the Charleston Library Society. Moreover, in 1822, he served on the town council of Moultrieville, on Sullivan's Island, which suggests that, like several other Charlestonians, he owned property there and used the island as a retreat. Ravenel was probably attracted to Ford because of the latter's interest in science, especially chemistry, which was a favorite subject of young Ravenel. Ford had been a charter member of the Literary and Philosophical Society, and in 1818 he presented a lecture before its members entitled "the Science of Chemistry." Ravenel also joined the Literary and Philosophical Society, and was one of the twenty-eight who held a life membership. In addition, he later served for a year as a curator of the Society's museum. In his will, Ford, who died in 1830, named his son-in-law as administrator of his considerable estate.[3]

Meanwhile, Ravenel had begun to study fishes, sharing a mutual interest with Stephen Elliott, who appears to have served as the young doctor's mentor in natural history. It is not known when Ravenel first met Bachman, though it was probably not before 1820. In a journal entry for 1821, Ravenel noted that he, Bachman, and Elliott examined together a specimen of the porbeagle shark, *Lamna nasus*, caught off the bar of Charleston Harbor. The entry continued for several pages and included excellent, detailed descriptions of the internal anatomy of the animal and a pen-and-ink drawing of the shark. Ravenel also recorded his observations of several other fishes, including sketches and morphological descriptions that he apparently intended to use in writing papers for publication. In 1822, he completed a description of *Chironectes ocellatus* (= *Antennarius ocellatus*), the ocellated frogfish, and submitted it for publication in the *Proceedings of the Academy of Natural Sciences of Philadelphia*. Unfortunately for the budding naturalist, the species had already been described in 1801, which Ravenel did not know, for he had limited access to the literature in ichthyology. The same would sometimes prove to be the case in the field of conchology as well, and later Ravenel complained, as had Bachman, of the inadequacy of libraries in Charleston.[4]

Perhaps this failed foray into ichthyology discouraged Ravenel from pursuing that subject further, or perhaps he simply became more interested in mollusks. Certainly, he did not cease to be interested in fishes, and, in fact, as late as 1830 he was sending specimens to Georges Cuvier and Achille Valenciennes in France. In some cases, he was evidently a joint collector with his friend John Edwards Holbrook, who had been a classmate in the Department of Medicine of the University of Pennsylvania. Valenciennes even named a species in honor of Ravenel, namely *Pomotis ravenelii*. Unfortunately for Ravenel, the fish had already been described by Linnaeus as *Perca gibbosa* (= *Lepomis gibbosus*). But the would-be ichthyologist could take pleasure in seeing his name in the renowned work of the great French naturalists, and he could delight in knowing that he had provided them with a new species, *Bryttus punctatus* (= *Lepomis punctatus*), described by Valenciennes in 1831. Meanwhile, by 1821 he had already begun to record in his journal some observations on mollusk shells and, in a few cases, on the nature of the animals living in them.[5]

Ravenel set up his medical office in Charleston on Tradd Street and began to see patients around the year 1820. As the highest ranked graduate of his class in medical school, he possessed a strong inclination toward scholarly work. Thus, when he learned of the plan of the Medical Society of South Carolina to establish a medical college in Charleston in 1824, he applied for, and was elected to, the chair of chemistry and pharmacy. Three months later, Holbrook was also elected to the faculty. The Medical College of South Carolina was the only medical school in South Carolina; indeed, it was one of only a few in the South at the time. As with other contemporary medical schools in the country, the professors did not receive a salary for their service but relied instead upon student fees. Ravenel assessed each student $20 a term for attending his lectures, which ran for a period of only sixteen weeks, the standard of the day. In addition, like other professors, he also continued his private practice.[6]

Meanwhile, on April 16, 1823, Ravenel had married Charlotte Ford. The happy marriage was blessed by the birth of a daughter, Mary Louisa, on October 8, 1826, but it came to a sudden, heartbreaking end only two days later, when his thirty-year-old wife died. Ravenel had previously experienced the sorrow of losing someone dear to him. His father had died when Edmund was only nine years of age, and his beloved brother James had died in 1817, at the age of twenty-seven. James's widow was left with only one of the three children born to them, the first two having died young. But the loss of his own companion threw Edmund Ravenel into a state of severe depression, and, after struggling unsuccessfully for six months to recover, he decided to travel to England in order to relieve some of his grief. Louisa Catherine Ford, the twenty-two-year-old half-sister of Charlotte, readily agreed to take care of the baby. On May 1, 1827, in the only surviving letter from Ravenel to Louisa during his sojourn abroad, the distraught widower referred to his "gloomy fit" of mind and of his "waking dreams" of home and the baby. He told Louisa that she was "the friend with whom I can converse most freely." Generally impressed by the English countryside, he spoke of its lovely cottages "covered with clinging ivy . . . and [the] tastefully laid out grounds of the Gentry," but the buildings in the city of Manchester struck him as "very much smoked and shabby in appearance."[7]

After his return to Charleston, Ravenel and Louisa fell in love, and they were married in 1829. Louisa gave birth to seven daughters and a son between 1830 and 1844, but, as with Bachman's family, tragedy struck often. The first daughter of their union, named for Edmund's first wife and Louisa's half-sister, was healthy and lived a long life. Their second daughter lived less than four months, however, and their third and fourth daughters, each named Theodosia, died within a few months of their births. Emma, the fifth daughter enjoyed a long life, but the sixth daughter, also named Theodosia, drowned in 1848, at the age of eleven. A son, named Edmund, born in 1840, and Caroline, born three years later, were the last of the Ravenel children.[8]

Despite those wrenching losses, Edmund Ravenel moved forward with his work as the professor of chemistry and pharmacy in the Medical College and was eventually selected to serve as the dean of the institution. While he enjoyed the esteem of both his students and his colleagues, Ravenel encountered difficulty with the trustees. The peculiar arrangement whereby the Medical Society of South Carolina governed the Medical College of South Carolina led to conflicts with the faculty over who had authority to set fees and to make faculty appointments. The Society's trustees grew increasingly hostile toward the faculty because they not only thought the Society should reap more profit from the college but also objected to the faculty's rejection of a number of their candidates for vacant chairs. In September 1831, the normally mild-mannered Ravenel, then serving as dean, openly objected to the decision of the Medical Society's trustees to place their own candidate in the vacant chair of surgery. At the next monthly meeting of the Society, his colleagues joined in the protest. Then, at the December meeting, the faculty notified the Society that they intended to petition the state legislature for a charter to establish another medical school in Charleston and to prohibit the appointment of trustees who did not possess the M.D. degree. The Medical Society issued a caustic rejoinder, referring to the "high-handed measure" of the faculty and the desire of the professors for "power and no responsibility" in appointing new faculty members. Indeed, the Medical Society declared that, despite a warning, the professors were not attending to their duties in the hospital and in giving clinical instruction. The state legislature sought a compromise, but eventually granted a charter for another medical college.[9]

The controversy in 1833 sharpened as the Medical Society sought to forbid the professors from offering external lectures on medicine, arguing that the faculty had no time for such and that they were shunning their official duties. Ravenel received a letter of reprimand from the vice president of the Medical Society. Incensed, he replied that, while the Society had appointed him to his post, he had never agreed to "bind" himself to the Society in any other way. Now, after many years of offering external lectures on medicine, the Society could not, he averred, pass an ex post facto rule. He had given the Society "no power," he exclaimed, "to control or regulate my conduct in any other respect, [and you have] no authority otherwise to interfere with me in any manner whatever." Other faculty members likewise responded with strong challenges to the authority of the Medical Society. Also at the heart of the matter was the faculty's disdain for the low standards of the Medical Society, a point made by Ravenel as early as December 1822, only a month after he joined the organization. He had complained then that members rarely discussed medical topics at the Society's meetings and that they had been unable to publish any transactions of the few papers read because they were so poor in quality. In short, said Ravenel, they had been apathetic in advancing medical work. Now he had a more serious complaint.[10]

Soon thereafter, the Medical Society declared all of the chairs vacant in the Medical College, claiming that the professors had accepted appointments in the newly chartered institution, named the Medical College of the State of South Carolina, but the controversy dragged on into the fall of 1834. On October 30 of that year the Medical Society informed Dean Ravenel that because the faculty "have violated their obligations," they must appear before the Society "for trial" on the following Wednesday. Because the Society had made the issue public, the highly regarded professor Samuel Henry Dickson decided to publish a rejoinder, in which he noted that the Society had become envious of the success of the faculty but wanted to exercise control without contributing any funds to the college. He refuted other claims of the Society and declared his resignation. The other professors also resigned and openly acknowledged their association with the new medical school. Now, Charleston had *two* medical schools. The best professors were at the new one, however, and by 1839 the original medical school had faded into oblivion. Ravenel never attended another meeting of the Society thereafter. The lengthy fight had taken a toll on him, however, and in March 1835, he announced his decision to resign later in the year. A resolution of the board of trustees expressed praise for "the zeal and labor" of Ravenel and referred to "his infirm health." No doubt Ravenel also wanted a change of scenery, for between June 1832 and August 1834, three of his infant daughters had died. Louisa had given birth in January 1835 to her fifth and Edmund's sixth child. Perhaps the couple thought that an estate in the country would be a good place to rear nine-year-old Mary Louisa, five-year-old Charlotte, and the baby Emma.[11]

Moreover, it appears that Ravenel wanted to devote more time to the pursuit of natural history, especially to conchology. The life of a planter, with slaves to do the hard work, would facilitate the efforts of Ravenel. Since the planting of rice requires substantial land, Ravenel had to look for a place large enough for the endeavor. But he desired a location not too far from Charleston and Sullivan's Island. He found what he wanted, a plantation on the Cooper River, about seventeen miles north of Charleston. After negotiating with the owner of "The Grove" throughout February 1835, Ravenel succeeded in striking a bargain. Unfortunately, the owner died soon after the verbal agreement, forcing Ravenel to negotiate with his heirs and placing him in an awkward position, for he had moved to the plantation immediately, before a written contract had been drafted. In May 1835, for $48,200, nearly $2,000 more than the original price, Ravenel completed an agreement with the heirs and thus got his plantation, which covered 3,364 acres of ground. The deal also included 104 slaves. The purchase was costly, however, and Ravenel had to borrow money to complete the transaction. Soon after his purchase of the Grove, he obtained a brick-making works, which enhanced his income, and later he established a timbering business. Since he had been acquiring rental houses in Charleston, he believed he could repay the loans in due course.[12]

It is clear that Ravenel had collected a large number of mollusk shells by 1829 and that he was busily engaged in writing about them. By then he was also corresponding with many other conchologists, both at home and abroad. Among his correspondents were the pioneering American conchologists Thomas Say, Isaac Lea, Timothy Conrad, and Amos Binney. His record book of exchanges, dating from 1829, and the extant correspondence for 1831–1839 reveal that Ravenel had found a field of natural history to which he was ideally suited, and in 1834 he published a catalogue of the shells in his cabinet, which included Recent and fossil and freshwater, terrestrial, and marine species. Ravenel's catalogue of shells consisted of only twenty-one pages, but it listed almost 750 species of Recent and fossil mollusks, of which nearly 160 were from the southern region. Ravenel apparently intended to produce an illustrated work on the shells of South Carolina, and it appears that he commissioned the able artist J. Sera to prepare the figures. In his manuscript journal, Ravenel referred to "33 Plates, [of] about 150 species," and the Sera drawings are dated either 1832 or 1833. The work was never completed, but Ravenel's catalogue proved to be popular with conchologists. It is also the source of the description of a new species of gastropod, namely the lettered olive, *Oliva sayana*, a beautiful, glossy, cylindrical shell of grayish background encircled by brownish-purple zigzag bands. The catalogue's description of *O. sayana* is quite brief, but modern malacologists give credit to Ravenel for first publishing a notice of its distinguishing characteristics.[13]

Ravenel was aware of his deficiencies in the taxonomy of mollusks, and for many years that reality seems to have kept him from trying to publish descriptions of new species. He did list several as *nova species* in his catalogue, but, unknown to him, most of them had been described previously. In large measure, his knowledge was limited because of the relative paucity of pertinent books and journals. In a letter to Binney in 1830, Ravenel said, "I may have erred in naming some of the Species [I sent to you] as my command of books is limited." Later, in 1834, after receiving a copy of Ravenel's catalogue, Binney urged Ravenel to submit a lengthy paper for publication in the *Journal of the Boston Society of Natural History*, but at that time Ravenel had just come away from the medical college controversy and was negotiating for the Grove. Moreover, he apparently did not think that he was ready to produce an article describing new species. The collecting of shells may have been merely a hobby for Ravenel in the beginning, as it was, and remains, for others who admire the beauty of those constructions of calcium carbonate and pigment.[14]

Like his contemporaries, Ravenel viewed mollusk shells as evidence of the hand of God in nature. In fact, he probably shared the view of many of his fellow conchologists that mollusk shells were specially important as representations of God's creative power. But Ravenel was far more than a collector of shells. He was also a dedicated naturalist, and his efforts to show natural relationships and to ascertain genuine species as opposed to varieties is evident throughout his writings. Like most of his peers, however, he was a conchologist, in the original sense of emphasis on the shell, rather than a malacologist, in the later sense

Edmund Ravenel. Courtesy of Waring Historical Library, Charleston, S.C.

that stressed the animal that secreted the shell. In any case, his task, as he viewed it, was to identify species and make sense of their natural relationships. As Neal Gillespie has noted, most contemporary conchologists tended to ignore the animals within the shell and based their classification upon the shell itself, which, because of intraspecific variation, often presented them with the dilemma of whether or not a variation was indicative of a new species. That phenomenon proved to be especially troublesome in the taxonomy of mollusks and led some conchologists to become "splitters," that is, naturalists who were inclined to make new species of subspecies or forms of a previously described mollusk. It also suggested the possibility of transmutation, but that idea found no favor in general. Ravenel wrestled with the problems of intraspecific variation as he tried to classify various gastropods and bivalves, but he was unsympathetic toward the splitters and, as far as extant evidence indicates, rejected the Lamarckian theory of evolution. After the Civil War, he discussed Darwin's theory of natural selection with John McCrady, but his views are not indicated in the account recorded by McCrady in his diary.[15]

Among those who soon came to recognize Ravenel as an authority on southern shells were two members of the Academy of Natural Sciences of Philadelphia. One was Isaac Lea, an expert on freshwater mollusks. Lea was very particular; he mainly wanted to obtain "fluviate shells" and "Uniones" from Ravenel, though he was quite willing to help Ravenel in identifying marine species. In fact, in 1834 Ravenel had first suggested to Lea that *Oliva sayana* was a new species, and Lea had agreed with him. Moreover, while Lea was encouraging his southern colleague to proceed with his work on the "treasures of the deep," he was often asking Ravenel to send him more shells. The other ANSP member was Timothy Conrad, a geologist and pioneer in American invertebrate paleontology. Especially interested in fossil mollusks, Conrad cultivated the friendship of Ravenel, for he viewed him not only as a conchologist but also as a paleontologist. In 1839, Conrad visited with Ravenel, who introduced him to the geology of the region. For his service, Conrad named a freshwater shell in honor of Ravenel. Thomas Say also sought information and specimens from Ravenel, and like Lea and Conrad, he encouraged Ravenel's interest in invertebrate fossils.[16]

Nothing is known about the origins of Ravenel's interest in fossils, but a comment that Ravenel made in a letter to Samuel G. Morton on February 13, 1833, makes clear that he had been collecting them for his friend and fellow Charleston naturalist Stephen Elliott for many years, and Elliott sent them on to Say. In any case, by the mid-1830s Ravenel was probably the most knowledgeable person in the South on invertebrate fossils, and by 1837 he was thinking of publishing some of his finds, which were being uncovered as his slaves cut roadbeds and dug for marl to use as fertilizer. Ravenel turned to Morton and the Academy of Natural Sciences of Philadelphia for support. At the behest of Morton, the Academy had elected Ravenel to corresponding membership in 1832. Morton, a prodigious collector of crania, especially human ones, had first called upon Ravenel to fill his request for an alligator skull, and later he asked for fossils. In 1837, Ravenel wrote a lengthy letter to Morton about two species of fossil echinoderms he had found, and he sketched a figure of each at the top of the letter. Ravenel hoped that Morton could tell him whether or not the species had been described already. Admitting his ignorance, he said, "my library is so deficient in these matters that I am obliged to be very cautious in forming an opinion and in consequence my recent and fossil shells have heretofore been described by others."[17]

For some reason, however, Ravenel did not submit a manuscript on the fossils until July 1841. Published a few months later in the *Proceedings of the* ANSP, the paper described the fossil echinoids *Scutella caroliniana* (= *Mellita caroliniana*) and *Scutella macrophora* (= *Encope macrophora*). It appeared again in 1842 in the *Journal of the* ANSP, accompanied by sketches of the two species. Two years later, encouraged by this successful effort, Ravenel published another paper on four fossils (two mollusks, a brachiopod, and an echinoderm) in the *Proceedings* and donated the specimens to the Academy's museum. He named one of the bivalves *Pecten mortoni* in honor of "our distinguished geologist Dr. Samuel George Morton," and the other, *Pecten holbrooki*, in honor of "our learned herpetologist Prof. Holbrook." More important to him than Charleston's now virtually neglected museum and the waning Literary and Philosophical Society was the esteem of Morton and the ANSP. Until Louis Agassiz came to Charleston in 1847 and several times thereafter, and encouraged rejuvenation of the Charleston Museum and the formation of a new scientific society, Ravenel and all the other Charleston naturalists considered Morton and the ANSP to be their closest allies in the pursuit of knowledge about the natural history of their region. By 1849, however, Bachman had begun to change his opinion of Morton because of the latter's contention that the human races were separate species. The other Charleston naturalists continued to view Morton as a superb naturalist, however, and when the Philadelphian died in 1851, all of them believed they had lost a great friend and co-laborer.[18]

Ravenel's main interest now seemed to be in locating and describing fossils, especially of the class Echinoidea. Perhaps his interest was whetted by the visit of the English geologist Charles Lyell in January 1842. Lyell specifically sought out Ravenel to serve as a guide. They journeyed northward up the Cooper River and along the Santee Canal to observe the geology of the region and to search for fossils. During their week's sojourn they found a number of specimens. Although Ravenel maintained an active interest in paleontology on into the 1850s, he published no further papers on fossils. Certainly, he was still uncovering fossils after 1845, and in 1848 he published a list of Recent and fossil echinoids in his cabinet. By then, however, Robert W. Gibbes, Francis S. Holmes, and Michael Tuomey had become the paramount paleontologists in the South. An entirely self-made naturalist and paleontologist, Holmes had also developed a fascination over the fossils he uncovered in marl pits on his plantation near Charleston, and he soon amassed a large collection. It was Holmes's good fortune that the State of South Carolina, in ordering a geological survey of the state, had employed the Irish immigrant and superb geologist Tuomey to direct the survey. In addition, R. W. Gibbes had begun to publish some excellent descriptions of vertebrate fossils in South Carolina. When Tuomey and Holmes teamed up later, however, they became the regional authorities on the invertebrate fossils of South Carolina and inadvertently pushed Gibbes and Ravenel aside. Moreover, of course, Gibbes lost favor with Bachman in 1847 over the Koch incident. Ravenel turned back to conchology, and eventually gave many of his fossil specimens to the ANSP and, later, to the Smithsonian Institution, the Elliott Society, and the Charleston Museum.[19]

During the decade he had devoted mostly to paleontology, Ravenel had continued to build his cabinet of shells. Aside from letters on the subject and exchanges of shells, however, he did not make a serious effort to publish on Recent species during the two decades from around 1837 to 1857. In part, Ravenel may have curtailed some of his writing after 1845 because of problems with his eyes. In fact, at one point, upon the advice of physicians, he confined himself to a darkened room for several weeks, believing that measure would halt the "opacity of the cornea" in each eye. That disorder does not account fully for the hiatus, however. Ravenel's friend Lewis R. Gibbes, a cousin of Robert W. Gibbes, helped to fill the gap during Ravenel's period of retrenchment. Indeed, it was Lewis Gibbes who played a major role in ultimately reviving Ravenel's interest in conchology.[20]

Although he set his scientific interests aside for a decade, Ravenel

Fossil echinoderms figured in M. Tuomey and F. S. Holmes, *Pleiocene Fossils of South Carolina* (1857). Of those in the above plate, Edmund Ravenel described No. 3, *Encope macrophora* (Ravenel, 1842), and No. 4, *Mellita caroliniana* (Ravenel, 1842). He also discovered the specimen figured as No. 5, which his young colleague John McCrady described as *Agassizia porifera*.

kept himself extremely busy in many other ways. Never completely giving up his interest in medicine, Ravenel published two papers on that subject, one in 1849 and the other in 1850. In the first paper he offered observations on the topography of the county in which the Grove was located and endeavored to show its association with diseases. Like his contemporaries, he believed that malaria resulted from the miasma of swamps. The second paper advocated the "pure air" of the seashore and bathing in the sea for the cure of several ailments. Of course, waves that run up on a beach must flow back when their energy is spent, and the undertow may be powerful in places, as Ravenel learned on July 14, 1848. On that day, his eleven-year-old daughter Theodosia Ford, her sister Emma, and three other children were playing in the surf at Sullivan's Island when a current suddenly swept them into deep water. Four of the children were rescued, but Theodosia sank beneath the waves. Her body was not found until the next day.[21]

Ravenel's other activities included acting as host to Louis Agassiz in the years around 1850, and, like Bachman and other local naturalists, he played a role in the Charleston meeting of the American Association for the Advancement of Science in March 1850, presenting a brief paper on his collection of the teeth of fossil sharks and a list of the Recent and fossil echinoderms he had collected in South Carolina. He also presented a paper illustrating other Recent and fossil species that he believed to be new to science, but nothing is known of these, as he did not submit the paper for publication in the proceedings of the meeting. Even more importantly, Ravenel became a mentor of the bright young Charlestonian John McCrady, who had developed an interest in invertebrates and would later study under Agassiz's direction at the Lawrence Scientific School, eventually becoming one of the great authorities on the hydrozoans of the North American Atlantic. From Ravenel he learned much about fossils, fishes, and echinoderms, and Ravenel even collected hydroids for his young friend.[22]

During the period from his move to the Grove until 1860, Ravenel also collected for many other naturalists: snakes and fishes for Holbrook; botanical specimens for Lewis Gibbes; mollusk shells, birds, and bird eggs for the Smithsonian Institution; soil samples for the state geologist, Oscar Lieber; and mollusk shells in abundance for dozens of conchologists. It is not surprising, then, that several naturalists named species in honor of Ravenel. For example, Holbrook, who had already paid tribute to Ravenel for collecting reptiles for him, acknowledged his old friend's assistance with fishes and named one for him, *Esox ravenelii* (= *Esox americanus*). In their *Pleiocene Fossils of South Carolina*, Tuomey and Holmes named a mollusk for Ravenel, called him "one of the pioneers of Natural History in South Carolina," and expressed appreciation to him for contributing a number of fine specimens. In the same work, McCrady named a fossil echinoderm species in Ravenel's honor, and in a later work, he named a genus after his special friend. When Holmes was working on *Pleiocene Fossils of South Carolina*, he called on Ravenel for assistance. Although he was not especially friendly toward Holmes, Ravenel nevertheless complied. He did so because, as he had said earlier, he preferred that South Carolinians describe native species and "not let us be always dependant [*sic*] upon Strangers to tell the world what South Carolina produces." When Holmes organized the Elliott Society of Natural History in 1853, he asked Ravenel to serve as one of its vice presidents. Ravenel could not attend the meetings regularly because of the distance from the Grove to Charleston, but he did prepare several papers for the Society, and he donated a number of specimens to its cabinet.[23]

Ravenel also devoted considerable time to business during the 1850s, adding more rental houses and a 1,400-acre plantation to his holdings, selling large quantities of board timber and bricks, and cultivating many acres of rice. By 1860, his real estate in Charleston, where he owned nine or ten houses, was valued at $27,000. In addition, of course, he owned the Grove and property on Sullivan's Island, and he still possessed around 125 slaves, down from the nearly 160 he had owned a decade earlier. Meanwhile, he continued to live on Sullivan's Island as often as he could, mainly during the summers. In 1846, the state legislature had appointed him as one of the commissioners to serve as the intendant of Fort Moultrie and as a physician to the island's inhabitants, especially to the indigent. On a single day in 1854, during an outbreak of malaria, Ravenel attended fifty patients in the Fort Moultrie hospital, and he was again overburdened with penniless patients during an outbreak of yellow fever in 1858, which, Ravenel insisted, occurred more among the "dissipated & reckless Soldier[s] of the Garrison" than among others.[24]

Praised by the wealthy inhabitants of the island for his "truly Samaritan labors . . . in behalf of the needy and the suffering," Ravenel had previously recommended the construction of a new and larger hospital to help him "meet the engagements which I have made for the Sick Poor." He was offended, however, by the objections of some of the leading citizens to the site he recommended for a new hospital, which, apparently, they considered to be too close to their homes. Later during the summer of 1858 the yellow fever epidemic spread, and the old hospital overflowed with patients. Ravenel had to find temporary quarters for them in homes near those of the gentry. Unfortunately for Ravenel and the sick, the island's aristocrats criticized the diligent physician for bringing the infected patients so close to them. After he issued a strong reply stating that he had no other choice and indicating dismay over their criticism, the prominent residents tried to assuage his feelings by saying that they had not intended to find fault with him but were merely trying to avoid a panic. The sixty-two-year-old Ravenel was so incensed over their complaint, however, that he resigned in April 1859 as intendant and as physician to the indigent.[25]

By that time Ravenel had become very active again as a conchologist. In fact, he published more on mollusks from 1857 to 1860 than he had previously done in the earlier period of interest. In a letter dated April 28, 1857, Isaac Lea said: "I was under the impression that you had given up Natural History entirely, which I very much regretted as you had done so much to promote a knowledge of it in your district." The correspondence between them thus resumed, and among the issues they discussed, and agreed upon, was the tendency of conchologists to become splitters. Earlier, in 1853, when a long-simmering feud between Lea and Conrad over taxonomic priorities came to a head, Ravenel had discreetly kept quiet, even when Lea apparently sought to drag him in on his side by making a point of Conrad's change of the name of a shell Ravenel had given to Lea. Ravenel had also begun to correspond with the British conchologists S. P. Woodward, Robert McAndrew, and Philip Carpenter; the bright, young marine zoologist William Stimpson, of the Smithsonian Institution; and the able naturalist Felipe Poey, of Havana, Cuba—all of whom he rightly viewed as authorities who could help him and also provide him with specimens from other coasts. All of them offered assistance and encouragement. No one did more than Lewis Gibbes, however, to persuade Ravenel to become active again in conchology. Although their homes were only a few miles apart, Ravenel and Gibbes often exchanged letters, a number of which are extant. Encouraged by Gibbes, Ravenel wrote papers for the Elliott Society. The first, presented in

May 1858, described two Recent and one fossil species of Mollusca, and the second, presented in December 1860, described fourteen species of the generally colorful clams in the family Tellinidae.[26]

Of course, Gibbes preferred that Ravenel publish all of his papers in the *Proceedings of the Elliott Society of Natural History*, for, like his peers in Charleston, he believed that southern naturalists had depended too long upon publications in the Northeast. Yet, both Gibbes and Ravenel understood the advantage held by the older publications in the Northeast, and to them, as to all other Charleston naturalists, both the Smithsonian Institution's publications and those of the Academy of Natural Sciences of Philadelphia (ANSP) ranked high in their esteem. Partly because of the continually friendly gestures of Spencer Baird, Joseph Henry, and, especially Stimpson, Ravenel decided late in 1860 to send a paper to Stimpson at the Smithsonian. In his paper, Ravenel described eight shells that he considered to be new species. Stimpson, who had spent time working in Charleston as an assistant to Agassiz, was particularly friendly with southerners. He liked Ravenel's descriptions and thought that they included many new species. Because the Smithsonian did not publish such short articles, however, Stimpson sent the paper on to the ANSP. The *Proceedings of the ANSP* was the more appropriate forum, and the article was published in its issue of February 1861. Ravenel was unaware of that fact, however, as sectional hostility was at a peak, and war was looming on the horizon. In June 1861, Ravenel told Lewis Gibbes that he had received no word regarding his paper, and he believed it was unlikely to be published because of his support of secession.[27]

In fact, scientists in the Northeast were generally not hostile toward their southern counterparts—first, because they viewed science as an apolitical, universal endeavor, and second, because some of them were not opposed to slavery, though they opposed the secession of the southern states. For example, W. G. Binney, son of Amos Binney, held Ravenel in highest esteem and corresponded with him between 1858 and 1860. As sectionalism increased, Binney told Ravenel that while war appeared to be inevitable, he would "always feel . . . under many obligations" to him. Ravenel strongly resented the efforts of abolitionists to grant freedom to the blacks, however, and late in 1860, he exclaimed in a letter to a newspaper editor that the "Black Republican Party" did not understand that "physical laws cannot be altered by human skill & resources." In his judgment, it was "the well-regulated

76 TREASURES OF EARTH AND SEA

system of Negro Slave labour" that was providing a civilizing effect upon blacks. By January 1861, he was convinced that war could not be avoided, and he decided that it was unsafe to leave his collection of shells at his home on Sullivan's Island. His home was less than 200 yards from Fort Moultrie and not far from Fort Sumter. By that time, Ravenel was feeble and nearly blind. Unable to do much for himself, he instructed his only son and one of his daughters to pack his shells for removal.[28]

Although many of Ravenel's taxa turned out to be previously described, largely the result of his limited access to the literature, several remain valid today. At least four of his taxa of Recent mollusks are among them: the lettered olive, *Oliva sayana*; the lineate dovesnail, *Anachis iontha*; the striate nassa, *Nassarius consensus*; and *Laevicardium pictum* (or *Lioadium picta*, as he named it). The last, sometimes called Ravenel's egg cockle but also the painted egg cockle, is an attractive little bivalve of the family Cardiidae, its smooth, creamy surface bearing zigzags of yellowish brown. Several of Ravenel's taxa of fossil mollusks and echinoderms are currently valid. While he had not earned a place among the greatest of American conchologists, Ravenel certainly belonged with that group who, as his contemporary George Tryon noted, had "done much to further the study" of conchology. Tributes to Ravenel, including a dozen taxa bearing his name, indicate the high regard in which he was held by his colleagues, both in his native region and elsewhere in the nation. In addition to fishes and fossils named for him, four species of Recent mollusks bear his name: the Carolina elktoe, *Alasmidonta raveneliana* (I. Lea, 1834), the Carolina spike, *Elliptio raveneli* (Conrad, 1834), a dovesnail, *Mitrella raveneli* (Dall, 1889), and the Ravenel scallop, *Pecten raveneli* (Dall, 1898). Those patronyms provide some indication of the significance of his role in advancing scientific knowledge.[29] Ravenel's magnificent collection of mollusk shells survives today in the Charleston Museum.

CHAPTER FIVE

A Low Class of Animals

D uring his first year in the medical department of the
University of Pennsylvania, Ravenel had developed
a friendship with John Edwards Holbrook. Al-
though Holbrook was three years older than Ravenel and ahead of him
in the medical program, he found that he shared a common temperament with the young southerner. Holbrook was born in the South but
had lived there only during the first three years of his life. His mother
never lost her affection for her native region, however, and she instilled
in her son a sense of vicarious attachment to the South. It is likely that
Ravenel strengthened the association and probably urged Holbrook to
settle in Charleston and practice medicine. The connection was no
doubt enhanced later when Holbrook's younger brother Silas Pinckney Holbrook, also born in the South, married Esther Gourdin, the
older sister of the widow of Edmund's brother James Ravenel.[1]

Holbrook's mother was Mary Edwards, the third child of John
Edwards, a Beaufort, South Carolina, merchant who had immigrated
from Wales. Mary had met the school teacher Silas Holbrook some
time around 1792 and married him during that year or the next. It is
unknown why Silas Holbrook, born in 1768, had left his home in

Wrentham, Massachusetts, to teach school in Beaufort, a small city on the South Carolina coast, about fifty miles southwest of Charleston. Nor is it known exactly when he arrived there. His name does not appear in the federal census of South Carolina for 1790. In any case, the first child of Silas and Mary was born on December 30, 1794, and named after his maternal grandfather. Their second child, Silas Pinckney, was born on June 1, 1796. Not long after the birth of their second son, Silas and Mary moved to the Holbrook home in Wrentham. Silas the elder died in March 1800, but Mary chose not to return to South Carolina. In October 1800, she married Daniel Holbrook, the forty-three-year-old brother of Silas. John Edwards Holbrook developed deep affection for his stepfather and for the three children born to Daniel and Mary.[2]

Only a meager amount of information is available on the early life of John Edwards Holbrook, but, since he had already acquired considerable knowledge of natural history when he went abroad in 1818, it would appear that he learned much as a child. His mother and stepfather enrolled him in Day's Academy, a local preparatory school. Afterwards he entered Brown University, from which he graduated in 1815. His activities during the next two years are unknown, but they apparently included a period of study with a physician in Boston. Holbrook then enrolled in the medical department of the University of Pennsylvania and received the M.D. degree in 1818. Soon after graduating, he moved to Boston, where he stayed for a few weeks in order to study with the noted physician William Ingalls. The young doctor aimed higher, however, for he wanted to go abroad for further study. Thus, he departed for Britain in June 1818.[3]

Holbrook desired first to attend the lectures of renowned professors of medicine at the University of Edinburgh and then to study with famous physicians and, especially, noted naturalists in Paris. Only a few of the letters he wrote to his mother and to his brother Silas during his years abroad are extant, and, unfortunately, only one of them is from his days in Paris. The lengthy letters he wrote to Silas from Britain reveal much, however, about Holbrook, his travels in Great Britain and Ireland, and his study at the University of Edinburgh. On the fifteenth day across the Atlantic Ocean, the ship passed along the coast of Ireland, which, Holbrook said, was justly named the "Emerald Isle." On the next day, he spied the coast of Wales, with its "wild and romantic shores" and "stupendous and bare rocks." Upon docking in Liverpool, Holbrook sought out the American consul, who showed him around the city. He liked the area and was impressed by its botanical garden and its libraries, hospital, and charitable institutions. He was less impressed, however, with the political oratory he heard and with a treatise on medicine that he read, the latter of which, in his opinion, revealed the high status of "quackery" in Britain.[4]

Holbrook then visited the city of Manchester, where he observed and praised the activities of the local scientific society. On he traveled to Lancaster and toward the Lake Country. Along the way to the region, he was especially struck by the gulf between the rich and poor and by the presence of "soldiers in every hamlet repressing the spirits and corrupting the morals of the people." Holbrook liked Glasgow, however, and called it "incomparably the best city in Great Britain." In Glasgow he listened to the eloquent oratory of the Scottish theologian and reformer Thomas Chalmers. Also of special interest to Holbrook was the Hunterian Museum, which, in his judgment, "contained some of the best anatomical preparations in the world" as well as good collections in natural history.

At last the happy traveler arrived in Edinburgh, where, in his opinion, "the common people know much less of the affairs of the nation than in America but are much more familiar with literature, especially the natural sciences." Holbrook soon formed judgments on the university's professors of science, calling John Playfair "a man of genius and great erudition" and noting that Robert Jameson was a learned geologist and leader of "the Wernerian school of geologists." But he declared that the scholars in "the literary and philosophical department are of no great name." Of the medical school professors, the fledgling physician found little to praise, complaining that the elderly James Gregory had "a mumbling manner of speaking" and lectured without notes. The latter, opined Holbrook, was bad because medical knowledge requires not only "a great number of facts" but also thoughtful selection of "the most applicable cases." He criticized several other professors and concluded that in general the lectures he heard were "quite inferior" to those he had attended at the University of Pennsylvania.[5]

Holbrook also journeyed to the Highlands of Scotland and then to the west coast of that country, taking time to collect plants and minerals along the way. Boarding a mail boat, he embarked for northern Ireland to visit the cities of Antrim and Belfast. Around Antrim, he paid special attention to the peat bogs and fossil remains uncovered from them. "We may imagine," he reflected, "what changes the surface of our earth has undergone. Many . . . curious remains are found in these bogs, some antedeluvian [*sic*] . . . and others doubtless preadamite [i.e., prior to the creation of Adam]." In addition, he commented upon the Neptunian and the Vulcanic theories of the formation of the earth. He was convinced, however, that more than one great deluge was necessary to explain the various strata and the fossils embedded in them. Quite clearly, Holbrook was at that time strongly interested in geology, and he was open to ideas that ran counter to the traditional interpretation of the biblical account of creation, including the notion of life on the earth before the creation of Adam.[6]

After exploring Northern Ireland, Holbrook returned to Edinburgh, and in June he wrote to Silas that the time had come for him to leave Scotland, which, on the whole, he had found very pleasant. During the next six months, he traveled about Wales, searching out the sister of his grandfather Edwards, who had immigrated from the Welsh town of Hollywell to Charleston sometime before 1773. He also observed topographical features of the land and visited a lead mining operation. In addition, Holbrook went to the home of Thomas Pennant, son of the naturalist-artist David Pennant, to whom he referred as "the Linnaeus of Great Britain." Pennant was away, but someone gave Holbrook "full range of the house." The ability of natural history artists seemed already to be of particular importance to Holbrook. Later he traveled again to Ireland, staying mostly in Dublin, where he talked with local naturalists, observed work in hospitals, visited the medical college, and toured the museum of the Dublin Society. Yet, he observed, "amid all this splendor and magnificence . . . I cannot walk without being shocked at the appearance of want and wretchedness. The streets swarm with beggars." Keen empathy for the suffering poor was deeply embedded in the character of John Edwards Holbrook.[7]

Back to Wales he went and then on to London, once again stopping often to study the local geology and to collect minerals. In September, after a few days in London, he traveled to Oxford and visited many of the colleges and libraries. Impressed by the libraries he saw, Holbrook noted that the Bodleian alone held "400,000 volumes." Heading westward from there, he moved on to the Cotswold, to Gloucester, Newport, and Bristol, all along the way paying close attention to matters of geology. The city of Bristol fascinated him, for "like its namesake in Rhode Island this city has deeply engaged in the slave trade." He also noted that many "unitarians and dissenters" lived in the city, but he offered no comment on them, though the context of his statement seems to suggest no displeasure over their views. Explorations of the cliffs and rocks around the Avon River led Holbrook to comment that "the rocks contain the remains of many animals—some extinct species—some inhabitants of the sea, and some of warmer climes." The "elegant and splendid houses" and the Sydney Gardens caught his eye as he sojourned in and around the ancient city of Bath. Soon after returning to Bristol, he set out for Salisbury, where he viewed with awe the grand spire of the cathedral of that city. He moved on to Stonehenge, then to Southampton, the Isle of Wight, Newport, Portsmouth, and Brighton. Finally, he returned to London and stayed until November or December 1819. To Silas he wrote that he would offer no account of the great city, for it "would fill volumes, as it is a world by itself." Holbrook now had his sights set on a long stay in Paris, followed by travel to Italy and through the German states.[8]

By January 1820, Holbrook was in Paris, and on February 2nd of that year he penned a letter to his mother. She had heard nothing from her peripatetic son since the previous November. Holbrook told her that he had spent the first several weeks trying to learn French, which, he lamented, he had "so miserably neglected in college." He complained that he had received no word from Silas for a year, and expressed concern that his mother had not forwarded the "nine long letters" he had written to his brother. It is very likely that Holbrook also wrote long letters to Silas about his life, studies, and travels in France, but none have been located. Thus, little is known about his contacts with the great naturalists of that country. Surviving evidence indicates, however, that he studied at the famous Jardin des Plantes and maintained an association with the renowned naturalists Georges Cuvier, Achille Valenciennes, Andre Duméril, and Gabriel Bibron. Indeed, it is quite likely that those men influenced him to focus upon the reptiles, amphibians, and fishes in his own country, which, they almost certainly noted, had received little systematic treatment. Perhaps the spark was already there and merely needed to be fanned, or perhaps the French naturalists generated Holbrook's interest in reptiles, amphibians, and fishes. The latter seems quite plausible since, during his travels in Britain, Holbrook had directed his attention mainly to geology and somewhat to botany, in addition to medicine. One of his former medical students, and later an associate, claimed that Holbrook was already a recognized naturalist when he got to France and that he knew enough to advise museum curators on the classification of some animals, but he likely confused that visit with the one Holbrook made two decades later. In any case, the experience with the French naturalists helped to set the course for Holbrook as a naturalist. As with the vast majority of his contemporaries, however, research in natural history for Holbrook had to be a part-time activity, to be pursued when he could eke out time from his occupation. Holbrook knew that he must first establish a successful medical practice. If he kept the plan he had conveyed to his mother in February 1820, he left Paris late in that year, or, more likely, early in 1821, and traveled to Italy and Germany, departing for home from Hamburg in the spring.[9]

Whether Holbrook pondered the prospects of a practice in Wrentham, Boston, or elsewhere in the Northeast is unknown. Also unknown is his reason for choosing Charleston, though, of course, his friend Edmund Ravenel likely touted the potential for building a successful practice there, the resurgence of scientific interest among Charleston's intellectuals, and the abundance of fauna in the region. In any case, Holbrook elected to return to his native state, arriving in Charleston in 1822. Since he did not apply for a license to practice in South Carolina until early in December of that year, he probably moved to Charleston only shortly before then. His practice apparently flourished, and Holbrook developed a reputation as a physician of gentle disposition and concern for his patients. Because he was keenly sensitive to suffering and pain, he avoided surgery if he could, and shunned obstetrics entirely. Indeed, a close colleague claimed that he never delivered a baby. As a bachelor until 1827, he had more free time than either Ravenel or Bachman for collecting specimens, a number of which, especially fishes, he sent to naturalists in Paris. In 1831, in their great work on the natural history of fishes, Cuvier and Valenciennes paid tribute to Holbrook for several specimens he sent to them.[10]

Like his friend Ravenel, Holbrook joined the Medical Society of South Carolina, although he waited over a year before doing so. When the Society's trustees met on April 12, 1824, to select faculty for the medical college in Charleston, they passed over applicant Holbrook and elected another medical doctor. That person resigned at the next meeting, however, and, when the trustees met again in June, they appointed Holbrook as professor of anatomy. Whether due to the devotion of much of his spare time to herpetology or simply to inability to keep an organized schedule, Holbrook developed a reputation for being absent-minded and disorganized. In fact, the Society levied several fines on him for failing to return books to the library on time. These traits seemed only to endear the kindly fellow to his acquaintances, but they are partially reflected in Holbrook's lack of systematic order in his volumes on herpetology and ichthyology. His mild disposition likely kept him from taking an aggressive role in the quarrel with the Medical Society in 1832–1834. Nonetheless, he obviously sided with his friend Ravenel and his other colleagues against the Medical Society trustees, for he also moved to the new medical school. Along with his colleagues, he was named in the Medical Society's letter of July 5, 1833, that declared all the faculty positions vacant.[11]

At some point, Holbrook joined the Literary and Philosophical Society, but he was never very active in the affairs of that organization. He was not among the group that founded the Elliott Society of Natural History in October 1853, though he joined it a month later. Holbrook never presented a paper before the Elliott Society, and the records show that he rarely attended its meetings. It appears that he thought time devoted to natural history could be better spent by quietly, but assiduously, working toward the completion of one major project at a time. Indeed, his first project, a systematic account of the reptiles and amphibians of North America, was a demanding task, and the first fruits of it did not appear until 1836. Given the nature of the endeavor, he must have begun the project soon after his arrival in Charleston. Although he was building a network of collectors and correspondents, carefully observing and dissecting specimens, writing descriptions, supervising the drawings of his animals, and making himself fully aware of all the pertinent studies he could get his hands on, Holbrook seems to have said little about the major work he intended to produce. The job was big enough, along with his practice and his professional duties, to keep him from participating in the activities of the Literary and Philosophical Society, although much later, when that organization devolved into the Conversation Club, he was slightly more active.[12]

Holbrook remained single until he was nearly thirty-three years old. On May 3, 1827, he married Harriott Pinckney Rutledge, the fourth of the eight children of Frederick Rutledge and Harriott Pinckney Horry. Born in 1802, Harriott Rutledge was descended from prominent

Charleston families and ultimately inherited property in Charleston, several slaves, and the 185-acre estate called Belmont, located on the Cooper River, less than five miles north of Charleston. Well-read and a stimulating conversationalist, Harriott Holbrook offered great encouragement and assistance to her husband. She proved to be the perfect hostess, as Louis Agassiz and, later, his wife Elizabeth, would attest after being guests in the Belmont home. Childless, she devoted her attention to her relatives, who lived with the Holbrooks for many years, and to nieces, especially Harriott Horry Rutledge, a daughter of her eldest brother Edward Cotesworth Rutledge, who was an officer in the U.S. Navy. When niece Harriott was nine years old, she spent much of the summer with her Aunt Harriott and Uncle John Edwards, including several weeks at Belmont, to which the Holbrooks retreated during the summers. Often called "Nunkey" by his niece, John Edwards Holbrook was very fond of Harriott, and even when she was grown, he continued to call her "Dear Child." The bond was further strengthened in 1851, when Harriott married Dr. St. Julien Ravenel, one of Holbrook's former students and then a colleague at the medical college. In letters to Harriott, Holbrook referred to her husband as "the Saint." A nephew of Holbrook's old friend Edmund Ravenel, St. Julien remained close to his uncle-in-law throughout Holbrook's life.[13]

By the early 1830s, Holbrook had built a considerable network of collectors. Among those in Charleston who sent him specimens and provided information on reptiles and amphibians were, of course, Edmund Ravenel and John Bachman. Thomas M. Logan, a medical doctor in Charleston who later moved to New Orleans, not only collected for Holbrook but also made "a number of beautiful drawings of Serpents, Salamanders, etc." for Holbrook's book. Another useful collector was the physician and naturalist J. G. F. Wurdemann. A truly able student of nature, Wurdemann also helped Holbrook with "many beautiful anatomical preparations." Later he sent fishes to Holbrook, but, at one point, when he sent to Lewis Gibbes a huge salamander— probably not a new species, as he guessed, but likely *Siren lacertina*— he urged Gibbes to describe it rather than send it on to Holbrook. While he strongly preferred "that a *Southern* man should have credit for describing it," he feared that the forgetful Holbrook would never return it to him. Especially helpful to Holbrook was Thomas L. Ogier, formerly a student and later a colleague on the faculty of the medical college. Indeed, Ogier, who often dissected specimens for Holbrook,

John Edwards Holbrook. Photograph from an 1857 portrait by Daniel Huntington, from the Andrew W. Mellon Collection, National Gallery of Art. Photograph © Board of Trustees, National Gallery of Art, Washington, D.C.

became a close friend of his mentor. No one helped more, however, than J. Sera. An immigrant from Italy whose first name remains unknown, Sera had been a successful painter of theater scenery before Holbrook discovered his rare ability to draw and color reptiles and amphibians. For at least five years, from 1832 until his death in 1837, Sera worked diligently to produce accurate drawings and coloring of specimens, often using, as Holbrook strongly preferred, live animals as his subjects. Another artist, J. H. Richard, continued Sera's work, but, while he was talented, he lacked the special gift of his predecessor. In addition, Holbrook turned on occasion to Bachman's sister-in-law Maria Martin. Doubtless Martin could have done more for Holbrook, as her later work for John James Audubon proved.[14]

A host of other local collectors assisted Holbrook, but he needed specimens from elsewhere in the South. Among the collectors outside the Charleston area who assisted him were Gerard Troost of Tennessee, who sent him several turtles; William C. Daniell, a physician in Savannah, Georgia; John Eatton LeConte, a native of New York, later a resident of Philadelphia, and a frequent visitor at his brother Louis's plantation in Liberty County, Georgia; and John M. B. Harden, a former Holbrook student who was married to a daughter of Louis LeConte. Holbrook owed a special debt to John Eatton LeConte, for the latter had done pioneering work on southeastern turtles, frogs, and salamanders. Holbrook was equally indebted to Harden for sending him a new species of salamander. Several specimens came to Holbrook from the prosperous planter and prodigious collector James Hamilton Couper, of St. Simons Island, Georgia. Also of considerable assistance was John P. Barratt, an able physician and excellent naturalist, in Greenwood, South Carolina. Essentially isolated in the up-country, Barratt, an immigrant from England, maintained a close relationship with the Charleston naturalists and furnished specimens for them, including a specimen of the alligator snapping turtle, *Chelonura temminckii* (= *Macroclemys temminckii*), which he collected in Alabama and sent to Holbrook.[15]

Specimens from other regions of the United States and information on their habits and distribution were necessary for Holbrook to do a complete work on North American reptiles and amphibians. Thus, he turned to naturalists outside the South, most of whom were in the Northeast. Like Bachman and Ravenel, Holbrook developed a strong relationship with naturalists in Philadelphia. Jacob Green, a professor at the Jefferson Medical College, in Philadelphia, provided Holbrook with useful information on the habits and distribution of salamanders. Although Green was mainly a chemist and not a taxonomist, he knew much about natural history, and he kept careful notes on the species he observed. Samuel S. Haldeman, a largely self-educated chemist, philologist, and naturalist, supplied Holbrook with information and specimens, and Richard Harlan served as an important source of information on reptiles. Holbrook had known Harlan since they were classmates in medical school. Though perhaps less well-known outside the field of herpetology, Edward Hallowell was among the most helpful to Holbrook of all the Philadelphians. William Blanding, a resident of South Carolina before moving to Philadelphia, furnished Holbrook with information on turtles in the South and in the West. Above all, however, Holbrook praised Charles Pickering for assisting him "with his accurate knowledge at every step of this work." Pickering was, of course, of great assistance to Bachman later on. Holbrook's esteem was well-placed, for Pickering, who settled in Philadelphia in 1827 to practice medicine, developed into an extraordinary naturalist. Another naturalist in Philadelphia who aided Holbrook was the widely traveled Englishman Thomas Nuttall, who sent specimens from Oregon to Holbrook, as he later did to Bachman. It appears that Holbrook had met all of these men during trips to Philadelphia.[16]

Other outstanding naturalists contributed to Holbrook's study. From Ohio, Jared P. Kirtland sent Holbrook a specimen of a rattlesnake, which Holbrook called a "beautiful reptile." In New York, Holbrook had an especially able ally in James E. DeKay, sometime editor of the *Annals of the New York Lyceum of Natural History* and scientist for the New York Natural History Survey, for which he later produced a five-volume report. Holbrook may have met DeKay at the University of Edinburgh, since DeKay earned his M.D. degree there in 1819. Finally, Holbrook had important contacts in Boston: Amos Binney and D. Humphreys Storer. Holbrook knew them well, for he visited with them and collected specimens in the Boston area on several occasions. Although Binney was primarily interested in mollusks, he was well acquainted with other areas of natural history, and he readily supplied Holbrook with specimens of reptiles from his region of the country, as did Storer. Like nearly all of those in Holbrook's network, Storer was a physician as well as a naturalist. He furnished Holbrook with a number of reptiles and referred to him in

1839 as a "distinguished herpetologist." Kirtland, DeKay, and Storer would themselves become well-known names in herpetology.[17]

Difficulty of access to books and journals discouraged Holbrook, as it did Ravenel and later Bachman, and on several occasions he asked friends to obtain some that he needed. Moreover, his project was enormous and had to take second place to his medical duties. In 1834, he told Samuel George Morton that he almost wished he had never begun the project, but his "love for the works of the Creator, even in this low class of animals," kept him going. Indeed, it did, and the first volume of *North American Herpetology* was published in 1836. Another appeared in 1838, and by 1840 the third and fourth volumes were off the press. Dissatisfied with the first two volumes, Holbrook enlarged or modified some of the descriptions and revised the plates in 1839, but he used the same title pages and failed to specify that the volumes constituted a revised edition. In fact, Holbrook was so displeased with the first two volumes that he decided to reorganize the work as a whole. As Kraig Adler has indicated, Holbrook's original volumes display no systematic arrangement of species; they appear in the order in which he either obtained the specimens or had them illustrated.[18]

Correspondence with Storer indicates Holbrook's dissatisfaction with the original volumes and reveals some of his frustrations, particularly over a complaint by Storer in late 1840 that Holbrook had ignored his published study of reptiles. In a letter to Storer in November, Holbrook reminded his friend that he had tried to talk with him about the study but that he was too "occupied to listen at the moment, and I forgot it afterwards." Holbrook ended his letter on a soft note, however, telling his Boston colleague that the governor of South Carolina had recommended that the state legislature authorize a subsidy for *North American Herpetology* and asking Storer to write in support of the recommendation. Storer wrote to Holbrook again, offering compliments but complaining further of Holbrook's neglect. Holbrook replied that he had not intended to produce more copies of the second volume "until I could revise the thing a little," and in a later letter to Storer, he said that around 1838 or 1839, he had become "troubled at the number of errors" in the first and second volumes and decided to recall them. In addition, he indicated that he wanted to "arrange the animals in their natural systematic position." Holbrook insisted that this would not be a new edition. He therefore told Storer that he could not have used his study because he was not issuing a new edition. Unfortunately, that was not the case with the second edition, published in 1842, though Holbrook told Storer in the later letter that he could not refer to Storer's work because he had made the revisions of it by 1840. Quite obviously, Holbrook had forgotten some of the facts, but he certainly had no intention of slighting Storer. As he told his friend-turned-critic, "No man ever entered into a work . . . with a more devoted love of truth and of doing justice to all."[19]

The alterations of the first two volumes had done little to rectify problems in *North American Herpetology*. Thus, Holbrook devoted considerable attention to making a better work, and in 1842, he issued a second edition, consisting of five volumes, with 147 plates colored by hand. Striving for more perfect plates, he continued to substitute new ones in later runs of the second edition. As a consequence, copies of the five-volume edition often vary somewhat. Nevertheless, even the initial volumes of Holbrook's *North American Herpetology* constituted a significant contribution to natural history in the United States. The illustrations alone represented a major achievement, not only because they were morphologically accurate in most cases but also because they were generally faithful in color. Sera and Richard strove to draw and color the animals as close to reality as possible, and they used living animals as models whenever they were available. Holbrook expected no less, and his desire to achieve perfection is evident in the continuing changes he made even after the initial printing of the second edition. His efforts set a standard in natural history publishing, not only in the United States but also in Europe, where *North American Herpetology* caught the eye of Louis Agassiz, who determined to seek out Holbrook when he came to the United States a few years later.[20]

Like Bachman and Ravenel, Holbrook often noted the inadequacy of libraries in Charleston and expressed "fear of describing animals as new that have long been known to European Naturalists." In addition, he complained that he had "only defective museums for comparison," presumably in reference to the herpetological collections in the museums of both the medical college and the Literary and Philosophical Society. Those wants bothered him especially because of the current "confusion . . . in Herpetology" in the United States. Nevertheless, as Kraig Adler has observed, Holbrook's work represented "the first great synthesis of information" on the amphibians and reptiles within the contemporary boundaries of the nation, which, of course, resulted in a focus upon species from the Atlantic Coast to the Mississippi River. The work also included twenty-four taxa new to science. Recognizing the significance of the contribution, the naturalist Charles Girard dubbed Holbrook "the father of American herpetology" in 1850. His contemporary and friend DeKay called *North American Herpetology* an "excellent and beautifully illustrated work" and predicted that it "will long remain a monument to his [Holbrook's] genius and his zeal." DeKay's assessment was correct, but, although many of Holbrook's contemporaries recognized the value of the work, it received comparatively few contemporary reviews. The early volumes of the original edition were noted in the *Southern Rose*, a Charleston children's magazine published by Charlotte Gilman, wife of the local Unitarian minister. Gilman praised the first volume as "highly creditable both to its accomplished author and to his country," and, later, she called the second and third volumes "splendid" and said they were certain "to elevate and sustain the reputation of the author, and the honour of his country."[21]

Longer, highly favorable reviews of the first two volumes appeared in the *North American Review* and the *American Journal of Science*. Though anonymous, the reviewer in the former was evidently well acquainted with the subject and knew that Holbrook had long been engaged in the study. He offered a few minor criticisms but generally praised the volumes, especially for the excellent quality of the illustrations and the synonymies. Added the reviewer: "Every species is minutely described, its geographical limits pointed out, and its habits elucidated, often times with great perspicuity." The last of those qualities, observed the reviewer, should arouse "uncommon interest in the mind of the reader." Indeed, Holbrook's descriptions reflect his views toward the creatures he studied. Of the green anole, *Anolis carolinensis*, for example, he stated that the lizard is "bold and daring" and the male "quarrelsome" and prone to "furious battle" during the spring, and he wrote that "the coloring of the animal has the liquid brilliancy of the emerald." Of the eastern diamondback rattlesnake, *Crotalus adamanteus*, Holbrook said, "a more disgusting and terrific animal cannot be imagined; . . . its dusky color, bloated body, and sinister eyes, of sparkling grey and yellow . . . combine to form an expression of sullen ferocity." Like other early naturalists, Holbrook sometimes gave anthropomorphic characters to the species he observed.[22]

Of Holbrook's twenty-four currently valid taxa, one is Blanding's turtle, *Cistuda blandingii* (= *Emydoidea blandingi*). Speckles on its carapace and bright yellow on its plastron and throat make it a very attractive turtle. Valid species of snakes (order Squamata, suborder Serpentes) described by Holbrook include the mud snake, *Coluber abacurus* (= *Farancia abacura*), which, despite its common name, is handsomely marked by shiny reddish to pinkish triangles on the lower half of its body, and the brown water snake, *Tropidonotus taxispilotus* (= *Nerodia taxispilota*), which, because of its size, color, and markings, is often confused with the poisonous cottonmouth, *Agkistrodon piscivorous*. Holbrook's taxa also included several species of amphibians in both the order Anura (toads and frogs) and the order Caudata (salamanders). The first includes the American toad, *Bufo americanus*, a very familiar species in most of the states east of the Mississippi River; the tiny oak toad, *Bufo quercicus*, which has black spots separated by a mid-dorsal line; and the ornate chorus frog, *Rana ornata* (= *Pseudacris ornata*), whose melange of bright colors justify its epithet and make it one of the most striking of creatures in the Southeast. The valid taxa of Caudata described by Holbrook include the mole salamander, *Salamandra talpoidea* (= *Ambystoma talpoideum*); the black-bellied salamander, *Salamandra quadrimaculata* (= *Desmognathus quadramaculatus*); and the dwarf salamander, *Salamandra quadridigitata* (= *Eurycea quadridigitata*). Unlike most salamanders, the last, one of the brook salamanders, has four rather than five toes on each of its hind feet.[23]

By 1842, Holbrook had attained rank among the best naturalists in America and in Europe, and when he traveled with his wife to Paris soon after the publication of the second edition, he was not only warmly received by Valenciennes and other naturalists but also prevailed upon to identify or confirm the labeling of several specimens of North American amphibians and reptiles in the collection of the Muséum d'Histoire Naturelle, to which Holbrook had sent a considerable number of American specimens. To Holbrook, that request represented singular recognition of his attainments. He had already received the honor of membership in the Royal Medical Society of Edinburgh, the American Philosophical Society, the Academy of Natural Sciences of Philadelphia, the New York Lyceum of Natural History, and the Boston Society of Natural History, but the plaudits of the Paris naturalists signaled international acclaim. In 1847, additional recognition

Blanding's turtle, *Emydoidea blandingi*, was described by John Edwards Holbrook in 1838 as *Cistuda blandingii*. Illustrated in the plate above is the type specimen, sent to Holbrook by the Philadelphia naturalist William Blanding. A northern animal, Blanding's turtle is related to the chicken turtle, *Deirochelys reticularia*, an inhabitant of southern marshes and ponds. The plate is reproduced from Holbrook, *North American Herpetology*, 1842 edition.

came to Holbrook when a Berlin society for natural history selected him as one of its foreign members.[24]

Meanwhile, Holbrook continued to teach medical students and to win their praise for his knowledge of anatomy, for his "easy and graceful" manner, and for "the charm with which he invest[s] his descriptions of a subject usually dry and uninviting." He also continued his private practice and, quietly, contributed to the cultural advancement of Charleston. In 1839, he had even found time to write a brief article on the need for educating farmers in the basics of botany, chemistry, geology, physics, mathematics, and zoology. Holbrook was a member of the Charleston Library Society, a patron of the Female Orphan Society, and a member of the New England Society. In addition, he helped to organize the South Carolina Historical Society. His membership in the latter two organizations reflected not only his familial and cultural connections but also an interest in historical matters. During the late 1840s and early 1850s Holbrook was a member of the Literary and Philosophical Society, better known informally by then as the Conversation Club, which included John Bachman; the College of Charleston professor of belle lettres Frederick A. Porcher; the banker Daniel Ravenel; the lawyer, writer, and artist Charles Fraser; the Presbyterian minister Thomas Smyth; the Unitarian minister Samuel Gilman; and the physicians James Moultrie, Samuel H. Dickson, Henry Frost, and Peter C. Gaillard. In his memoir on the club, Porcher observed that Holbrook rarely said anything at the few meetings he attended. Bachman, on the other hand, spoke often—and sometimes vehemently, as on the occasion in 1847 when the guest was Louis Agassiz and the topic of discussion was the origin of human races. Holbrook was likely present for that meeting, but whether or not he joined in the discussion is unknown. That his views sided with those of the pluralist, or polygenist, school is certain, however.[25]

Such activities were of no great importance to Holbrook, for he was already deeply engaged in another major project: a description of the fishes in southern waters. His interest in ichthyology ran back at least to the late 1820s, and, of course, by the early 1830s he had sent several specimens of fishes to Valenciennes. He had maintained contact with Valenciennes during the 1830s, and visited with him during his trip to Paris in 1842. Holbrook's project on all the southern fishes proved to be even more demanding and frustrating than his first one, for the subject was enormous in scope. Moreover, while his network of correspondents and collectors was still in place, it could not function as effectively as it had on the reptiles and amphibians because it was confined to a region in which there were no real experts. Nevertheless, Holbrook set to work seriously on the project in the early 1840s. Once again, he had to call upon friends in Philadelphia and Boston to lend him some works dealing with ichthyology. Many of his descriptions would be new, however, for little literature on southern fishes existed.[26]

In 1847, Holbrook published the first of his volumes on fishes, titled *Southern Ichthyology; or a Description of the Fishes Inhabiting the Waters of South Carolina, Georgia, and Florida*, and a year later he published a second volume. Strangely, he labeled the volumes as Part II and Part III, respectively. Holbrook later said that two parts had been published in 1845. Nothing has been found to confirm that any parts appeared in 1845, nor have any copies of Part I or Part IV (to which he referred in Part III) been located, confirming the statement of the noted ichthyologist Theodore Gill, in his biographical sketch of Holbrook in 1903, that Holbrook published no other parts of *Southern Ichthyology*. As before, the forgetful and disorganized medical professor got matters confused. In any case, the first volume (Part II) contained forty-seven pages, and the second (Part III), sixty pages. J. H. Richard drew the fishes illustrated in both parts. In those parts Holbrook described two new species: *Umbrina littoralis* (= *Menticirrhus littoralis*) and *Otolithus nothus* (= *Cynoscion nothus*). The first, commonly called the gulf kingfish, runs to eighteen inches in length, and the silvery body of this surf-zone fish is rather unremarkable. The second, the silver sea trout, resembles the common name it bears. Both belong to the drum family (Sciaenidae).[27]

Except for a list of the fishes of Georgia, published in *Statistics of the State of Georgia* in 1849, Holbrook published nothing in ichthyology again until 1855. Probably, as Gill later suggested, he must have come to an understanding of "how impossible it would be . . . to realize his desire for describing and painting his subjects only from life." The task was insuperable, at least during the active years he had left to him. Holbrook therefore narrowed the scope of his study to the fishes inhabiting the waters of South Carolina. Moreover, based upon the several requests for loan of books or articles, Holbrook continued to need sources unavailable to him in Charleston libraries. Of Spencer Baird he made several requests. Baird, the assistant secretary of the Smithsonian Institution, was especially interested in Holbrook's study of southern fishes, for so many of them remained to be collected and described. Holbrook was warmed by both the professional and personal concern of Baird, and he urged him to attend the AAAS meeting in Charleston in 1850. During the previous year, he had planned to visit Baird, but the sudden illness of Holbrook's mother-in-law forced him to cancel his plan. Later in that year or very early in 1850, Holbrook sent a letter to Baird, listing some of the "best fishes for the table," but his note about culinary delights also contained the bad news that he had been the victim of an accident. He had fallen from a wagon and received "some very serious bruises." The misfortune, however, would not prevent him, he believed, from seeing Baird at the AAAS meeting in March. Unfortunately, shortly before the meeting, Baird informed Holbrook that he would be unable to attend, but he offered to send "specimens of such of our fishes as you may desire for purposes of comparison with yours." Ever the collector, Baird added that he would like to receive duplicates of all the southern fishes Holbrook had collected, especially "anything rare," for the Smithsonian Institution. Baird was also mindful that Holbrook remained interested in reptiles and amphibians, and he asked him to send specimens of those inhabiting the southern region. Holbrook did send some, but he had likely disposed of many of the specimens used for his *North American Herpetology*. Apparently unimpressed with the Charleston Museum, particularly because it was in one of its periodic declines around 1840, Holbrook had offered his collection to the Boston Society of Natural History at the price of $300, even though, he said, it had cost him $1,000, not counting his own labors.[28]

Meanwhile, in the early spring of 1846, Holbrook had received a letter from the German naturalist Edward Rüppell, a resident of Frankfurt. The letter suggests that Rüppell and Holbrook had corresponded earlier, though the former had only recently obtained a copy of Holbrook's "beautiful American Herpetology and some proof plates of the unpublished figures of fishes, which are remarcably [*sic*] fine." No doubt Holbrook relished the praise of his German correspondent, but almost certainly he was even more pleased to hear that "Professor Agassiz, who is here on his way to the United States," wanted to meet the author of *North American Herpetology*, which Agassiz later called a book "far above any previous work on the . . . subject." Indeed, said Agassiz, until the appearance of that work, European naturalists had generally viewed American publications in natural history as inferior. In particular, added Agassiz, "the accuracy and originality of his [Holbrook's] investigations" made a strong impression upon European scientists. Given his esteem for *North American Herpetology*, it is no wonder that Agassiz would want to meet Holbrook during his planned sojourn of two years in the United States.[29]

Accorded an enthusiastic welcome by American scientists after settling in Cambridge, Massachusetts, in 1846, Agassiz launched a lecture tour in 1847 through several cities, including New York, Philadelphia, and Charleston. In Philadelphia, Agassiz met and conversed at length with Samuel George Morton, whose collection of human crania especially fascinated him. It was in Philadelphia that Agassiz saw a black person for the first time in his life, and, in a letter to his mother, expressed a feeling of disgust and repugnance over the skin color and facial features of the man. Soon he would see black people in great numbers and, unlike the individual in Philadelphia, they would be subjects of bondage, representing, in the views of most southern writers and in the view of the Swiss visitor, an inferiority of physiognomy, character, and mental capacity that warranted their status as slaves. Agassiz arrived in Charleston at the end of November 1847, and on the third night of December he delivered the first of a series of lectures that would not only endear him to the naturalists of Charleston but also foster a feeling of their importance to the scientific community and, eventually, provide a sort of scientific sanction of their views on the races—that is, the views of all the naturalists but John Bachman.[30]

As expected, Agassiz sought out Holbrook, and the self-effacing physician and naturalist and his gracious and gregarious wife entertained the internationally prominent scientist as a guest in their home in Charleston and at Belmont. Agassiz found his host to be "a man of singularly modest nature . . . and lovable personal qualities." Moreover, he found Holbrook to be a man who could calmly and rationally discuss the basic character of the slaves at Belmont. Charmed as well by the well-read and readily conversant hostess Harriott, Agassiz knew that he had found friends in the South.[31]

But the friendship was not confined to the Holbrooks. Agassiz found others eager to associate with him and to show him some of the riches of their personal collections and, despite the present state of the Charleston Museum, some of the marvelous objects donated to it during the previous half century. Immediately, Agassiz recognized the potential of Charleston for development as an even more important center of activity in natural history. The area abounded in fauna, flora, and fossils. He must return to this city and encourage the able naturalists he found there and foster the general spirit of scientific patronage that was evident to him. Invitations came swiftly, with medical college officials urging Agassiz to return the next year. Indeed, Agassiz did return, but not until 1849, as he had much business to attend to in Cambridge, by which time he had given up any idea of returning to Europe.[32]

Instrumental in persuading the American Association for the Advancement of Science to hold its third meeting in Charleston, Agassiz knew that all of the major naturalists and many others in the city would not only welcome the meeting in the culturally most southern of southern cities but would also prepare papers for the sessions. Indeed, they did. As already noted, Bachman, Ravenel, and Holbrook presented papers; so too did Lewis Gibbes and Frances S. Holmes. In addition, a few other Charleston residents and scientists from elsewhere in the region took advantage of this singular opportunity to show that science in the South was flourishing. Holbrook's paper was titled "On the air-bladder of the Drum-fish, Pogonias fasciatus, and the mechanism by which the sound is produced." Unfortunately, nothing is known of the contents of the paper, for Holbrook never gave a copy to the secretary. For him, the great task was to get on with his important volume on South Carolina fishes.[33]

In 1855, Holbrook published an article titled "An account of several species of Fish observed in Florida, Georgia, &c.," in which he described two new species: *Pomotis marginatus* (= *Lepomis marginatus*) and *Brytus gloriosus* (= *Enneacanthus gloriosus*), both of which are colorful sunfishes. Although he might have presented the paper before the Elliott Society of Natural History (ESNH), which had been founded partly in response to a suggestion by Agassiz, Holbrook elected to submit the paper to the ANSP for publication in its *Journal*. He had not been active in the affairs of the Elliott Society and was probably aware that the ANSP would get the article in print sooner than the ESNH could do so. It is likely that Holbrook collected some of the fishes described in that article while he was on an expedition in 1853 along the coast of Georgia and Florida. Agassiz had originally intended to be on the schooner with Holbrook but had backed out at the last moment. Holbrook had become so ill in February of that year that he relinquished his duties at the medical college for the rest of the term. By April, however, he felt strong enough to make the trip, which, he believed, would hasten his complete recovery and allow him to collect numerous specimens. With him on the schooner was the young Charlestonian Gabriel Manigault, who would complete his studies at the medical college the following year, go to Paris to study medicine and natural history, develop skill as an expert osteologist, and later become curator of the Charleston Museum. Manigault considered Holbrook to be one of the three best professors in the medical college.[34]

By 1855, Holbrook had completed *Ichthyology of South Carolina*. Unlike his *North American Herpetology* and *Southern Ichthyology*, he had this volume published in Charleston. The work contained 182 pages of descriptions and twenty-seven plates done by Richard. Information on the size, color, habits, and distribution of the species added to the quality of the work, which included four new species and two new genera. Meanwhile, as with his work on reptiles and amphibians, Holbrook was dissatisfied with some of the drawings, and only a few copies of the book were run. Just after some new plates had been prepared in Philadelphia, the building in which they were stored caught fire and burned to the ground. In due course, Holbrook turned this discouraging event to his advantage, however, for he set about to correct some errors and to make some of the drawings "more accurate and highly finished." Thus, he substituted some new drawings, done by Agassiz's gifted artist, Jacques Burkhardt, and the superbly talented Auguste Sonrel. The expense of the plates was high, however, and Holbrook called upon the South Carolina legislature to grant a subvention to assist in producing the new work. That aid, noted Holbrook in the preface to the 1860 edition, was "freely given." Mindful of the achievements of the naturalists in the state, especially as sectional hostility worsened, state leaders had already granted support to Francis S. Holmes for his second book on South Carolina fossils and to the Elliott Society to publish a journal. Works of natural history, they proudly proclaimed by their financial support, could be done in the South and did not have to be sent to a publisher in Philadelphia or New York.[35]

The second edition of *Ichthyology of South Carolina* was published in 1860. Beautifully done, it consisted of 205 pages of descriptions and twenty-eight color plates. Unfortunately, Plates 27 and 28 were reversed, and, as in *North American Herpetology*, the arrangement of species was somewhat haphazard. The young ichthyologist Theodore Gill noted those problems, and he criticized Holbrook for using some outdated names. Nevertheless, Gill praised Holbrook for "his great zeal . . . and his laborious and pains-taking endeavors to perfect" the second edition. Indeed, Gill declared that he hoped Holbrook would produce another volume to include even more species. Unfortunately, Gill's review did not appear until 1864. The Civil War had broken out soon after the publication of *Ichthyology of South Carolina*, and, except for the belated publication of Gill's review, the book received little notice until much later. Holbrook had proven, however, that he was a keen student of fishes. Although he described as new several previously known species, altogether he had added eight new species and two new genera of fishes.[36]

As the second edition of *Ichthyology* was in press, Holbrook learned of the illness of John Eatton LeConte, and in a letter to Joseph Leidy on February 12, 1860, he told his Philadelphia colleague that he was grieved to learn that "the old major" was suffering and would not be able to make his annual visit to the South. A month later he learned of the death of his "old and trusty friend" Hallowell, but he rejoiced to hear that LeConte was on the mend. Soon thereafter, however, Holbrook himself was seriously ill. Bothered by intestinal bleeding, Holbrook grew so weak that "great fears are entertained concerning him," said young John McCrady. By late April, however, Holbrook was on the road to recovery, though, Harriott noted, he "has not regained his usual strength." Unable to attend fully to his duties at the medical college and nearing his sixty-sixth birthday, Holbrook decided to notify the president and trustees of his intention to retire at the end of the current term.[37]

Although he had never earned much as a professor of medicine or in private practice and while the expenses of his collecting trips and publishing efforts had consumed much of what he had earned, Holbrook could retire in relative comfort, for, although he did not personally own much, his wife was heir to considerable wealth. Devoted to her husband, Harriott Pinckney Rutledge Holbrook freely shared her possessions with him. By that time, she owned the house in Charleston, the Belmont estate, and thirty-four slaves, and had several thousand dollars in her bank account.[38]

Financial security seemed certain for the Holbrooks as the year of 1860 came to a close, but the clouds of secession were darkening the nation's skies. As the threat increased, Holbrook counseled caution. Sectional differences could be settled amicably, he believed. After all, moderation and a generous spirit had carried him through more than six decades of life, and he enjoyed the admiration of northerners and southerners alike. By then, however, he was deeply rooted in the South, though his ties with friends in the Northeast remained strong. Indeed, many of his northern friends were not persuaded that a war should be fought to abolish the institution of slavery, but they did believe that the separation of the Union by armed rebellion was an unacceptable action. Once the storm broke, Holbrook elected to stay with, and serve, the South.[39]

From Alpha to Omega

T o escape the sweltering heat of the summer months in Charleston, some families fled to Sullivan's Island to savor the effects of the cooling ocean water and the gentle breeze that swept the largely uncultivated isle. While adults basked on the broad beach, children splashed in the surf. In addition to Edmund Ravenel, those attracted to Sullivan's Island around 1820 included Lewis Ladson Gibbes, his wife Maria Henrietta Drayton, and their five young sons and a daughter. For their firstborn, ten-year-old Lewis Reeve, however, the excursion was an opportunity to explore nature. Insatiably curious, the small and studious lad wandered inland among the thickets of shrubbery to collect leaves and seeds and roamed over the beach to pick up mollusk shells and catch crustaceans. His interests included other subjects as well: mathematics, astronomy, physics, chemistry, mineralogy, music, poetry, and languages. Lewis learned to play the piano, the organ, the violin, and the flute, and eventually he mastered Latin, Greek, and French and could read Spanish, German, and Hebrew.[1]

Both of Lewis's parents were descended from prominent Charleston families, but, while they enjoyed a comfortable living, they were not wealthy. They did place great stock in education, however, and found the means to send Lewis to good schools. The elder Gibbes had been a student at Eton and later at the Sorbonne, and his wife was a student of literature and an accomplished amateur botanist. In 1821, they enrolled Lewis in the grammar school of the University of Pennsylvania. Their reason for choosing that school is unknown, but evidence indicates that Lewis Ladson Gibbes and his family lived in Philadelphia during the years 1821 and 1822. At some point thereafter, the senior Gibbes bought 767 acres of land near the village of Pendleton, South Carolina, located in the northwest corner of the state. Whether he made the choice because Pendleton Academy was close or because he thought the location was a good place for a plantation is uncertain. In any case, the elder Gibbes built a large house on the estate, and purchased thirty-two slaves to run the plantation. Young Lewis enrolled in Pendleton Academy in 1823. An excellent student, he gained particular notice for his achievements in mathematics, French, Greek, Latin, and the classics. In his spare time, he studied local flora, conducted chemical experiments, and, with a good telescope given to him by a cousin, observed the stars and planets.[2]

The birth of another brother in 1821 and a second sister in 1823, had swelled the Gibbes brood to eight. In addition, the plantation had cost so much that young Lewis was likely beginning to wonder whether his parents could afford to send him to college. A serious blow struck the family in April 1826, when the forty-two-year-old wife and mother Maria died. The elder Lewis persisted in his goal of getting his son Lewis through the Academy and into a college, however. After graduating in 1827 from Pendleton, with the highest honor, Lewis Reeve was admitted to the junior class of South Carolina College, in Columbia. He began his studies there in January 1828 and quickly impressed his professors, especially in mathematics, natural philosophy (physics), and chemistry. In December 1829, Gibbes delivered "the Salutatory Oration" and received the baccalaureate. Then, he returned to Pendleton and commenced the study of medicine under the direction of his cousin Arthur S. Gibbes, a local physician.[3]

His plan to enter a medical school seemed to be in doubt, however, when his fifty-seven-year-old father died near the end of Lewis Reeve's first year at South Carolina College. Along with his brothers Charles and John, ages seventeen and sixteen, respectively, Lewis Gibbes suddenly found himself thrust into the role of guardian, with two of his seven siblings under the age of eight. Apparently, financial assistance came from relatives, especially from an aunt, Henrietta Drayton. The situation improved further in mid-1830, when, upon the sudden death of the principal of Pendleton Academy, the trustees appointed Gibbes as the acting principal and instructor in classics and mathematics. Determined, however, to pursue the study of medicine, Gibbes left the position in November 1830, moved to Charleston, and enrolled in the Medical College of the State of South Carolina. He left his brother Charles to run the plantation and care for the younger children.[4]

No doubt Gibbes made the decision to attend the medical college in Charleston rather than one in the American Northeast because his father had left little money to his children. As with an increasing number of southerners, however, regional pride also played a role in his decision not to study medicine in Pennsylvania or New York. In addition, by the early 1830s, many southern physicians were touting the notion of the "medical singularity" of their region, and they argued that "the particularity of southern medical practice" required training that could best be obtained in their own medical schools. After the 1830s, some southern families continued to send their sons to the older, more reputable medical colleges in the Northeast, but the number who did so was declining. The decision to return to Charleston proved to be fortuitous for Lewis Gibbes, however, for it whetted his interest in natural history. Of course, at the medical college he received instruction from the naturalists Elliott, Ravenel, and Holbrook, and he became acquainted with Bachman as well. Near the end of his first year of medical study in 1831, however, Gibbes learned that South Carolina College intended to fill the position of tutor in mathematics and that the trustees of Pendleton Academy and several of his former professors were recommending him for the post. As a salaried position would help him to "assist a family of seven brothers and sisters" and because he loved mathematics, Gibbes decided to become a candidate. With the support of Thomas Cooper, the president and professor of chemistry, and of his cousin Robert W. Gibbes, assistant professor of mineralogy and chemistry, Lewis Gibbes easily won out over five other candidates.[5]

A prodigious worker, Gibbes not only tutored students in mathematics but also continued to study medicine as a student of the local physician Thomas Wells. In addition, he commenced a study of local flora, which led to the publication of his *Catalogue of the Phoenoga-*

mous Plants of Columbia, S.C. and Its Vicinity in 1835. Although it contained no descriptions, that list of approximately 775 phanerogams, or flowering plants, would eventually establish his reputation as one of the most knowledgeable authorities on southern flora and bring him to the notice of the nation's leading botanists, John Torrey and Asa Gray. Meanwhile, he continued to study astronomy and, in a newspaper article on December 17, 1834, gave a report on the solar eclipse that occurred on November 30 of the same year. He also drafted a notice on the value of calculating the "resistance of fluids," as formulated by his colleague James Wallace, the professor of mathematics at South Carolina College. Published in the *American Journal of Science* in 1835, the brief article revealed Gibbes's knowledge of physics and mathematics.[6]

By late November 1834, however, Gibbes was feeling somewhat uncertain about his future at South Carolina College. The source of the problem was President Cooper. Appointed to the faculty in 1819 and elevated to the office of president a year later, Cooper was a man of extraordinary ability, but he held unorthodox views on religion. Those views might have elicited no complaint or special notice had Cooper kept them to himself or at least within a close circle of friends. But, combative and outspoken by temperament, Cooper resisted every effort to keep his views under his vest. The board of trustees defended him nobly in December 1831 and again in late November 1834, but the public outcry was so great that the trustees eventually buckled under the pressure; on December 9th, following the typical pattern in southern colleges, they called for the resignation of the president and the entire faculty. Gibbes need not have worried, however, for soon thereafter the trustees appointed him as temporary professor of mathematics, a post he filled until June 1835, when he elected to return to Charleston to complete the program in medicine. Had he chosen to be a candidate for the permanent post, he would certainly have received strong support. Indeed, in recommending him for the position, one trustee declared that "he has no superior of his own age" and praised "his zeal and steady habits of application." Five months after his departure, Gibbes heard from Cousin Robert Gibbes that the tutorship would likely be restored and that Lewis would unquestionably receive the unanimous support of the trustees if he merely let it be known that he would accept. "The kindest feelings towards you prevail in the

Board," said Robert Gibbes. But since Lewis Gibbes wanted to earn the M.D. and to study in Paris, he declined to return to Columbia.[7]

Gibbes finished his medical studies in March 1836, after completing a thesis, written in French, for which he received the highest honor. Meanwhile, he had been trying to raise money. His brother Charles had experienced little success in making the plantation profitable and in late 1834 had moved to Texas, leaving his brother John in charge of the estate. John was no more successful and even borrowed $700 from Lewis. In 1837, with the consent of Lewis, he sold the plantation and also left for Texas. Even though Lewis Gibbes had saved some money, he had to borrow more from his Aunt Henrietta Drayton in order to make the trip to Paris, for which he set out soon after graduating. With him he carried collections of South Carolina plants and marine and freshwater mollusk shells from the Southeast to exchange with French and German naturalists. In his valise he had letters of introduction from John Edwards Holbrook and the Charlestonian and American statesman Joel Poinsett.[8]

Traveling first to Philadelphia and then on to New York, Gibbes collected plants in the vicinity of each city. He also visited Niagara Falls and was intrigued by the possibility of working out a mathematical method of determining the rate of recession of the bed of the great cataract. In June of 1836, Gibbes set sail from New York for the port of Le Havre and by August was attending lectures at the French Academy of Medicine. Within a few weeks, he had become acquainted with many of the noted professors of medicine and naturalists in Paris. In early October, he presented the collection of plants and shells he had brought to the Muséum d'Histoire Naturelle and received a number of specimens in exchange. In his spare time, the twenty-six-year-old American physician and naturalist attended meetings of the French Academy of Science to observe "instructive and animated discussions" by illustrious figures. The "dark dingy . . . [and] absolutely dirty appearance of the Cathedrals, churches, Palaces" and other buildings in Paris initially disappointed the young traveler, but later he came to appreciate "the elaborate and exquisite architecture" of some of them. Indeed, he said, "I really felt their magnificence." News from home proved to be depressing, however, for both his twelve-year-old-sister Esther Marie and his eleven-year-old-brother Nathaniel died during the first few months he was abroad, and by mid-1837, Gibbes had to ask his Aunt Henrietta for an additional loan. During the evenings and on weekends, the impecunious visitor read commentaries on the Bible and made notes on points that impressed him. Thwarted by lack of funds, he had to forgo his intention to travel elsewhere in Europe, confining his sojourns to nearby regions to collect plants. Gibbes left France during the fall of 1837 and was back in Charleston by November of that year, thinking it might be the best location for establishing a medical practice.[9]

As he was pondering his options, however, a fortunate circumstance occurred: the chair of mathematics at the College of Charleston fell vacant. Gibbes quickly applied for the position. With an impressive background of experience and training and with recommendations from supporters in Columbia, he was almost a certain choice. Indeed, the trustees notified him on February 3, 1838, that he had been "duly elected Professor of Mathematics." Perhaps he anticipated an enduring connection with the college, for, within a few weeks after the appointment, he informed a friend in Paris that he had "left medicine . . . perhaps forever." Nor could he be persuaded to leave Charleston, even when, three months later, and again in 1845 and 1846, he received word that South Carolina College was searching for a professor of mathematics and that Gibbes would readily be selected for the post if he applied. Obviously enjoying his affiliation with the College of Charleston, he declined to do so on each occasion.[10]

During the first decade after he was appointed to the College of Charleston faculty, Gibbes devoted himself to instruction, but he did find time for some scientific work and minor publishing. Within a year after his appointment, the trustees added natural philosophy to his duties. No doubt Gibbes liked the subject because of its close association with mathematics and astronomy, but the task was rendered more demanding by the paucity of equipment for demonstrating physical principles to his students. To a friend in Paris he wrote that he had to design and make his own apparatus, which left him little time for other activities. His ability to explain principles to his students soon became known, but so too did his reputation for exacting work from them. Although he could occasionally reveal to his students that he possessed a sense of humor, and although, outside the classroom, he displayed warmth toward his friends and relatives, especially to his young sister Louisa, Gibbes often revealed, as one former student said, a manner that was "austere, if not harsh, . . . [for] he was too dreadfully in earnest." The former student also recalled that the physically small professor "threatened to personally eject any student from his lecture-room who was talkative or frivolous." As a long-time colleague penned in his autobiographical memoirs years later, Gibbes was "accurate and methodical in everything he undertook," and, he added, considered the reading of light literature as a waste of time. Indeed, in 1838, Gibbes had counseled a young cousin that she should use every moment for serious study, for dedication is "the key to success." On another occasion, he advised the cousin to attend concerts and never spend time in going to balls.[11]

It is clear, however, that Gibbes possessed a forward-looking view of the purpose of education. He believed that students should learn principles and acquire "mental furniture for future use." Indeed, Gibbes complained that too many students "conceive their chief duty . . . to be 'to say lessons' and these from the 'textbook,' so that instruction given by me in lecture[s] . . . is considered important *only for the moment.*" He gave further notice of his dislike of rote learning in 1859, when, upon evaluating the teaching of Joseph LeConte, a professor of geology and chemistry at South Carolina College, he reported that LeConte strove "to impress upon his class . . . the principal cases related in the books, without calling on the class to examine in the field, cases actually occurring in nature." Although Gibbes never won the esteem later bestowed upon LeConte as a teacher, he acquired a reputation for his devotion to duty as an instructor. To his responsibility for teaching mathematics and physics, college officials later added courses in chemistry and duty as secretary to the faculty. Compounding his heavy load were the grossly insufficient funds allocated for equipment, supplies, and books for an already inadequate library. Under such conditions Gibbes's successful efforts are especially notable.[12]

As with many contemporary Americans, Gibbes collected data on daily weather conditions. Unlike most of the weather watchers, however, he joined company with a small group that sought to determine meteorological patterns, especially as related to the phenomena of storms. Although he published his reports and speculations in local newspapers, he hoped to bring them to national attention. By 1841, his interests had come to the notice of the well-known American meteorologist James P. Espy. Particularly fascinated by storms, Espy sought to develop a theory to account for them, believing, as did most contemporary meteorologists, that all storms had a single cause. Gibbes and Espy exchanged information several times between 1841 and 1844. Unfortunately for Gibbes, little came of his efforts, and he gained no significant standing because of his observations on the nature of storms.[13]

More rewarding was his work in astronomy, which he pursued diligently between 1841 and 1850. During that decade, he sent a score of articles on astronomical phenomena to local newspapers, including discussions of a solar eclipse, the transit of Mercury, comets, and the rings of Saturn. Gibbes published another thirty articles on astronomy in the local newspapers during the next fifteen years. Unfortunately, although his treatment of those topics was excellent, Gibbes made no effort to publish his articles in a scientific journal. He did, however, send copies of some of them to James Dwight Dana, editor of the *American Journal of Science*, for notice. He also sent copies to, and corresponded with, two of the leading contemporary figures in American science, namely, Joseph Henry and Alexander Dallas Bache. Gibbes had visited Henry at his home in Princeton, New Jersey, in 1839, and he corresponded with him several times thereafter, mostly on matters pertaining to physics. In December 1843, Henry solicited from Gibbes a letter of support for the appointment of Bache as superintendent of the United States Coast Survey. Henry complimented Gibbes by including him among "some of the best names in science" and referred to the "ingenious instrument [Gibbes] had invented for the determination of the height of a cloud," though, he added, Bache had indicated that it was not "intirely [*sic*] new." Bache won the appointment and by 1847 had asked Gibbes to establish an observatory in Charleston, under the auspices of the Coast Survey, for the purpose of making observations, eventually to be telegraphed to Washington. Bache promised Gibbes compensation of $300 "for a year's observation of [the] moon's culmination and occultations."[14]

Gibbes envisioned the construction of a good building for the observatory, but, while the College of Charleston trustees were sympathetic, they did not possess funds for such a project. Gibbes decided then to erect a small building in his own backyard and sent a plan to Bache, who liked it but said the building "should be wider to give you room to move about." Moreover, said Bache, he could allot a sum to help if the costs were higher. Gibbes drew mainly upon his own meager funds for the project, however. Meanwhile, in February 1848, Bache sent him some of the necessary instruments, to complement those Gibbes already owned. Gibbes then petitioned Bache to call his station the "Charleston Observatory," to which the superintendent readily assented. To Gibbes, the designation apparently signified the scientific standing of his city. The observatory was, after all, only the third one in the lower southern states, the others being at the University of Alabama and the University of Mississippi. As Gibbes told Elias Loomis in October 1854, he built the observatory, a wooden building of twelve by fifteen feet, with a sliding roof, and he proudly declared that he had been the "sole observer" since 1848. The house in which he lived, however, was in bad condition, and, in 1853, when the landlord decided to raze it, Gibbes was compelled to move. The property owner allowed him to continue to use his little observatory until June of 1854, after which he apparently ceased his work for the Coast Survey, although he continued to use his telescope from the upper floor of a college building for many years thereafter.[15]

From 1848 to 1850, Gibbes often labored into the night to carry out his duties, and on June 4, 1849, he wrote to the eminent Yale University naturalist James Dwight Dana that he had been so "much occupied with astronomical observations" that he had been unable to reply sooner to a query Dana had sent him. His observations were in addition, of course, to his teaching duties. Furthermore, he was receiving so many letters of request for zoological and botanical specimens and for information that, he confided to Dana, "I feel almost in despair. . . . I sometimes think I lend myself too readily to the requests of friends but cannot bring myself to reject their applications. . . . I too readily undertake more than it is possible for me to execute." That Gibbes did in fact often take on more obligations than he could manage without hampering his own scientific inquiries is evident in the abundant letters that have survived. Turning away neither novice nor accomplished scientist, he dutifully took time to reply to queries and to collect specimens they requested. Perhaps no contemporary scientist in America other than Spencer Baird, and almost certainly none in the South, spent as much time as did Gibbes in corresponding with novices. Among them was a distant cousin, Thomas Parker, of the Abbeville District in western South Carolina, who told Gibbes that "the want of a companion in . . . pursuit . . . [of scientific matters] and your residence . . . in a large sea-port City where all things requisite could be so easily obtained" had prompted him to elicit information and encouragement from him. Indeed, the situation described by Parker characterized the antebellum South as a whole and explains much about the comparative paucity of scientific pursuits outside the larger cities in the region. Lewis Gibbes seemed to recognize those difficulties, and thus encouraged many would-be scientists in the sparsely populated areas of the South.[16]

Gibbes did not neglect the requests of the advanced students of science, however, and frequently set aside his own work to honor their calls for assistance. The nature of the queries varied, indicating that Gibbes's interests and knowledge ranged widely. John Bachman had sought information from Gibbes about botanical specimens when Gibbes was a student at the medical college in Charleston. Gibbes also possessed some knowledge of mammals, and, in January 1842, he sent to Bachman a small mouse he had taken at Ashepoo, South Carolina, thinking it might be a new species, but, as Bachman noted, it was a form already described by John Eatton LeConte. Bachman also called on Gibbes to assist him with Latin and Greek names of genera and species but had to plead with his learned friend not to write the Greek names in Greek characters, one of the meticulous habits that irritated some of the admirers of the fastidious Gibbes. In the *Quadrupeds*, Audubon and Bachman thanked Gibbes "for several specimens of rare quadrupeds, and for his kindness in imparting to us much information and scientific knowledge."[17]

Gibbes's familiarity with mollusks, both living and fossil, was also remarkable, and the polymathic professor furnished specimens to a number of conchologists, including the little-known R. T. Brumby, professor of natural history at the University of Alabama, and the noted authority Isaac Lea of Philadelphia. In addition, he assisted his fellow Charleston naturalist Edmund Ravenel and furnished many specimens for Michael Tuomey during the latter's geological survey of South Carolina. For the assistance given by Gibbes, Lea named a freshwater bivalve in his honor. Gibbes's own collection of mollusk shells included approximately 1,100 specimens and nearly 300 species. Among them were specimens sent by the able naturalist James Hamilton Couper, owner of large plantations in coastal Georgia. Gibbes also sent terrestrial shells to the noted Boston conchologist Augustus A. Gould, to whom he confessed in March 1845 that he could not always fill his wants because he had "access to very few books" and had therefore "never seen descriptions" of some of them. When the Virginian Edmund Ruffin completed his report on the preliminary agri-

Lewis Reeve Gibbes, ca. 1874. Courtesy of Special Collections, Robert Scott Small Library, College of Charleston, Charleston, S.C.

cultural and geological survey of South Carolina in 1843, he acknowledged receipt of a number of fossil shells from Gibbes, and Tuomey, upon completing his survey in 1848, noted that Gibbes had prepared the "first list of any extent, of the [invertebrate] fossils" of the Charleston region. In addition, Tuomey noted Gibbes's "assistance in comparing recent and fossil shells" and in providing a "synopsis of Fauna of the State." In a separate article, describing the fossil skull of a primitive, toothed whale from Charleston, Tuomey indicated that a portion of it had been found by the Charleston naturalist Francis S. Holmes but that Gibbes had visited the site of the find, located "the rest of the scull [*sic*]," and carefully "picked up the fragments." The skull was, as Tuomey observed, a "valuable relic."[18]

The noted Harvard College astronomer Benjamin A. Gould understood something of the burdens borne by his correspondent Gibbes, and on November 12, 1849, wrote him that "the duties of American professors are usually so severe that they have felt themselves for the most part unable to devote their leisure hours to research and abstract investigations, [for] they have needed all the time for relaxation." He added, however, that Americans were increasingly inclined to consider scientists as "too valuable an article to be thrown away upon mechanical routine." Gould's estimate was correct, but improvement of the situation came more slowly to the South. Obviously, Gould held Gibbes in high esteem, for he had written mainly to inform Gibbes that he was planning to launch a journal of studies in astronomy, and he expressed a desire for his southern colleague to contribute articles to it. It is time, he said, for work to be "devoted to the advancement of science rather than [to] its diffusion," and, in a statement probably aimed at the able Charleston astronomer, he advised that, while "popular and rhetorical articles" have a place, they should be left to "other journals." Although Gibbes valued the opinion of his Harvard friend, and although he enjoyed the esteem of other noted American astronomers, including Ormbsy M. Mitchell, Sears Walker, and Stephen Alexander, Gibbes persisted in believing that he had a mission to edify Charlestonians about astronomical phenomena. His view was shared by the southern publishers J. D. DeBow and William Gilmore Simms, each of whom solicited contributions from Gibbes. One of the requests from DeBow was for "a few pages from your ready pen and elaborate studies" on the subject of tides. He asked Gibbes to explain "their theory, their action, various hypotheses, etc., etc.," for he believed that most readers knew little about the topic. Simms was especially solicitous of Gibbes, urging him to write for the *Southern Quarterly Review* in May 1849, boasting that his journal was "the only organ of letters in the South." Simms renewed his request in 1850, telling Gibbes that he knew his "mind is quite full of his topic" and that "just at this moment the scientific subjects are . . . provocative." In 1849, Gibbes had also received an invitation to write articles for the *Charleston Medical Journal and Review.* The editors were seeking contributions in all areas of "medical and natural science" from "men of ability."[19]

Gibbes published more than sixty articles in the Charleston newspapers. Many of them, as previously noted, dealt with astronomical phenomena, but the others covered a wide range of topics, including meteorological information, an explanation of the phenomenon of dew, a fire plan for Charleston, and the importance of an agricultural and geological survey of the state. Obviously convinced that some of his newspaper articles constituted contributions to science, he sent them to the *American Journal of Science* to be published as notices. Editor James Dwight Dana at first ignored them, for they were not cast in the form required for publication in the *AJS*, though he said nothing to Gibbes about his reason for not publishing them. In a letter to Dana on June 4, 1849, responding to a query about his collection of Crustacea, Gibbes seized the opportunity to insert an aside that he had "sent newspapers with [my] articles," but had "never seen them acknowledged in the Journal, though . . . newspapers from all parts of New England, from Ohio, etc. are acknowledged." Dana admitted that he had not personally seen Gibbes's newspaper articles until prompted to read them. After all, he observed, he could not read all of the newspaper articles sent to him. Now that he had looked them over, however, he had "observed that your articles were of real value, and exceedingly well drawn up." He praised Gibbes for "doing good service to science and the community," but, he added, "such articles are necessarily drawn out with more detailed explanations than would be desirable." He asked Gibbes to "pardon the past, and mark off what you wish inserted in the Journal, [and] I will try to do better in the future."[20]

Many months later, Gibbes told Dana, "I know that many persons consider 'a man who writes for the newspapers' as one employing a low and easy means of notoriety, and the articles as of less value than if they appeared in a journal of a more elevated character." He was aware, he added, that Bache considered such writers to be "pretenders to science," but, he said, "I regard the endeavor to instruct the minds and exalt the characters of those around us by articles written with honesty of purpose and truthfulness of statement . . . [in] a daily press as a most laudable purpose." Gibbes noted that he had published many such articles and would continue to do so. He promised, however, to follow the procedures, and later sent his newspaper article on "an Aurora seen here [Charleston] Sept. 29 [1851]." On November 21 of the same year, he told Dana he had hoped he would publish a notice of it and reminded him that he had sent other such articles, "but none have ever been copied in the Journal." As revealed in his later comments, Gibbes viewed the matter, first, as discrimination against articles in natural history and, second, as discrimination against southern scientists. Neither was a correct assessment. Dana published the notice in the *AJS* issue of May 1852.[21]

Dana recognized that Gibbes was an authority on the subphylum Crustacea (phylum Arthropoda), particularly the class Malacostraca, and since the task of identifying and classifying the large number of crustaceans collected on the Wilkes Exploring Expedition of 1838–1842 had fallen to Dana, he sought some information from Gibbes. As Louis Agassiz noted in 1852, in a letter to Dana regarding the larger decapods, "There is nothing to be found after Professor Gibbes has gone over the ground." Gibbes had been collecting specimens of crabs and shrimps for a number of years and had been receiving others from the Cuban naturalist Felipe Poey and, especially, from John G. F. Wurdemann, who collected a wide variety of specimens from Florida and Cuba for the naturalists in his native city. He sent Gibbes several decapods from Key West.[22]

Gibbes had made his mark as an authority on the Crustacea, however, by studying and identifying specimens in the collections of the Boston Society of Natural History (BSNH) and the Academy of Natural Sciences of Philadelphia. He journeyed to Boston in mid-August 1845 and examined all of the specimens in the BSNH Cabinet, working from 7:30 A.M. until dark for several days, even getting locked in once when he stayed too long. Soon after he completed a list of the species in the BSNH collection, the Society published it in its *Proceedings*. Two years later, he went to Philadelphia and studied the ANSP crustacean collection for a week. He also worked up a list of the species in that collection and appended some valuable notes. Because of the demands of his work for the Coast Survey, however, he did not have it ready to submit until 1849. It was published in the *Proceedings* of the ANSP in March 1850. Gibbes also studied the crustacean collection of the New York Lyceum of Natural History but, for reasons unknown, did not publish a list of the species there. Meanwhile, in 1849, he published a list of Crustacea in *Statistics of the State of Georgia*. Unlike the first two, that list included isopods, amphipods, and copepods. It also included the horseshoe crab, *Limulus polyphemus*, but that ancient species is now known to be a chelicerate, not a true crab at all.[23]

Those studies, while important and useful, did not constitute new contributions to science. They did, however, indicate that Gibbes was intimately familiar with crustacean taxonomy and that, by 1849, he was one of the leading American authorities in the field. Moreover, by that year he was planning more ambitious works, and he mentioned to his friend Augustus A. Gould of Boston that he was preparing four monographs on crustaceans. Before he finished those, however, he found a unique opportunity to combine his lists from the BSNH and the ANSP and to present descriptions of species in his own cabinet, including many that he believed to be new to science. He noted that his collection possibly included other undescribed species but explained, "the want of works of reference, and particularly want of [additional specimens] deter me from describing them at present." The third meeting of the American Association for the Advancement of Science, held in Charleston in March 1850, gave Gibbes a special incentive to present a paper on the malacostracan Crustacea of the United States. In his "On the Carcinological Collections of the Cabinets of Natural History in the United States" Gibbes listed all of the species he had seen in the three northern collections and in his own cabinet, the last of which he noted was "the largest . . . in the South," ultimately including more than 350 species. In addition to giving the location of each species, he provided commentaries on "the most remarkable" of the nearly 240 species listed and descriptions of twenty-four new species. The vast majority of species on the list were decapods, but it also included thirteen stomatopods, or mantis shrimps, noted for their powerful, snapping, raptorial claws in place of the pincers in decapods. Gibbes noted that he had "paid less attention to the lower orders . . . [for] few of them [were] found in the collections."[24]

Of Gibbes's twenty-four new taxa, at least thirteen are still valid: eight crabs, four shrimps, and a stomatopod. In the first group is *Pisa mutica* (= *Pelia mutica*), a small spider crab that attaches bright-colored sponges to its carapace for camouflage. Another of Gibbes's crabs is the gulf weed crab or sargassum swimming crab, *Lupa sayi* (= *Portunus sayi*), so called because the carapace of the adult, which is about two inches in width, resembles in color the sargassum weed in which it lives. Gibbes also described the tiny grapsoid crab *Grapsus transversus* (= *Pachygrapsus transversus*), whose carapace tapers posteriorly to resemble the upper half of an hour glass, and is lined with rows of "fine transverse plications." A fourth species described by Gibbes is the pentagon crab, *Cryptopodia granulata* (= *Heterocrypta granulata*), which is triangular in shape and possesses enormous chelipeds for its size. In addition, Gibbes described *Pagurus tricolor* (= *Clibinarius tricolor*), *Ilia armata* (= *Petrolisthes armatus*), *Ilyas aculeata* (= *Pitho aculeata*), and *Chlorodius floridanus* (= *Cataleptodius floridanus*). Four shrimps are also among Gibbes's valid taxa: the dorsally humped *Pontonia domestica*, commonly called the Atlantic pen shrimp, which is commensal with lamellibranch bivalves; the beautiful peppermint shrimp, *Hippolyte wurdemanni* (= *Lysmata wurdemanni*), whose transparent exoskeleton is banded in bright red; the riverine grass shrimp, *Hippolyte paludosa* (= *Palaemonetes paludosus*); and the striped snapping shrimp, *Alpheus formosus*, which bears a light mid-dorsal stripe and orange speckles on its carapace. Gibbes also described a rare species of the order Stomatopoda, *Gibbesia neglecta*, which he described as *Squilla neglecta*. This species was not reported again until 1933. Following their recent study of that stomatopod, Raymond B. Manning and Richard W. Heard have placed it in the genus *Gibbesia*, a name honoring its discoverer.[25]

Dana was impressed by Gibbes's contribution and, in a synopsis of it written for the *American Journal of Science*, he praised the paper as "enriched with many valuable notes . . . and descriptions of several new species." However, he had declined to publish the entire article in the *Journal*, he told Gibbes, "because of the dryness of the subject," adding that he did not even publish his own similar studies in the *Journal*. The contributions of Gibbes to carcinology also captured the attention of Joseph Leidy, a member of the ANSP and a splendid naturalist who was especially interested in paleontology. On March 1, 1851, he wrote to Gibbes to say that the Academy had recently received more than 100 species of Crustacea "from all parts of the world" and to offer him "an opportunity of studying . . . and describing them in the Journal of the Academy." The busy and overworked Gibbes obviously wanted to examine the new collection, but he could not get away from Charleston. He waited over a year before responding to Leidy's letter and then, after noting that he would not be able to travel to Philadelphia for at least a year, hinted for Leidy to send them to him. The specimens, he said, "are too fragile to risk their transportation here, unless with very careful packing and by steamer," adding that if anyone else wished to describe them Leidy should not hold them merely "to gratify me." Leidy ignored the hint to send them to Gibbes, and the opportunity passed.[26]

Continuing his interest in the Crustacea, Gibbes presented three papers on them before the Elliott Society of Natural History (ESNH). In the first paper, presented in April 1854, he described six species in the family Porcellanidae, and two years later he read a paper on the genus *Cryptopodia*, though his taxon *Cryptopodia granulata* has since been referred to *Heterocrypta*. Gibbes's last paper on the Crustacea was a complement to the description of *Ranilia muricata* (Milne Edwards, 1837). After the formation of the ESNH, Gibbes published most of his articles in its *Proceedings*, quite clearly as a consequence of his increasing sectionalist sympathies and pride in a southern scientific society. His decision worked somewhat to his disadvantage, not because the *Proceedings of the* ESNH was an inferior publication but because the first volume did not appear in print until 1859 and because its circulation was more limited than the journals of the older scientific societies. Other reasons account for Gibbes's decline as one of the leading American carcinologists, not the least of which were the heavy load he carried as a professor in the College of Charleston and his refusal to concentrate upon a more limited field of scientific inquiry. Added to those factors was his increasing sensitivity to presumed slights by northern publishers. Still, his work on crustaceans remains as a major contribution, and four species of crabs and a mantis shrimp genus now bear his name: *Percnon gibbesi* (Milne Edwards, 1853); *Albunea gibbesii* (Stimpson, 1859); *Portunus gibbesii* (Stimpson, 1859); *Polyonyx gibbesi* (Haig, 1956); and *Gibbesia* (Manning and Heard, 1997).[27]

Indeed, as the president of the College of Charleston would later say in tribute to the man who gave fifty-four years of service to the institution, Gibbes "came as near as possible to the realization of the Baconian dream, the taking of all scientific knowledge for his province." In addition to his contributions to the fields already mentioned, Gibbes described a new salamander: *Menobranchus punctatus* (= *Necturus punctatus*), the dwarf waterdog. He also presented this description at the AAAS meeting in Charleston in 1850, but he published a fuller account of it in the *Boston Journal of Natural History* in 1853. Like other species of the family Proteidae, the dwarf waterdog is a neotenic salamander, one that retains its larval form of external gills

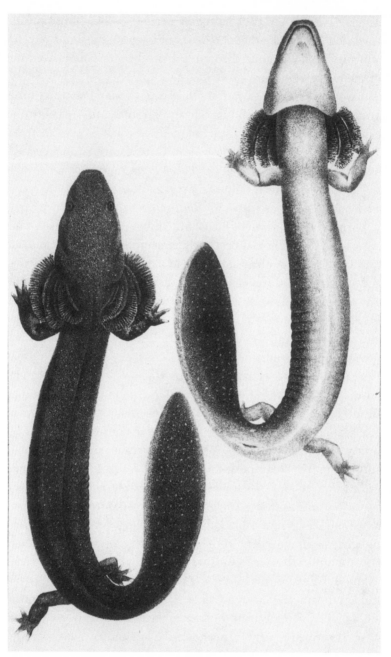

The dwarf waterdog, *Menobranchus punctatus* (= *Necturus punctatus*), was described by Lewis Reeve Gibbes in the *Boston Journal of Natural History* 6 (1853): 369–73. The above figure of the salamander is from a lithograph by A. Sonrel, from the original drawing by J. H. Richard. Courtesy of The Charleston Museum, Charleston, S.C.

throughout life. An inhabitant of slow-moving streams, it ranges from central Georgia to the Atlantic Coast and northward on the Coastal Plain to southern Virginia. Nine years earlier Gibbes had described another salamander, *Salamandra melanosticta*, which he had collected in the vicinity of Abbeville, South Carolina, while visiting Dr. John Perkins Barratt, the remarkably good naturalist who also assisted Bachman and Holbrook. *Salamandra melanosticta* turned out to be a synonym of *Hemidactylium scutatum*, the four-toed salamander. Thus, Gibbes added only one taxon to the class Amphibia, but it was a significant one nevertheless and a remarkable accomplishment, considering all of his other pursuits.[28]

Among Gibbes's interests was geology, especially mineralogy. In fact, he built an extensive collection of minerals that filled "20 shelves, with 30 to 36 specimens on each," or somewhere around 700 altogether. During the summer of 1849, while examining some of the new gold mines near the north Georgia town of Clarkesville, Gibbes discovered "an abundant precipitate" of xenotime crystals, a sample of which he sent to Yale University for analysis. The analysis was probably done by Charles Upham Shepard, an authority on minerals. Shepard had studied with Yale's pioneering chemist and mineralogist Benjamin Silliman and was married to an adopted daughter of the Sillimans. Shepard was also closely associated with the nation's leading mineralogist in the nineteenth century, James Dwight Dana, who had been one of his students when Shepard was a lecturer at Yale and who later married the third daughter of the Sillimans. In the winter of 1835, Shepard had become a lecturer at the medical college in Charleston, and he taught there every winter until 1861 and from 1865 to the early 1870s. He joined the faculty of Amherst College in 1845, an association he continued until his retirement in 1877. Gibbes sent Dana a letter on his discovery on November 21, 1851. Meanwhile, in the same year, Shepard had discovered xenotime in McDowell County, North Carolina, and sent his comments to his brother-in-law. Dana informed Gibbes on December 10, 1851, that because Gibbes had made his discovery two years previously, he was placing his letter in the *AJS* ahead of Shepard's notice. Such generosity should have shamed Gibbes for his complaints about Dana's failure to publish notices of his newspaper articles. Perhaps it did, for, in an immediate reply to Dana, Gibbes humbly noted that, since Shepard may have "obtained" xenotime earlier, "his announcement . . . ought not to give me any trouble."

After all, he added, "these objects . . . are not mine, nor his, nor yours, but our Maker's, and he appointed whom he chooses to bring them to light. All *contests* about priority I desire to avoid." A second notice of the discovery of xenotime could only help to establish a fact, he added, because "two . . . witnesses have concurred." Gibbes was not done, however, and sought to reinforce his own claim to priority by noting that he had earlier talked with Shepard and "mentioned my crystals." Perhaps that mention, he suggested, "induced Shepard to examine his [own] collections or perhaps he was already doing so." Clearly, contrary to his proffered largesse, Gibbes believed he deserved priority of claim.[29]

In 1836, Gibbes had visited Niagara Falls on his way to Europe, and ever since then he had been intrigued by the erosion of the bed of "the mighty cataract." In 1856, after reading yet another estimate of the erosive effects of the falls, he penned his own views on the subject and sent them to the noted geologist James Hall, of Albany, New York. Although Hall did not accept Gibbes's views, he believed the paper worthy of presentation before the tenth meeting of the AAAS, in Albany, New York, in August 1856. Gibbes also read the paper before the ESNH in January 1857 and later published it in that society's *Proceedings*. Agreeing with Sir Charles Lyell's estimate that the process of erosion had taken 35,000 years, Gibbes maintained that the recent argument of the Swiss naturalist Edouard Desor of only forty yards of erosion since 1678 would equal "less than a foot per annum" and thus require "*hundreds of thousands of years* for the present state of the falls." By his own calculations, Gibbes found the fall on the Canada side was eroding its bluff at a rate three to four times faster than that on the American side. Admitting that his calculations were based on approximations of the volume of water, the rate of flow, type of bedrock, and other factors, all of which he based upon information contained in other geological studies, Gibbes endeavored to derive a rate of erosion. His estimate hardly broke new ground, but it did show the kinds of data needed and a workable method of calculating the rate of erosion. His comments seem to suggest, moreover, that he objected to any notion of a really ancient earth. They do not necessarily imply, however, that he opposed the contemporary speculations of an earth created long before the appearance of humans.[30]

Earthquakes also interested Gibbes, and on September 1, 1858, he presented before the ESNH a paper on the phenomena associated with a minor earthquake in the Charleston region on December 19, 1857. He also wrote a newspaper article on the event. His interest in observing the effects of an earthquake was especially notable when a major shock rocked and heavily damaged Charleston in 1886, and his report of that event later proved to be useful to geologists. Gibbes also possessed some knowledge of paleontology, as indicated by his interest in fossil crabs, his discovery of the remainder of the fossil whale skull, and various of his comments, especially his newspaper article on "Geology of the Sidewalk," in which he discussed the "fossiliferous rocks" in and around Charleston. While these diversions in geology reveal yet another aspect of Gibbes's scientific universalism, they also indicate what the man could have done had he concentrated more upon the fields in which he made his greatest contributions. Such universalism was not uncommon in his era, but several of his contemporaries were beginning to recognize a need to specialize or at least to limit themselves to two or three areas of scientific inquiry, as they became aware of how much was to be learned, for example, just in vertebrate zoology or about the mammals or the reptiles alone.[31]

As Gibbes told Asa Gray in March 9, 1852, whenever he was "in the country and especially in the mountains," his interest in botany was renewed "with all the emotions of a first love." He had said essentially the same thing in a letter to Dana a few months earlier, indicating that he believed he had discovered a new plant in the genus *Rhus* and hinting that "some Journal" might like to publish it. Dana did not take the bait, but Gibbes was undeterred and spent considerable time during the next few years in pursuit of his "first love," all the while keeping busy with other duties and inquiries as well. A prodigious collector of botanical specimens, he eventually built a large herbarium, containing an estimated 4,000 species of native plants and some 900 from other lands. In addition to the botanists Torrey and Gray, he exchanged specimens with J. W. Bailey, a specialist on the algae. Other than the catalogue of the flowering plants of Columbia, South Carolina, that he published in 1835, however, Gibbes wrote only two other brief papers in that field, namely "Botany of Edings' Bay" and "On the representatives of the genus Cactus in this State [South Carolina]," both of which he presented before the ESNH, the first in October 1857, the second in January 1858. In the former, Gibbes listed all the flora he found on the small island called Edings' Bay, located at the mouth of the Edisto River. Organized by family, the list included seventy species of trees, shrubs, and vines and contained comments on several of them. Although it included no descriptions of new species, it is useful as a record. The paper on the cacti listed four species and offered a few comments on them. Especially interested in the trees of the genus *Pinus*, which then included the hemlock and other related trees, Gibbes sent a number of specimens to Gray. Thus, although he devoted considerable time to botany and was extremely well versed on the subject, he made comparatively few original contributions to the field.[32]

In October 1851, Joseph Henry invited Gibbes, as one of the "gentlemen distinguished for their attainments in different branches of knowledge," to present a series of lectures under the aegis of the Smithsonian. He issued another invitation to Gibbes in June 1852 and also invited him to "give us a paper on some branch of *physiques*." Henry noted that the Smithsonian had a two-year backlog of papers in natural history. Gibbes offered no paper in physics, however, and in December 1853 Henry sent him yet another invitation: "We shall be happy to receive from you at any time, any paper which you may prepare, whether it be on natural history or on any other branch of science." He stated his preference, however, for "a memoir on physics." Gibbes declined, though he honored Henry's request in June 1854 to referee a paper on a topic in mathematics. Henry did not give up in his attempt to draw the able southern scientist into the circuit of Smithsonian lectures, and in August 1857 he again extended an invitation to Gibbes, who declined once more.[33]

Yet, late in the summer of 1859, Gibbes complained to Henry that the Smithsonian was neglecting him because he was a southerner. Henry denied "that there is any design to 'shove you out' from any privilege extended by the establishment to scientific men." Indeed, quite the opposite was true. Henry had warmly befriended Gibbes and strongly urged him to present lectures for the Smithsonian and to contribute to the institution's noted *Contributions to Knowledge*. Moreover, Henry had neither shown bias toward southerners nor offered any objections to the institution of slavery. By then, of course, sectionalist feelings were running high, and Gibbes was quick to perceive a slight, real or imaginary. Furthermore, he was likely miffed that, although Henry had left the door open to a contribution in natural history, he had stressed his preference for one in physics. That preference was not directed solely to Gibbes, however; it was merely a part of Henry's well-known intention to make the Smithsonian a center of research. That Gibbes declined the invitations to present lectures for the institution was probably due in large measure to his onerous load at the College of Charleston, his service as the president of the Elliott Society of Natural History, his multifarious scientific pursuits, and his duties as the father of a large family, but his decision in part was almost certainly a consequence of his sectionalist propensity.[34]

Gibbes was an able physicist, but unlike Bachman and like most of his fellow naturalists in Charleston, he was not inclined to conduct experiments, either in physics or in chemistry. Although the Baconian philosophy of scientific inquiry, which stressed the avoidance of hypotheses and theories and emphasized the inductive method, had not lost its hold upon the minds of all northern scientists, it continued to carry considerable weight in the mind of the southern scientist. The relative paucity of interest among southerners in sciences other than natural history seems to reflect their entrenchment in Baconianism. Gibbes was no doubt aware of the ideas of Alexander von Humboldt and the Humboldtian model of scientific inquiry, which stressed the importance of interrelationships among, rather than the independence of, scientific phenomena. Since Bache was a leading advocate of the Humboldtian scheme, Gibbes was almost certainly familiar with the ideal promoted by his friend. By that model, Gibbes could retain his interest in physics or natural philosophy, viewing it as a means of determining the underlying system of order and laws. Perhaps that was the reason why mathematics, another of his teaching fields, also appealed to him. In any case, by early 1859 he had sent a paper on a topic in mathematics to J. D. Runkel, editor of the *Mathematical Monthly*. Runkel liked Gibbes's "Note on the Cycloid" and published it in the June issue. Encouraged, Gibbes soon sent a second paper to Runkel, titled "Note on Maxima and Minima," which Runkel published in the July issue. So impressed was Runkel that in November he told Gibbes, "any articles you may send me I will imprint as fast as possible." He added, "I do not get many contributions from your part of the country." Meanwhile, Gibbes had sent a third article, titled "An Easy Mode of Approximating to the Time of Vibration in a Circular Arc," which Runkel published in the issue for November 1860.[35]

As with Bachman, Gibbes possessed an exemplary commitment to accuracy in scientific inquiry, which was no doubt the reason why the clergyman had called upon him to assist in demolishing the mermaid hoax in 1843. Indeed, the dispassionate and coolly rational arguments in Gibbes's letters to the Charleston newspapers were perhaps even more telling against Richard Yeadon than was Bachman's blunderbuss of criticism. Yet Gibbes could also carry his commitment to perfection to the point of annoyance, particularly when language and accentuation were involved. His meticulosity resulted in a command of seven languages, but he could be petulant with those who made even a slight error. For example, in a letter to A. A. Gould on September 12, 1845, Gibbes noted that he was preparing a catalogue on accentuation of terms in natural history, and he noted that he disagreed with "the rule [Asa] Gray and yourself separately gave me, viz. That in compound words the last syllable of [the] first part or word is to bear the accent." He had, he added, "examined the principles of accentuation in Latin and Greek languages . . . and hesitate not to say that in Latin it is utterly untenable." Gibbes said he would therefore follow the correct Latin accentuation, since Latin "is the language of naturalists," but he declared that his intention was not to cavil, only "to be accurate." Gould responded six weeks later. Accentuation, he agreed, is important, and Gibbes's catalogue should be useful. He had, he noted, consulted with scholars at Harvard and confirmed that Gibbes was correct. Gibbes could not let the matter rest, however, and replied soon thereafter, saying " 'the rule' then is no rule." Furthermore, he said that when he had been in Boston and the so-called rule came up, he had kept quiet because no one asked his opinion, though at that very moment he had with him "*all* the rules requisite for accenting the latin [*sic*] words Naturalists use." He had not thought it proper, he added, "to instruct 'the Heads of Science,' " especially since he was a visitor in the "land of nations."[36]

Gould waited seven months before replying. Sensing chastisement by his Charleston friend, he said, "I could not help feeling that there was something churlish in relation to the accentuation matter." After all, he added, he never "pretended to be an authority" on accentuation. Gibbes tried to make light of the subject in his response on August 3, 1846. He stated that Gould had taken his comments too seriously, and, to show that he thought his criticism was offered in a jovial spirit, Gibbes related a game played by him, Bachman, and others when they were botanizing together. Anyone who could not immediately name a plant had to confess that he was "stumped." The person with the greatest number of "stumps" on the trip then became the object of derision. He had more than once been the victim, he added, and Bachman had "rubbed me hard" each time. It was in the same spirit, he declared, that he had rubbed Gould. But Gibbes was not quite willing to play the game either, for he complained, "I hardly think it quite just to called me 'churlish.'" He asked Gould, however, to pardon him for any unintended offense.[37]

Gibbes displayed a similar spirit of petulance in a letter to Spencer Baird, in March 1851. He had received a copy of the Smithsonian's publication of J. W. Bailey's microscopical studies and "was mortified . . . to find in more places than one *ovae* used as the plural of *ovum*." The mistake, he said, "is really a serious etymological error, the true plural being *ova*." Baird replied that he was embarrassed over the error and, ever the diplomat, assured Gibbes that the mistake was "not likely to occur again." Gibbes's colleagues in the ESNH were impressed by their president's mastery of accentuation and published his *Rules for the Accentuation of Names in Natural History, with Examples Zoological and Botanical* in 1860. The thirteen-page catalogue gained little notice, though in part that may have been due to its appearance only shortly before the Civil War began. Certainly, Gibbes was not wrong in wanting naturalists to accentuate words correctly or to use the proper plural of words, but his fussiness over them suggested that he was sometimes willing to allow form to take precedence over substance.[38]

Still, Gibbes enjoyed the esteem of his northern counterparts. For example, in 1840, two years after U.S. Secretary of War Joel Poinsett had established the National Institute for the Promotion of Science, Gibbes was elected to membership. One of Poinsett's aims was to create a national museum of natural history, and in May 1843 Gibbes sent some specimens to the corresponding secretary, Francis Markoe, for deposit in the museum: twenty-one species of mollusks and sixty-eight species of plants. He promised to send specimens of annelids, crustaceans, echinoderms, and more mollusk shells and plants. The life of the National Institute was brief, however, and the Association of American Geologists and Naturalists (AAGN), also founded in 1840, then became the only national organization in the United States devoted solely to science. Invited by Benjamin Silliman to attend the meeting of the AAGN in 1845 and again in 1846, Gibbes declined the first time but decided to attend the latter. He did not find it to his liking, however, because zoology and botany took second place to geology and because, in his hypersensitive view, too many of the participants were more eager to "exalt self" than to "advance science."[39]

Meanwhile, in 1844 the ANSP had elected him as a corresponding member, and three years later the BSNH had done the same. Thus, the two leading organizations in natural history recognized Gibbes as a significant naturalist. Samuel George Morton urged him to attend the first meeting of the American Association for the Advancement of Science (AAAS), formed from the AAGN in 1848, but as Gibbes had traveled to the North three times during the past few summers, he probably could not afford the expense. Nor did he attend the second meeting, held in Cambridge, Massachusetts, in August 1849. The third meeting of the AAAS, in March 1850, was held in Charleston, of course, and Gibbes played a very active role in it, serving on three committees, formally commenting on a paper, presenting five of his own papers, serving as its secretary, and, later, editing the proceedings for publication. At its next meeting, the Association voted its thanks to him "for the industry and fidelity with which he performed the duties of Secretary and for the promptness of the publication of the Proceedings of the meeting." Gibbes continued to stand in high regard among the leaders of the AAAS, and in 1859 was appointed as a member of the Committee on the Coast Survey.[40]

Like Bachman and Ravenel, Gibbes experienced the tragedy of losing children. Of the nine borne by his wife Anna Barnwell, whom he married in 1848, four or five died in infancy or childhood, including the first of his only two sons, who died in 1858, just short of his seventh birthday. Relatively few details of the personal and family life of Gibbes are available, for, although many of his letters are extant, the more personal ones have been lost. The views of Gibbes upon slavery and the argument over the races as separate species must be inferred from various sources. There can be no doubt, however, that the devout Episcopalian supported slavery, for the record indicates that he legally shared ownership of several with his brother Charles, and the census records show that he kept six in Charleston, apparently using two as servants and likely hiring out the others in order to enhance his relatively low income. Gibbes did not enter the great controversy between Bachman and the proponents of pluralism. That he was an ardent supporter of the South and its culture is clear enough, however, and, when South Carolina chose to secede from the Union, Lewis Gibbes was among those who celebrated the occasion.[41]

Ancient Animals

I n 1839, as he directed slaves to dig for the marl formed from the sediment of ancient seas, a young planter in St. Andrew's Parish, a few miles west of Charleston, eagerly extracted mollusk shells of all shapes and sizes; shell-like tests, or calcareous skeletons, of plump sea urchins; tiny mats of bryozoan exoskeleton; and thin little cuticles that once housed the feathery feeding and respiratory lophophore of brachiopods. The planter, Francis Simmons Holmes, had no training in paleontology and, in fact, had left school in 1829 at the age of fourteen. Native intelligence and a driving curiosity about those traces of the past were sufficient, however, to make up for his incomplete education. His interest in natural history had likely been whetted during his boyhood days in Charleston, when he viewed the collections of the old Charleston museum.[1]

Born in Charleston on December 9, 1815, to John Holmes and Helen Boomer, Francis was the fourth of seven children in the family. Although John Holmes was not wealthy, he had prospered as a merchant and thus earned a place of respect in Charleston society. He died when Francis was only eight years of age, but, because he had been a member of the charitable organization known as the South Carolina

Society, his children were entitled to a free education in the Society's school. Francis later enrolled in the academy founded by John England, the Roman Catholic bishop of the Charleston diocese. Illness forced the boy to withdraw in 1829, however, and his mother sent him to the English city of Liverpool to live with his uncle George William Lee, who was apparently an agent of John Fraser and Company, a Charleston shipping firm. When Francis returned a few months later, he went to work in a mercantile enterprise, probably for the same company with which his brother-in-law George Alfred Trenholm was associated. Later, he operated a dry goods business. In 1837, he married Elizabeth S. Toomer, of Charleston.[2]

In 1839, Francis Holmes moved to adjacent St. Andrew's Parish to operate the 811-acre plantation known as Springfield, which Trenholm had purchased. Although he had no experience in farming, the young man succeeded in the new venture. Local newspapers were noting his attainments by 1840, and two years later Holmes enhanced his standing by publishing a little book titled *The Southern Farmer and Market Gardener*. He had first sent the manuscript to the Horticultural Society of Charleston, and a committee of that organization had strongly praised it, one prominent member saying that Holmes had adapted "many valuable principles . . . to the peculiarities of the Southern latitude." The book sold so well that Holmes reissued it in 1852. Meanwhile, in 1844, he presented to the State Agricultural Society a report of a series of agricultural experiments he had been conducting. The carefully controlled experiments showed that marl was an effective fertilizer. The Society awarded him "the Ruffin premium . . . for his well-conducted experiments." The prize was named for the Virginian Edmund Ruffin, who had published *An Essay on Calcareous Manures* in 1832. In fact, Ruffin had visited Springfield in 1842 while he was conducting a preliminary agricultural and geological survey for the State of South Carolina, and Holmes had shown him not only nodules containing phosphate but also a collection of the fossils he had unearthed on the plantation.[3]

By then, according to his own recollection, Holmes had built a cabinet of "thousands of remarkable specimens . . . of interesting and beautifully preserved forms of shells, teeth, . . . bones [and] rocks filled with the casts of shells, coral and corallines." No doubt his collection was larger than any other in the state, indeed, probably in the South. In 1843 he sent several fossil shells to Lewis Gibbes and around the

Francis Simmons Holmes, ca. 1850. Courtesy of Special Collections, Robert Scott Small Library, College of Charleston, Charleston, S.C.

same time made the acquaintance of Edmund Ravenel, who, of course, had built an impressive invertebrate fossil collection of his own. Like John Bachman, Holmes also met a man who altered the course of his career. The man was Michael Tuomey. Then residing in Virginia, the Irish immigrant was continuing the survey of South Carolina's agricultural and geological resources in 1844 when he saw a fossil on display in a Charleston bookstore, probably the noted enterprise and gathering place for local literati run by John Russell. Tuomey wrote to Gibbes, whom he had met not long before, to ask if he knew the owner of the fossil. Of course, since Holmes had sent fossils to Gibbes, Gibbes was able to put Tuomey in touch with the "lover of nature," as one knowledgeable Charlestonian referred to Holmes. Before long, the two paleontologists were planning to produce a book on the invertebrate fossils of the Charleston region.[4]

In 1845, Holmes began to broaden his contacts, first by approaching the Philadelphian Samuel G. Morton, who was already well known to Bachman, Ravenel, Holbrook, and Gibbes. "I have found a number of shells, casts, bones, and other fossils, and I feel much interest in reading on the subject," he told Morton in his introductory letter. He wanted to know how he could obtain a copy of Morton's *Synopsis of the Organic Remains of the Cretaceous Group of North America*. Soon thereafter, Holmes was offering to exchange specimens with Morton, and in January 1846 he shipped several fossils to him. Later in the year, after Bachman had sent a letter of introduction to J. E. Gray, Holmes exchanged mollusk shells and fossils with that British naturalist. Gaining confidence through these contacts, especially with Tuomey, Holmes began to publish scientific papers. As with other contemporary naturalists, his interests ranged widely, and his first paper was not on fossils at all. Rather it was a brief description of a bezoar stone, or gastric concretion, taken from the stomach of a white-tailed deer, *Odocoileus virginianus*. Published in Charleston's *Southern Journal of Medicine and Pharmacy*, the report gave the weight, color, and features of the concretion, including its inner layers, which Holmes had observed by cutting the object with a watchmaker's tool. Though hardly a major contribution, the brief account indicated Holmes's ability to write a scientific report, and it opened a door for the unheralded naturalist. Holmes took a more important step in 1848, when he published a paper titled "Notes on the Geology of Charleston." He placed it with the same journal, though its name had been changed to the *Charleston Medical Journal*. A year later, the paper was published in the *American Journal of Science*, its topic no doubt appealing to the editor, James Dwight Dana, whom Lewis Gibbes would soon unfairly accuse of neglecting papers from southerners. The article clearly showed that Holmes had learned quite a bit about geology, no doubt largely from his association with Tuomey. While the article dealt primarily with the topographic features of the city, it also contained some useful information on the stratigraphy and fossils of the area, including a list of 147 species that had been unearthed there.[5]

Although Holmes was much better acquainted with invertebrate fossils than with vertebrate fossils, he was interested in and collected specimens of the latter as well. As noted previously, one of his most important finds of vertebrate fossils was part of the skull of a primitive odontocete, or toothed whale, later described as *Agorophius pygmaeus*. Soon afterwards, Lewis Gibbes visited the site, on the west bank of the Ashley River, and found the remaining fragments of the skull. Holmes might have collaborated with, or at least sought assistance from, Lewis Gibbes's cousin, paleontologist Robert W. Gibbes, since the latter was certainly knowledgeable about a number of extinct vertebrates, but the touchy temperaments of the two men were too similar to expect a truly harmonious relationship. Moreover, evidence indicates that both sought the attention of Tuomey, a contest that favored Holmes, likely because of his larger collection of fossils. That Gibbes considered it improper for a man with no college education, let alone no doctorate in medicine, to be vying for scientific recognition seems to be apparent also. Holmes did send a portion of a fossilized placoid fish to Gibbes to describe, and in 1850 Gibbes described the ray as *Myliobates holmesii* in honor of its discoverer. The truce was temporary, however, and ill feelings would surface again. Holmes had in the meantime found a friend who was well known for his knowledge of fossil vertebrates, namely Joseph Leidy, a Philadelphia physician-turned-naturalist. Their friendship flourished during the 1850s.[6]

Upon the recommendation of Lewis Gibbes and others in October 1847, the trustees of the College of Charleston had designated a basement room in the college to house some of Holmes's collection of fossils, thus giving recognition to the planter-naturalist without an official appointment. The decision was fortuitous, for Louis Agassiz made his first visit to Charleston in late November and early December of that year. Agassiz was impressed with the collection and viewed it as the core of a rejuvenated museum for the city. As noted earlier, Agassiz had deliberately sought out Holbrook, with whose work he had been familiar ever since *North American Herpetology* appeared in Europe. Agassiz had met Bachman in Freiburg in 1838, and he visited with him also, though the clergyman-mammalogist had come away from the meeting with an unfavorable opinion of Agassiz as a mammalogist. Edmund Ravenel came to the attention of Agassiz as well, and so did Lewis Gibbes. To that group of acquaintances and co-laborers in natural history, Agassiz added Holmes. After seeing the impressive fossil collection at the college, Agassiz traveled over to the Springfield plantation to visit with the man who had built it. Apparently, Agassiz suggested to Holmes that the Charleston Museum could be given new life and become the center of scientific activity in the South. Certainly, he brought up the idea of forming a new scientific society in Charleston. That visit and annual ones by Agassiz in 1849 to 1853 strengthened the standing of Holmes as one of the major naturalists in the region.[7]

In 1850, Holmes joined with the other naturalists of Charleston in the program of the AAAS when it met in the city during March, presenting a paper and commenting on another. In his paper, titled "Observations on the Geology of Ashley River, South-Carolina," he described the stratum of marl that ran along the river and noted that it contained only a few bones of the larger vertebrates but abounded in marine vertebrates, which, he added, Agassiz called the richest deposit of fossils he had seen. Holmes observed that he had personally collected from it many thousands of shark teeth. He also noted the absence of the Miocene and Pliocene formations along the river and commented on the more recent strata and their fossiliferous contents. Although the presentation was quite general in nature, it served to show how familiar Holmes was with the fossil deposits of the region, and Agassiz touted the significance of the paper. Holmes's comment, also published in the proceedings of the meeting, pertained to a paper by Robert Gibbes, "Remarks on the Fossil Equus," in which he described some fossil teeth he had found in a bed of marl on the Ashley River. Gibbes suggested they were of a new species of *Equus* and that the marl was of the Eocene epoch. Holmes did not dispute the claim of a new species of the horses and asses, but he asserted that, in all of his searches, he had "never discovered a single specimen of the remains of *Equus*" in any of the pits he had dug, one of which was "seventy feet in length, twenty [feet] wide, and *twenty feet deep*." He added that fossil remains of *Equus* would be found only in "a still more recent formation." His assertion was right, for the horses of that genus did not evolve until the Oligocene epoch. Yet, Holmes later published a paper in which he placed *Equus* in the Eocene. Of course, he had not flatly stated that

Gibbes was wrong, only that he had personally never found any remains of *Equus* in the Eocene marl. His comment was sufficient to irritate Gibbes, however, though it was perfectly in harmony with the spirit of criticism essential to scientific inquiry. As it turned out, both men were in error, but on different counts. The marl bed in question, long thought to be of Eocene age, is now known to have been deposited during the Late Oligocene. Gibbes's specimens were almost certainly of Pleistocene age but were probably found lying on top of the marl, having been eroded from overlying Pleistocene beds.[8]

The meeting of the AAAS in Charleston in March 1850 brought much praise to the city's naturalists. The noted astronomer B. A. Gould said, "we are sometimes disposed to imagine ourselves [in the North] the pioneers in all matters of science and education, . . . [but] the claims to similar distinction must not be overlooked. Few states have preceded . . . [South Carolina] in scientific efforts." Such tributes, along with a recommendation by Agassiz, prompted the College of Charleston trustees to talk of establishing a museum at their institution. Agassiz noted that specimens from the cabinets of Holmes, Tuomey, Holbrook, and others would quickly make the museum an important center of natural history studies. In July 1850, then, the trustees authorized the establishment of a museum. At the same time, the Literary and Philosophical Society voted to turn its much-neglected collections, many from the old Charleston Museum, over to the college. So too did the medical college in Charleston, which had a small natural history collection of its own, in addition to its anatomical specimens. By that time, the Literary and Philosophical Society had devolved informally to the Conversation Club and had no curator to care for its collections. In 1850, the medical college was preparing to raze the building in which its natural history specimens were housed in order to erect a hospital on the site. As one observer had noted only five years earlier, the medical college collections consisted "principally of minerals and a few mounted birds and quadrupeds." The birds and mammals, he added, "were so infested by insects that only a small number could be retained." By November 1850, then, the College of Charleston had a museum, and its trustees opted to retain the old name, The Charleston Museum. To fill the post of curator they appointed Holmes, and a month later they also named him professor of geology and paleontology.[9]

The College of Charleston trustees wanted to open the new museum in January 1852. Thus, Holmes had much work to do before then. He initially petitioned the trustees for an allotment of $500 for cases and galleries but exceeded that amount by $107. Then he sought another appropriation of $587 to place panes of glass in the cases. The *Charleston Mercury* kept an eye on Holmes, not for his expenditures, but for his enthusiasm. The new curator, it reported early in 1851, was working hard "to build up at the South an institution similar to those found in every large city North." The *Mercury* also noted that heretofore "numerous valuable specimens of the utmost importance to our own Naturalists" had been sent "to the museums of the Northern States." The museum that Holmes was developing, opined the newspaper, now provides "the great desideratum." Throughout the year Holmes solicited specimens for the museum. Holbrook donated a collection of fishes; Tuomey deposited a series of the fossils he had collected during the geological survey; Shepard contributed several specimens of minerals; and Thomas Burden, an obscure but able collector who lived just outside Charleston, presented a collection of Tertiary fossils.[10]

The museum was opened to the public in January 1852. It occupied the entire second floor of the main college building, some 10,000 square feet. In a note to Spencer Baird on July 25, 1852, Holmes proudly observed that the new museum had 7,000 square feet of shelves in wall cases with glass fronts and sides, and he boasted that the space included "a suit of rooms called the *Zoological Laboratory* . . . [and a] work room for the taxidermist and his assistant." Holmes also declared that the collections already contained "about three thousand specimens," and the library held "a number of valuable books on Natural History." On hand for the opening celebration was Agassiz, who presented a brief address in which he praised Holmes for "re-arranging the museum" so well. The *Charleston Courier* also lauded Holmes, calling him an "able, scientific and devoted curator" and noting that in only fourteen months he had gotten the museum ready for the opening. The newspaper exulted, moreover, over Agassiz's comment that only the Academy of Natural Sciences of Philadelphia now exceeded the Charleston Museum in the natural history collections of the United States. The claim was somewhat exaggerated, but, as the *Courier* noted, the museum was indeed "an honor to our City."[11]

To Francis Holmes belonged a large share of the credit for getting the new museum under way and for the excellent progress it made during its first decade of existence. In his annual report for 1854, Holmes listed the names of more than 300 contributors of specimens and books to the museum, including Edmund Ruffin, who donated a number of fossils; Henry W. Ravenel, of Aiken, South Carolina, who added more mycological specimens to those he had presented earlier; and Dr. Lingard A. Frampton, a local resident, who presented "a valuable Library of several thousand volumes." Donations also came from naturalists in Boston, New York, and Philadelphia, including the ANSP, which contributed "a case of mounted birds." In addition, more than 200 residents in the Charleston area, including thirty women, donated specimens to the new museum, reflecting, no doubt, the pride expressed by the *Courier*. The trustees of the college were obviously pleased with the work of the curator, not only for his success in building up the collections of the museum but also for his effective promotion of the institution. Holmes certainly believed in the value of a good museum, and in his annual report for 1854 he stated that he had heard "expressions of surprise and gratification" from patrons when viewing objects in the display cases and declared that "many have been . . . led to examine for themselves . . . these mysterious creatures of the Great Creator."[12]

Meanwhile, in 1853, Holmes and Tuomey launched their collaborative project to produce a book on the invertebrate fossils of lower South Carolina. Publishing it originally in parts, the authors had completed six of them by 1855, and they published the next eight at various times during 1856. The final part appeared in a compilation of the whole volume, published in 1857. Unfortunately, Tuomey, who had been serving as professor of geology at the University of Alabama and director of the geological survey of Alabama, died on March 30, 1857, shortly before the completed work appeared in print. Holmes included a warm tribute to his coauthor in the finished volume, reflecting his high regard for him. For Holmes, the loss was incalculable, but he could find some comfort in the masterly work that he and his late friend had produced. Partially subsidized by a grant of $2,000 from the South Carolina legislature, *Pleiocene Fossils of South Carolina* contained 152 pages of descriptions and plates of two sea stars, two brachiopods, nine echinoderms, nine bryozoans, and 181 mollusks, or a total of 203 species. Although the strata from which some of the specimens were extracted were later interpreted differently and even though the book contained some taxonomic errors, *Pleiocene Fossils* described a significant number of new taxa and was a major contribution to paleontology. The work became a classic and remains valuable to systematists.[13]

In a circular designed to promote the book, Holmes said that friends had originally advised him and Tuomey that it was "impossible to publish such a work at the South in a style to compare even with the ordinary productions of Northern cities." Then he proudly proclaimed: "It is the first successful attempt to publish such a work among ourselves, and it has signally developed the ability to accomplish *at home* what we have been dependent for upon the *North*." He boasted, moreover, that the book "appeals . . . strongly to the pride, not only of our State and city, but of the whole South." *Pleiocene Fossils* was in fact the first of the major works of the Charleston circle published in the South, for Holbrook had turned to the North for the production of his *North American Herpetology* and his *Southern Ichthyology*, though later he had his *Ichthyology of South Carolina* published in Charleston. The work on North American mammals by Audubon and Bachman was published in New York. But in 1857, when *Pleiocene Fossils* was published, sectionalist sympathy and pride were at a peak in the South, and Holmes, a passionate champion of his native region, was delighted over the production of the book "at home." The volume was also a matter of great pride to South Carolina legislators, a committee of which called it "a highly scientific work, beautifully executed, and calculated to reflect great credit upon the state." No doubt, the satisfaction of the legislators with *Pleiocene Fossils* and their knowledge of the necessity for assisting the naturalists in order to demonstrate the state's scientific independence prompted them later to grant a subsidy to Holbrook for his *Ichthyology of South Carolina* and to the Elliott Society of Natural History for the publication of a journal.[14]

A number of prominent scientists also praised *Pleiocene Fossils*, among them James Dwight Dana, who told Holmes that he "admired the perfection of your plates and the whole style of typography and description." Then he paid Holmes an even greater compliment by saying the volume "is certainly equal to the best European Palaeontological Works, and surpasses nearly all that this country has produced." For Dana, as with most of his American contemporaries, the competition in science was not between the South and the North but between the United States and Europe. Thus, the distinguished scien-

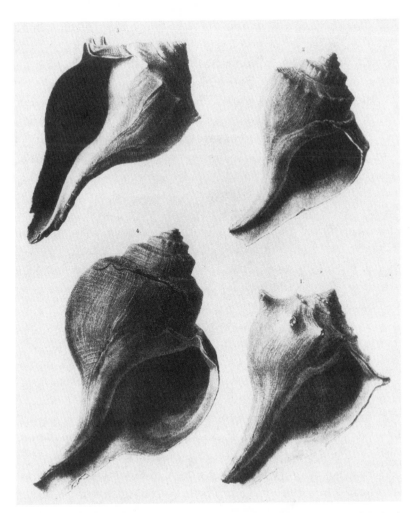

Fossil gastropod shells figured for M. Tuomey and F. S. Holmes, *Pleiocene Fossils of South Carolina*. The four specimens of fulgur whelks in the plate above lived millions of years ago, but each species thrives into the present.

tist was not congratulating Holmes so much for a southern publication as he was for an American publication. Similarly, Joseph Leidy wrote to tell his friend Holmes that *Pleiocene Fossils* "hardly has a parallel in our country," and Isaac Lea lauded the "perfect manner" of the plates and the "accuracy of the descriptions." With the praise of Agassiz added to those favorable comments, Holmes could bask in the limelight cast on his achievement and begin work on his next major project, a volume on the fossils of what he and his contemporaries called the "Post-Pliocene" epoch.[15]

Extraordinarily active during the year of 1853, Holmes completed a study of the harbor of Charleston, in connection with a project to create a new channel, called the Maffitt Channel. Among his observations, he noted a process that continues into the present, namely that currents erode sand from one portion of a shore but deposit it a little farther away. These "modifications and changes," he said, "are still progressing." In addition, Holmes examined the fossil contents of borings to twelve feet beneath the floor of Charleston Harbor. In the meantime, he was taking another important step toward the professionalization of science: the creation of a society for the naturalists of the city. Of course, the old Literary and Philosophical Society had encouraged the pursuit of natural history and other sciences, and it had once had a museum under its care. Despite the best efforts of the first president, Stephen Elliott, and of Edmund Ravenel and John Bachman, the Society had not been able to sustain a program of regular meetings, though some good papers had been presented before its membership. Furthermore, the Society, as its name suggests, was not interested exclusively in the sciences. Indeed, by 1842 it was commonly called the Literary Society, the Conversation Club, or, simply the Club, and only three of its members were active in the study of natural history. The Literary Society, or Conversation Club, continued until 1860, and its members were treated to some interesting presentations, especially between 1847 and 1849, when a recurring topic was the specific unity or plurality of the races. Meanwhile, however, Agassiz had suggested to Holmes that a natural history society in Charleston would advance the field much better. It is likely that he mentioned the same to other Charleston naturalists, who certainly longed to see in their fair city a natural history society like the ANSP or the BSNH. Holmes took the initiative in 1853 and formed the Elliott Society of Natural History. The ESNH began to flourish at once and soon became the focal point of scientific activity in the Southeast. Holmes became the secretary of the organization, and John Bachman was elected as the first president. In 1859, the Society entertained the notion of expanding its interests to include other sciences and changing its name to the Elliott Society for the Advancement of Science, noting that "the limited circumspection of knowledge . . . induces us to think [incorrectly] that Natural History is an exclusive department." It took no action on the proposal, however, for it concluded that all sciences "have been harmonized in one grand unit—nature." Quite obviously, the scientists of Charleston continued to view natural history as the queen of the sciences.[16]

During the first three years of the ESNH, Holmes was a central figure in its activities. He presented three papers before meetings of the Society, the first of which was a brief description of fossil barnacles of the genus *Balanus*, the acorn barnacles, of which a number of species are living today. Holmes discovered them in what he identified as the Eocene (now, correctly, the Oligocene) marl of the Ashley River. In another paper he offered a short description of a living oyster, but, contrary to his belief, it was not a new species. Holmes also described the "American Devil Fish," as he called it, but it was in fact the giant devil ray, or manta, *Manta birostris*. Nevertheless, as modern authorities have noted, Holmes recorded information useful to later ichthyologists, and his efforts helped to advance the interests of the ESNH. In 1857, however, Holmes resigned from the organization he had founded because of disagreements with two decisions of the majority. He did not rejoin the Society until after the American Civil War.[17]

Somewhat wanting in ability to compromise and viewing both the ESNH and the museum as his possessions, Holmes soon created another controversy. Although his faculty colleagues respected his ability and his admirable work, they disapproved of his frequent absences from faculty meetings and disliked his propensity to place the museum above the interests of the college. Thus, they objected when the dedicated and aggressive curator sought to get the trustees to allocate more space for additional museum collections, which, of course, would have reduced the space for other purposes. In 1859, the trustees noted that the "plans hitherto proposed for the accommodations of Professor Holmes have been exceedingly distasteful to the members of the Faculty." A year later, in 1860, the trustees were still struggling to find the "best method of preventing any encroachments [of the museum] on the other departments [of the college]." Even then, Holmes's colleagues remained unhappy about a decision he had made three years earlier to restrict access to the collections of the museum. Under Holmes's rule, they must obtain his permission to study museum specimens, even during their meetings, which were held in the college building that housed the museum.[18]

On behalf of the Society, Gibbes petitioned the trustees to establish rules for free access to the collections. "It is rather difficult I fear," said the chairman of the board of trustees, "to harmonize the various views." Certainly, the trustees did not wish to displease Gibbes, and, aware of the standing of Holmes in the community and of his prodigious labors for the museum, they could not easily reprimand him. Besides, they understood why Holmes wanted to restrict access when the curator was not on duty, but they also understood why ESNH members objected to "being regarded as intruders." In his petition, Gibbes said that "we . . . feel that we have no inconsiderable claim to the privilege of using, under proper restrictions, the museum . . . which we are called on to cherish and support." Surely, Gibbes must have been pained somewhat by the action of a man he had helped so much. In any case, the trustees, not wishing to let the controversy continue, agreed that rules must be drawn up, and instructed Holmes to do so. Adopted in June 1858, the rules specified that anyone engaged in the study of natural history could "at all times, obtain access . . . on application to the Curator," but they also stipulated that "the Curator's room shall be considered strictly his private apartment." Neither provision was unreasonable, and each allowed Holmes to maintain a high degree of control. The question of housing the collections of the ESNH in the museum remained unresolved, however, as did the question of ownership of the specimens brought in by Holmes, the latter of which created another problem later on. In effect, Holmes had won the battle, but it ultimately proved to be a Pyrrhic victory. When he needed their support later, he could muster no sympathy from those he had rankled.[19]

In the meantime, Holmes continued to study and to write descriptions of the "Post-Pliocene" fossils of Charleston. His friendship with Leidy grew even closer, and he sent several fossil specimens to him for description. Leidy even came to Charleston in 1857 to visit with Holmes and to conduct field study for himself. Holmes stopped sending specimens to Agassiz, however, for he came to realize that the busy naturalist never returned them. Indeed, Holmes claimed that Agassiz had failed to return several hundred specimens he had sent to him for study, and he told Leidy that Agassiz had held some for at least five years. Both statements are believable, given the experience of others who at first thought it was an honor to assist Agassiz but eventually learned that he rarely returned borrowed specimens. Lewis Gibbes also sought the return of specimens from Agassiz, and, in desperation, once urged young John McCrady, while he was a student of Agassiz, to ask his mentor for their return. McCrady declined, knowing that no mere student should dare to approach the authoritarian master with such a message.[20]

Holmes published his *Post-Pleiocene Fossils of South Carolina* in two parts, beginning in 1858. The completed volume of 122 pages and twenty-eight plates appeared in 1860. Again, the typography and the plates were excellent. The first of the two parts of the volume contain descriptions of about 150 species of invertebrates, primarily mollusks but also echinoderms, crustaceans, and other marine animals. As indicated by the designation "Cabinet of Francis S. Holmes," a large number of the specimens had been collected by the author. Most of the other specimens belonged to the Charleston Museum, but several of those had been donated by Holmes. Actually, most of the specimens were of the Pleistocene epoch, but at the time anything younger than Pliocene was called "Post-Pliocene." In any case, Holmes referred to it as "the epoch just antecedent to the advent of man upon this earth." Certainly, he thought of the age of the earth as measurable in at least a few million years and thus probably shared the notion of some of his contemporaries that the Mosaic, or biblical, account of creation could not be taken literally. Perhaps Holmes even subscribed to the Pre-Adamite theory, which postulated that inferior types of people had existed before God made the forefather of present humans. In any case, Holmes found no conflict between the biblical story and the solid evidence of animals that lived in distant ages.[21]

His view was no doubt like that of his friend and coauthor Tuomey, who, in his report on his geological survey of South Carolina in 1848, had asserted that the remains of ancient animals could not be explained if "the age of the earth was only 6,000 years" as some supposed. "The interpretation of the Mosaic narrative of the creation that is most in accordance with the discoveries of Geology," he added, "is that which supposes the 'beginning,' mentioned in the first verse [of the Book of Genesis], to be a time immeasurably distant from the 'first day,' mentioned in the fifth verse, and that in the interval . . . all the phenomena of Geology were brought about." His argument was not original, for it had been popularized nearly two decades earlier in Scotland by the theologian Thomas Chalmers and in the United States by the Amherst College professor and geologist Edward Hitchcock, who based much of his argument on the geological theory of Cuvier. Tuomey also declared that "those Divines who have examined both sides of the question" accepted the view he had given. Certainly, he stretched his claim somewhat, but the evidence is clear that a considerable number of contemporary theologians had no difficulty with the liberal, as opposed to literal, interpretation of the Mosaic account, though John Bachman was not among that number. The able South Carolina paleontologist Robert Gibbes also subscribed to the theory of an ancient earth, and in 1849 he said, "The researches of the geologist oblige us to assign millions, rather than thousands of years, as the age of the earth." In his *Post-Pleiocene Fossils*, Holmes referred to the "Post-Pliocene" as the "period in which . . . the earth had been finally prepared and made ready for him who was to be formed in the likeness of the Creator." It was thus the period that "will ever be distinguished as the *grand connecting link* between the past and the present." His view did not, of course, suggest any notion of evolution, and Holmes likely shared with Agassiz the view that the appearance of new species in different strata represented separate acts of creation. At the time of the publication of *Post-Pleiocene Fossils*, Charles Darwin's *On the Origin of Species by Means of Natural Selection* had just appeared in the United States. Later, Holmes could not have avoided thinking about Darwin's theory since it contained observations relevant to his work, but he never addressed it, at least not in his extant papers.[22]

The second part of *Post-Pleiocene Fossils* consists of twenty-three pages of descriptions and fifteen plates of vertebrate fossils, mostly found in the Charleston area. Although he had discovered most of the thirty-two species described in Part II around the Ashley River, Holmes asked Leidy to write the accounts. He did so for good reason. Previously, Holmes had published a paper in which he identified the remains of horses, oxen, sheep, and dogs as being from the Eocene. Leidy had disagreed, maintaining that it was more likely that they were the bones of domestic animals of the Recent era that had become "intermingled with the true fossils of the Post-Pleiocene and the Eocene," possibly excluding *Equus*. The remains of giant ground sloths, mammoths, mastodons, and other extinct vertebrates found by Holmes, Leidy argued, almost certainly belonged to an earlier period than those of Recent animals. Moreover, noted Leidy, Holmes had been correct when he had questioned the claim of Robert Gibbes at the AAAS meeting in 1850. Among Leidy's significant contributions in Part II was a description of a new Pleistocene tapir, *Tapirus haysii*.[23]

Because *Post-Pleiocene Fossils* appeared in print shortly before the

Civil War began, it received less notice than it deserved. An important review did appear in the *American Journal of Science*. Signed "W. S.," it was the work of William Stimpson, who, though still very young, had established himself as an outstanding authority on marine invertebrates. Stimpson had studied informally with Agassiz, but because he refused to accept the authoritarian manner of his mentor, he had moved on to the Smithsonian Institution. A friend of the Charleston naturalists both before and after the Civil War, he had collected specimens in Charleston Harbor and in other southern waters. Paleontology was not his special field, but he was well acquainted with it. Certainly, he was an authority on the taxonomy of many phyla of marine invertebrates, especially the phylum Mollusca. Stimpson called *Post-Pleiocene Fossils* an "important work" and praised the "excellent plates" in the volume. He did criticize Holmes, however, for using "generic nomenclature" that had become outdated. In addition, he noted several misidentifications, some of which Holmes had based upon the only remaining portion, the tip, or protoconch. A man of merriment, Stimpson punned that "a little *tipical* knowledge is quite necessary to an investigator of fossil shells." Whether or not Holmes knew that Stimpson loved to jest is unknown, but in all likelihood the Charleston paleontologist saw no humor in the comment. He had to be pleased, however, that Stimpson called the volume "an important work," noted that he believed fourteen of Holmes's taxa were new, and indicated that the book provided "much valuable information as to the distribution of our shells, both in a recent and fossil state." Fortunately for Holmes, he did not know that Stimpson had told Edmund Ravenel that the book contained "many bad mistakes."[24]

Holmes no longer had Tuomey to help him, of course, but he could perhaps have avoided some of the errors if he had called upon Ravenel, John McCrady, and Lewis Gibbes. Pride and independence, however, made it difficult for Holmes to call upon those who had criticized him. Of course, he had called upon Leidy, the best person he could have gotten for the task. Although *Post-Pleiocene Fossils* was somewhat marred, it was nevertheless a very good book, and it remains useful even today for its descriptions of Pleistocene fossils in South Carolina. That it contained fewer new species than Holmes claimed is hardly surprising inasmuch as most of the species have survived into the present, and many had already been described as living species.[25]

Of the mollusks described by Tuomey and Holmes in *Pleiocene Fossils*, seventeen taxa are currently valid. Among them is the bivalve *Lucina multilineata* (= *Parvilucina multilineata*), which is, in fact, a Recent species commonly called the many-line lucine. Two of the bivalves and two of the gastropods described by Holmes in his *Post-Pleiocene Fossils* are valid taxa of living mollusks. The bivalves are the gray pygmy venus, *Chione grus*, and the concentric ervilia, *Ervilia concentrica*; the gastropods are the crenate pyram, *Pyramidella crenulata*, and the waxy miter, *Vexillum wandoense*. As the decade of the 1860s opened, Holmes could justly take pride in his accomplishments and in the recognition he had received, but he could not ignore the rift between the South and the North that had widened into a nearly impassable gulf. The Charleston Museum was continuing to prosper under his direction, and, thanks to a highly favorable series of articles on the museum by the *Charleston Courier* in late 1858 and early 1859, Holmes was enjoying deserved acclaim. The *Courier* aptly stated that the curator was "well and widely known for his zealous, devoted and indefatigable pursuit of the Natural Sciences generally, and of the Palaeontological branches of zoology especially." A few months later, the *Courier* reprinted an article from the *Country Gentleman*, in which the reputable naturalist S. B. Buckley said that he had recently visited the museum and was surprised to find that "the collection is the best in the United States, except that of the Academy of Natural Sciences of Philadelphia." He added: "The Smithsonian Institution has a larger collection . . . , especially in reptiles and perhaps birds, but the [Charleston] Museum . . . has certainly more mounted birds . . . and is more complete in the other branches of Natural History." Although Buckley's comparison of the Charleston Museum with the Smithsonian was perhaps a bit exaggerated, it was nevertheless true that the southern institution had made remarkable progress and had earned a place as one of the best of the American museums of natural history, largely through the efforts of Francis S. Holmes.[26]

Tragedy struck at the same time, however, for Holmes's wife Elizabeth died after giving birth to her eighth child on February 24, 1859. Meanwhile, like Bachman, Holmes was fretting over the abolitionist movement. Firmly committed to the institution of slavery and himself the owner of five bondsmen, he feared that a war over the issue was likely. He was certain, however, that the South would triumph if war came. "Our young men are the best horsemen on this continent," he wrote to Leidy on January 7, 1861. "We will," he added, "never flinch before a '*Lincoln* Force' three times our number." Holmes informed his Philadelphia friend that "we are literally in camp and *armed to the teeth*," but, he lamented, "these are sad times for our happy country." The Union must be dissolved, he said, and "*there must be a Southern Confederacy*." Holmes volunteered for a local militia troop, but as a nearly forty-six-year-old father of eight surviving children, he was rejected. He would have to stay at home and wait for the unflinching young men to defeat the enemy. The museum could go on in the meantime, though his scientific work had to be placed in abeyance for awhile.[27]

Passionate Pursuits

P ulsating rhythmically near the surface in Charleston
Harbor, the thimble-like medusa would have been
missed by most observers, but it was soon caught up
in the dip net of twenty-five-year-old John McCrady. After obtaining
several more specimens of the medusa during the summer and the fall
of 1856, and after carefully studying the gelatinous little creatures
under his microscope and counting the annulated tentacles on the
fringe of their bell, McCrady began to prepare a description and
figures of *Turritopsis nutricula*, a new genus and species of hydro-
medusa. Read before a meeting of the Elliott Society of Natural History on December 1, 1856, the lengthy paper and the detailed figures,
drawn by McCrady himself, clearly indicated the superior ability of the
young Charleston native.[1]

The third of fourteen children of the prominent Charleston attorney and devout Episcopal layman Edward McCrady and his wife
Louisa Rebecca Lane, John was born on October 15, 1831. Enrolled in
the school run by Samuel Burns, who was noted for his "inhuman
floggings" and his mastery of the classics, John was a highly self-
disciplined and precocious student. As a former classmate later re-

called, by the age of eight he was impressing his peers because of his "superior strength and talent." McCrady entered the College of Charleston in 1846 and quickly manifested his excellence as a student. A classmate of the poet Paul Hayne, McCrady possessed something of a romantic streak and even wrote some verse, which is not especially memorable, though, as he said later, some of it was "magnificently flaming." In mathematics and the sciences, however, he excelled, impressing Professor Lewis Gibbes in particular. Perhaps he also studied natural history in the class taught by John Bachman, for the latter had joined the faculty in 1848, but in that field he considered Edmund Ravenel and John Edwards Holbrook to be his informal mentors.[2]

McCrady was particularly attracted to Louis Agassiz, whose lectures in Charleston in 1847 and 1849–1853 he no doubt heard with great enthusiasm, particularly those dealing with the phylum known then as the Radiata, which lumped together the poriferans, cnidarians, ctenophores, and echinoderms, or, respectively, the sponges; the jellyfishes (or medusae) and allied classes; the comb jellies; and the sea urchins, starfishes (or sea stars), and allied groups. It is probable, moreover, that young McCrady read the lectures of Agassiz published in 1847 as *Introduction to the Study of Natural History*, which contained considerable information on one of Agassiz's favorite subjects, the coelenterates (a taxonomic category that places the cnidarians and ctenophores in the same phylum and at the time included the sponges). John McCrady determined that he would study with Agassiz at Harvard's Lawrence Scientific School in Cambridge, Massachusetts. The official record shows that he was enrolled only in 1853 and 1854, but his son later claimed that he was with Agassiz during each of the summers of 1851–1854. A sketch written by McCrady and some of his letters establish that he was definitely there for awhile in 1852 and from February through October of 1853 and January through March of 1855. It is likely, then, that he studied with Agassiz off and on during the period from 1852 to mid-1855. In any case, it is certain that he mastered knowledge of the phyla Cnidaria (at least the classes Hydrozoa and Scyphozoa), Ctenophora, and Echinodermata (at least the classes Echinoidea and Asteroidea).[3]

While he was in Cambridge, McCrady completed descriptions of five species of fossil echinoderms for the volume by Tuomey and Holmes, after first seeking approval from Agassiz to do the work. A letter he wrote to Ravenel on August 30, 1853, indicates that McCrady had already learned a great deal about echinoderms, and the descriptions he wrote for *Pleiocene Fossils* further demonstrated his command of the subject. Indeed, at that point he expressed an intention to study the living and fossil echinoderms of America. He never returned to that subject, however, for he developed a passion for pursuing the study of the hydrozoan animals. Although he relished the association with his idol Agassiz, McCrady did not like the culture he encountered in the North. He first visited New York briefly, but he was not favorably impressed. In "these crowded streets," he wrote to his father, " 'every man for himself' is the law . . . and what with the hundreds of vehicles and their reckless drivers, who hardly ever stop for foot-passengers, I am always wondering that the women and children are not run over." Even more troubling to the conservative southerner was that "everything [in New York] seems to say 'away with the old things—down with the old building.' Novelty is the only food to fill this greedy multitude."[4]

The city of Boston also displeased him, and he complained of its dirty and winding streets. He reserved his strongest criticism, however, for the city of Cambridge. While he expressed pleasure over being in "a society of gentility and education," McCrady abhorred the abolitionist sentiments of some of the citizens. Fortunately, he told a sister, in a letter dated May 8, 1853, his landlady was strongly opposed to abolition and to the woman's rights movement. During the early winter of 1855, he described Cambridge to his sister as "the muddiest, sloppiest, most uncomfortable hole in which it was ever the lot of a poor student to be buried. . . . Then will come a snow storm, when the only way to get along [on the streets] is ploughing literally." Indeed, it was so bad in the month of January, he moaned, that "frogs and toads are the only conspicuous offspring of nature in Cambridge." Worse to the southern gentleman, however, was the behavior of women in the city. It is, he told his sister, "not until you come to observe the doings of the Womankind, that you can form any just conception of the depth to which this town has sunk, or rather of the height of enormity to which its subversion of the natural order of things has risen." For McCrady, it was an especially "bad sign when women undertake to wear *men's boots*." Elsewhere in the world, he added, women have "occasionally on great occasions stepped into their husband's *shoes*," but, in his opinion, "Providence has reserved it for the northern women to cap the climax and assert an equal right with man."[5]

A temporary rift occurred between McCrady and his mentor in

January 1855, apparently because McCrady did not wish to remain at Harvard beyond the end of the current term. If that were the first time Agassiz asserted his notion that a student was not ready to function independently, it certainly was not the last. In later years Agassiz became even more assertive about a lengthy apprenticeship. McCrady believed he would soon be ready to be "an independent man" and did not back away. "It's all settled now," he wrote to his mother, "and he [Agassiz] has promised to put me in the way of being ready for a professorship within the year." McCrady had little reason to worry, for Lewis Gibbes, burdened by a beastly load, was urging the College of Charleston trustees to create an assistant professorship in mathematics, and he was happy to have McCrady appointed to the position. The trustees readily complied, for, after all, McCrady was an excellent mathematician, a student of the much revered Agassiz, an alumnus of their institution, and a native and champion of the Queen City. In fact, McCrady was set on returning to Charleston, for he viewed the city as the center of civilization. Indeed, he once penned an ode to his beloved home:

Upon a tongue of land . . .
A Southland city sits beside the wave
And points to Heav'n from many a flame tipp'd spire.
On either side the rivers kiss her feet
And bring far-faring ships, white-wing'd that bear
All wealth and tidings out of other lands
To deck and glad her Queenliness therewith.[6]

He must return to his beloved home, not because of the muddy streets and liberated women in Cambridge but because Charleston was the place where culture was infinitely superior. Indeed, he was already thinking of drafting essays that compared the two cultures of America.

Meanwhile, the young naturalist began to conduct research on marine invertebrates, especially the hydrozoans. After his superb study of *Turritopsis nutricula*, McCrady set out to describe all the species of hydromedusae he could collect in Charleston Harbor. During the next few months he devoted most of his time and energy to the effort. Prior to McCrady's work, the South could claim no naturalist whose main interest was marine invertebrates. Edmund Ravenel studied mollusk shells, but he paid less attention to the living animals. He did make a few drawings of marine gastropods, but the task was not easy since the animal usually retreats into its shell. Lewis Gibbes had done pioneering work on the decapod crustaceans, but that was only one of his many interests. In fact, the South had not produced a naturalist of the caliber of James Dwight Dana, whose work on the Wilkes Exploring Expedition from 1838 to 1842 had qualified him as one of the world's authorities on marine invertebrates. McCrady held the potential for coming close to Dana, but he gave no thought to joining an exploring expedition or even to doing research outside his region. Certainly, an abundance of species flourished in the South, and McCrady would do his part to study those known as medusae. But, like his fellow Charleston naturalists, McCrady lacked the adventuresome spirit of Dana or of his contemporary William Stimpson, a spirit that would carry the latter to other oceans and seas.[7]

Given Charleston's proximity to the ocean, it is somewhat surprising that its naturalists, excepting McCrady, generally showed little desire to investigate the vast hordes of invertebrate species there. In part, of course, the problem was that the free, white population of Charleston was simply too small to produce a sizable number of scientists, and thus a Charleston naturalist who concentrated upon marine invertebrates other than mollusks and crustaceans would be rare. The student of nature who would take an interest in sea jellies would be even rarer, for the general view of those animals, both in the South and in the North, was one of revulsion. Even today, many surf bathers flee to shore at the sight of any sea jelly, assuming that it must be a vicious stinger. All cnidarians (the root word of which, *cnido*, means "nettle") possess nematocysts or batteries of stinging cells, but in the North American Atlantic only a relatively small number of scyphozoans and cubozoans are capable of inflicting a painful sting, though the sting of one of the hydrozoans, the Portuguese man-of-war, *Physalia physalis*, can be extremely painful, even life-threatening in rare cases. In fact, however, the sea jellies, whether the generally small hydromedusae or the larger scyphomedusae, as McCrady recognized, are both fascinating and beautiful. A zeal for studying them is essential, and John McCrady possessed such a passion.[8]

By April 1857, McCrady had completed his remarkable study, and he presented it before the ESNH, though, because of its length (165 pages of text), he likely presented it in abstract or read it at more than one session. Titled "Gymnopthalmata of Charleston Harbor," the paper presented a taxonomic account of thirty-one nominal genera and

John McCrady, ca. mid-1850s. Photograph in the McCrady Family Papers, Sewanee, Tenn. Courtesy of Edith D. McCrady, Sewanee, Tenn.

thirty-three nominal species, including three genera and species of siphonophores, which constitute a separate order of animals that possess a swimming bell or float, like the Portuguese man-of-war. McCrady placed the taxa in two suborders that closely correspond to the present-day orders: the Leptothecatae, whose hydroids have a theca or cuticle to protect the hydranth or feeding tentacles, and the Antho-athecatae, which have no protective theca. Of course, he recognized that most of the hydromedusae (Leptomedusae and Anthomedusae, respectively) were the sexual stage of those hydroids, having budded off from a sessile polyp and developed from a larval stage to a free-swimming creature, male or female. The mingling of eggs and sperm from the two sexes then results in new larvae or planulae, which eventually settle on a substrate and develop into a new polyp or hydroid. That process he recognized as an alternation of generations. In the scyphomedusae the process is essentially reversed, the medusa being the dominant stage in terms of size and complexity, and the scyphistoma the secondary or polyp stage. McCrady also recognized that the sea float *Velella velella*, or the by-the-wind sailor, while roughly resembling the siphonophores, was in fact distinct.[9]

In this classic paper, McCrady described nine genera and twenty-five species of hydrozoans as new to science, and he established four new family names. Of those, a substantial number are currently valid: two familial taxa, five generic taxa, and thirteen specific taxa. The paper established McCrady as a pioneering student of North America hydrozoans. His descriptions and figures of medusae were generally so clear that later authorities rarely encountered difficulty in identifying the species first made known by McCrady. Unfortunately, Mc-Crady provided a drawing of only one hydroid and generally gave fewer details on the hydroids. In May 1860, William Stimpson published a review of McCrady's two papers on hydromedusae, in the *American Journal of Science*. Calling them "the most valuable contribution to the history of our American *Hydroidea* that has appeared since Agassiz'[s] papers in the Memoirs of the American Academy [of Arts and Sciences]," Stimpson lauded McCrady for his new views on classification and his writing style. It was well-deserved praise.[10]

Among the new species, McCrady named one *Corynitis agassizii* (= *Sphaerocoryne agassizii*) for "my former Master in Science, Professor Agassiz, to whom America owes the only special publication on her Medusae." Another he named *Nemopsis gibbesii* (= *Margelopsis gibbesii*) for his former teacher and present colleague Lewis Gibbes. "The appearance of this Medusa," he wrote, "is at once singular and beautiful. The conspicuous crescentic outline of the pale, orange-colored sexual ribbons, the vivacious movements of the mouth and its appendages, the graceful, waving outline of the flapping dish . . .

make it an unusually remarkable object." Gibbes had, he noted, pointed out the first specimens he saw, washed up on Sullivan's Island. A third he named in honor of his native state, *Hippocrene carolinensis* (= *Bougainvillia carolinensis*). That species, he noted, is very common during summer, but he had found only one small polyp of the hydroid stage (on a piece of water-logged wood), which he had placed in a jar, hoping it would produce a medusa.[11]

Although he was truly enamored of the hydrozoans, McCrady also investigated other natural phenomena as well, reporting on them before meetings of the ESNH. In one paper, he described two species of Ctenophora, *Bolina littoralis* and *Beroe punctata*, but neither proved to be a valid taxon. He did make the perceptive observation that the premature form of *Bolina* is considerably different from the adult stage. His excellent efforts later came to the attention of the great American authority on the ctenophores, Alfred G. Mayer, and Mayer named a new species in his honor, *Mnemiopsis mccradyi*, a "greenish-amber color[ed] and opalescent" comb jelly that Mayer collected first from Charleston Harbor. McCrady also presented a paper on graptolites, which are somewhat mysterious little fossils. He speculated that some of them could be hydroid polyps, but he thought it more likely they were "parts of the skeletons of Echinoderm Larvae." His first guess was closer to fact, for it is now generally recognized that the graptolites were a form of hemichordates. In addition, he described "an embryo unlike anything known" that he thought was the "young of [the brachiopod] *Lingula pyramidata*"; he presented a paper on the annelids (i.e., segmented worms) of Charleston Harbor; and he described a fossil echinoderm, which he placed in a new genus that he called *Ravenelia*, after his friend Edmund Ravenel, "an observer to whose zeal our State is almost entirely indebted for the knowledge of its Echinoderms." McCrady also commented on a species of *Oculina* (likely *Oculina arbuscula*, a shallow-water scleractinian or stony coral) that he had found growing in two feet of water near Sullivan's Island. The Civil War broke out before some of those papers were published, however, and, when the bulk of his manuscripts went up in the flames that consumed his research journal and other items in February 1865, McCrady lost the papers on the brachiopod embryo, the annelids, and the species of corals.[12]

The subject of transmutation of species had arisen at the ESNH meeting of May 15, 1857. Contrary to the popular belief that the idea of

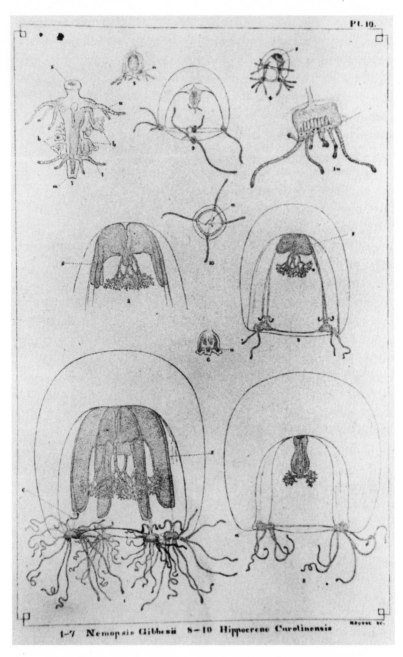

Pt. 10.

1-7 Nemopsis Gibbesii   8-10 Hippocrene Carolinensis

Two species of hydromedusae described and figured by John McCrady in his "Gymnopthalmata of Charleston Harbor," *Proceedings of the Elliott Society of Natural History* 1 (1859): 103–221. McCrady named *Nemopsis gibbesii* (= *Margelopsis gibbesii*) for his colleague Lewis R. Gibbes, and *Hippocrene carolinensis* (= *Bougainvillia carolinensis*) after his beloved state of South Carolina.

organic evolution was first introduced by Charles Darwin in 1859, various theories of transmutation had appeared long before, including the theory offered by the French naturalist Jean Baptiste de Lamarck that species were transformed by an inherent tendency toward greater complexity and by responding to environmental changes in ways that were passed on to their offspring. In the discussion, McCrady asserted that most naturalists believed that specific forms (i.e., species) remain "absolutely unchanged" from the time of their origins. He expressed doubt that one species could be transformed into another, but he thought it was "consistent with the analogy of nature to suppose that each specific form . . . exhibit[ed] in the course of its history a cycle of changes belonging to itself and included in its original conception." He was referring to the idea of stages of embryonic development espoused by Agassiz and others. But McCrady believed there was a "law of development" governing all life, and he hoped to discover that law.[13]

For such theoretical subjects as the law of development and for lengthy monographs, the Society should publish a journal, McCrady suggested to his fellow society members in 1857. They listened to him, for he was, along with Lewis Gibbes, the most active and productive member of the ESNH. A journal required funds, however, and the Society, with fewer than sixty members and fewer than a dozen faithful participants, had accumulated barely enough money to publish the first volume of its *Proceedings*. The Society had already issued brief parts of the *Proceedings,* but it could not see its way clear until May 1859 to publish the full volume for the period 1853 through December 1858. McCrady viewed the publishing of a journal as a matter of state and regional pride, however, and thus believed the state legislature would come through with a subsidy, as it had for the Tuomey and Holmes book and for Holbrook's volume on South Carolina ichthyology. Indeed, he was right, and on January 7, 1859, he wrote to Edmund Ravenel that the South Carolina legislature had granted $500 to the ESNH for the publication of the first volume of a journal and for other matters. He also expressed optimism that a grant for a second volume would be forthcoming. All of McCrady's ESNH colleagues shared his belief in the importance of advancing science in the South and of raising the ESNH to equality with the ANSP and the BSNH in scientific publishing, but none pursued the matter with the passion of McCrady.[14]

However, McCrady recognized many deficiencies in southern science, and in 1857, before the College of Charleston's alumni society, he had presented an address titled "A System of Independent Research, the Chief Educational Want of the South." It was the first in a series of addresses and papers by McCrady dealing with his views on the deterioration of culture and civilization in the American North and the superiority of culture in the American South. Indeed, for him the matter was somewhat analogous to the alternation of generations in some hydrozoans, with the South having budded off from northern culture and presently developing into a fully formed medusa that could swim freely. The South was thus a higher civilization that was superseding the earlier stages in Europe and the northern United States. "We are," he told his eager audience, "a peculiar people—a people separated from the rest of Christendom by a process of segregation, which has been working for nearly two centuries." The South, he averred, has been touched by "the finger of that Providence which guides the progress of man's history, and . . . has written character by character, a strange and diverse destiny upon our social being." McCrady argued that slavery had been forced upon the South and ultimately became "a necessity of our social system, an essential to our prosperity, and . . . as righteous and legitimate as it is useful." He added, "a social philosophy . . . supporting itself upon Science, Revelation, and Common Sense solves the problem of the union of Happiness and Labor, points out the distinction between a stable Republic and the rule of the mob, and shows how the races of men . . . may yet be combined into new social unities, to the benefit of all."[15] Just what McCrady meant by "new social unities" is unclear, but he probably envisioned that the "inferior" races would accept their status and work in harmony with the "superior" race.

In order to reach the highest level of civilization, however, McCrady argued, the South must develop a better system of higher education. On the one hand, then, McCrady contended that the South had progressed beyond the North, but on the other he admitted that it was behind. More specifically, he noted a paucity of institutions that must be "the centers of intellectual progress." Southern scientists, he declared, ought to recognize the need for a "system of independent research . . . and no longer be compelled to look at science through the eyes of other men." McCrady then proceeded to indicate what was necessary to achieve the system he advocated. His points were a perceptive assessment of the conditions that kept the Old South from attaining the same general level of scientific success achieved in the Northeast.[16]

"We need in this vast Southern country," asserted McCrady, "at least two or three principal and central institutions . . . designed for purposes which our present College systems not only do not accomplish, but to which they seldom even pretend." Such institutions, he declared, would include each specialty in science and a professor of each who would not only teach but also conduct research in that specialty. The South must also develop more learned societies for every discipline. Too many of the organizations in the South, he added, are merely "clubs, whose object is the amusement of men of leisure." The region, he continued, has too many political organizations, whose members fail to recognize that the most important problem for the South is the lack of a strong national system of higher education.[17]

A good system of education and research, added McCrady, also necessitates excellent museums that collect every kind of specimen, both from the region and from around the world. Chemical laboratories were likewise essential to the research system advocated by McCrady. Equally important were "libraries so great that they will adequately represent the intellectual activity of man in all ages of time, all parts of the world, and in every department of thought." Doubtless, McCrady, like Bachman, Gibbes, Holbrook, Holmes, and Ravenel, appreciated the notable deficiencies of Charleston's libraries, and he, more than any of them, knew the strength of Harvard's library. "It is vain to open the doors of opportunity to research," he astutely observed, "when the very material for its work is not to be found within." Especially troubling to McCrady was the paucity of scientific journals in the South. "So much of the results of Southern Science" he noted, has to be published in "the Journals of Northern Societies." Moreover, it was especially unfortunate that his own state could not claim "within her borders a single Astronomical Observatory," since the little Charleston Observatory of Lewis Gibbes had been razed. McCrady ended his address with an appeal for the South to make "a valid declaration of independence, not from the political, but from the *intellectual* domination of the Old World."[18]

A year later, beginning in June 1857, the young scientist cum philosopher and sociologist, was penning a four-part series titled "A Few Thoughts on Southern Civilization" for *Russell's Magazine*, a Charleston publication edited by his former college classmate Paul Hayne. In those essays, McCrady once again sought to demonstrate the superiority of southern culture. Indeed, he viewed the South as a distinct and peculiar civilization that was reaching its highest form. His idea amounted to an embryogenesis of southern civilization. By then McCrady had come to believe that the contemporary theory of developmentalism was really the basis of the law of development he was trying to formulate. As he viewed the matter, he was applying science to sociology, but he did not hesitate to make assertions based upon the opinions of authorities, something he studiously avoided when writing about hydrozoans. His cultural obsessions were so strong, however, that he readily adopted any idea designed to achieve his end. Unquestionably, he thought he was writing from a scientific frame of reference.[19]

McCrady devoted considerable discussion to the ideas of the French historian François Guizot, in his history of civilization in Europe; the German philologist Baron Wilhelm von Humboldt, in his work on language and civilization; and, especially, the French writer Joseph Arthur de Gobineau, in his work on race and civilization. Those authorities agreed, he said, that civilization is a state of "progressive intellectual development" that leads, in its highest form, to ideal social and moral conditions. The initial, or embryonic, stage occurs when a people establish a distinctive "intellectual unity." As in the development of any animal embryo, the growing civilization begins to form specialized parts in accordance with the law of development, which, while based on Divine precept, is not the direct action of God but his mode of operation. By "specialized parts" McCrady obviously meant such organs as the gastrozoid (feeding polyp) and the gonophore (sexual polyp) in the case of a hydroid, and limbs, internal organs, and integumentary derivatives in a mammal. In the embryonic civilization, the specialized parts were individuals with various types of skills, ranging from low to high degrees of usefulness to the collective whole. "The whole population," wrote McCrady, "becomes divided into the laborers and capitalists, brain-workers and hand-workers, the refined and the coarse." The differentiation, he added, also occurs in morals, and "good morals . . . gradually become the peculiar property and heritage of a class." Moreover, degrees of distinction among the races were, to McCrady, an inevitable part of the law of development, with, quite naturally to him, blacks at the bottom, Asians at an intermediate stage, and whites at the top, though there were distinctions or grades among the last. The hierarchy was, by McCrady's scheme, further divided by degrees of success in the process of civilization: the northern United States had budded off from the European, and the southern civilization was budding off the North and developing into the highest form yet achieved—or potentially so, as it had to remedy certain short-comings first. McCrady also believed that gradation characterized the intellectual realm, with science being the highest type of knowledge. The notion of the South as a superior culture should have offended some of McCrady's scientific acquaintances in the North, even Louis Agassiz, but chances are they never read the essays, or, if they did, dismissed them as mere eccentricities of a brilliant man.[20]

In his scheme of the South as an ideal state, McCrady also defended the slaveholders as men of practical wisdom, lauding them as the only ones qualified to offer a realistic view on "the intellectual diversities of the races." A poet may rhapsodize about the fine qualities of a donkey, said McCrady, but, if he has never driven a horse, he has no basis for comparing the lowly ass with a fine thoroughbred. Excepting "Irish laborers at the North" and "the Negroes of the South," all other Americans, he asserted, stem from the Teutonic embryo, and he believed the Germanic germ was flourishing to its fullest in the South at the time he was writing. Clearly, McCrady failed to acknowledge the great achievements of Americans in the northern states, and he ignored the fact that some of his northern colleagues shared his views on race. Blinded to such considerations by his zeal for the culture of the South, he believed that he was offering a genuinely scientific justification of slavery.[21]

McCrady was not done with his thoughts on southern civilization. As the sectionalist crisis worsened, he expanded upon his theme—first, in an address before the Chrestomathic Society of the College of Charleston on March 2, 1860, and second, in an article in the May–June 1860 issue of *DeBow's Review*. In the address, he reiterated his ideas on civilization and protested even more directly against the arguments of abolitionists. The notion of liberty and equality for blacks was based on "a false creed," and the abolitionists' claim of scriptural justification for their actions was blasphemous, McCrady charged. Moreover, he insisted, the abolitionists had repudiated the American Constitution and "joined a league with hell and a covenant with the devil." Southerners have examined "the subject of slavery calmly, thoughtfully, and as much without prejudice as men have ever examined a subject in which they were personally so interested," he told the members of the Chrestomathic Society. Zoology, he added, has shown that the "physical structure of the Negro . . . [is] so very inferior, so decidedly lower in type" that thinking men even doubt that "the Negro and the white man could possibly have descended from the same origin." Obviously, McCrady agreed with the view of Agassiz and other pluralists that each race originated in its own geographic region and represented a distinct species.[22]

The South must be guided by knowledge, he continued, and knowledge indicates that "the Negro is fit only to be a slave, incapable of originating a civilization himself, and incapable even of keeping one up, when it has been delivered to him by the white." Northern critics of slavery, asserted McCrady, have confused liberty with license. Their call for a democracy in place of a republic, he added, ignored the need to provide for "a check upon the unbridled freedom of the press, the tongue and the pen." Thus, argued the passionate southerner, in the North "fanaticism" and "the wildest theories, amounting to monomania in some cases," are allowed to run rampant. Lest his audience assume an attitude of self-righteousness, however, McCrady added that the South, because of its patriarchal nature, had failed to educate its citizens "for philosophical thought." Nevertheless, in his view, the South had made "the only real advance . . . in philosophic sociology," for it had developed a proper conception "of the relation of the races to each other as contributors to the grand civilization of Christendom." Much more must be done, however, said McCrady, and he returned to the points he had made in 1856: "I ask, where are the great libraries, the laboratories of Science, the galleries of Art, which are needed for men?" McCrady noted that the absence of those institutions was mainly due to the region's relatively small population, but he remained optimistic that southern civilization was in ascendance. The South, he opined, is "like one of those vast coral reefs which skirt our Southern shores, made up of millions of individuals, all working together for a common end, and together spreading . . . and growing upwards, despite the floods of fanaticism . . . at last to rise above that tempestuous sea—a spiritual continent!—firmly and lastingly fixed—the intellectual New World!"[23]

The flourish and feeling of McCrady's analogy no doubt impressed his audience, but an analogy from zoology was a far cry from scientific evidence or a scientific theory. Although he may have thought otherwise, McCrady was speaking from a heart pulsating with provincialism, not from a mind that addressed the question of race on the principles of scientific inquiry. Among the Charleston naturalists, none quite possessed the passion of young John McCrady for his beloved region, though Bachman and Holmes came close. To censure McCrady for his zealous defense of the South and slavery or to fault him for his effort to use science in a mistaken way, however, is to ignore the cultural context of his era. His views were wrong, of course, but even many of his northern peers considered blacks to be inferior. Unlike McCrady and those sharing his sympathies, however, they had never developed a special sense of regional peculiarity. Certainly, his northern compatriots possessed a sense of intellectual rivalry with Europe, and their patriotism to America is unquestionable. Conditions made them more temperate and open, however.[24]

In his last essay on the superiority of southern culture, McCrady again wrote with enthusiastic feeling, but he resorted less to harsh criticism of the North. Furthermore, he endeavored to draw upon science more than in his previous addresses and essays, though not with telling results. Published in *DeBow's Review* in 1861 as "The Study of Nature and the Arts of Civilized Life," the essay included many references to God and referred very directly to the law of development he was trying to formulate. Using more analogies from natural history in that essay, McCrady endeavored to expand upon his idea of the embryogenesis of southern civilization. In an extensive discussion of the nature of cells, he referred to "a *nucleus*" that, in the case of a tree, grows its roots and branches by "obeying the great law of development by specialization." More important for his argument, however, was the budding-off process, which, in the case of southern civilization meant separation from the Union. "The great Confederate Republic," he argued, "is about to break up into two or more confederacies," and each would "specialize into a multitude of different parts." McCrady predicted that it would be a "peaceable separation," but nonetheless, "like those vast convulsions of geological time, it will be a convulsion of development—a pang and throe of the birth-time of great nations which are yet to be." To McCrady, it would be a scientific process, not a revolution. Nor would it be an evolution, for, in his judgment, a nation could no more be transmuted into a new species than could any animal be transformed into another.[25]

Believing that civilizations were no different from animal life, Mc-

Crady contended that the budding off of the Confederate nation was simply a stage in a natural process. Or, as in another analogy he used, the new nation was like a seed falling from a plant and taking root in fertile soil. "Every true civilization is a native growth," he said, but once more he urged the South to "build up a system of *home education*." The region must not try to graft another culture upon itself, for such a culture "is the egg of an exotic not adapted to our climate . . . [and] it may hatch out into a viper." Concluded McCrady, "these are not visions; they are sober principles." In the South, he rhapsodized, "the palm tree uplifts its queenly height amid the powerful and outspreading arms of the oak; . . . here the apple tree and the orange tree cast their fruit upon the same ground together; here the vine and the olive, and the fig tree, and the pomegranate bring forth their fruit in due season; here the roses bloom forever; and lastly, here King Cotton has fixed his royal throne, and all the world must pay him tribute." So John McCrady romanticized the South, the land he loved so dearly that he would lay his life on the line to preserve it and its peculiar culture if he had to.[26]

Meanwhile, at meetings of the ESNH on February 15, May 1, and August 1, 1860, McCrady had offered comments on Darwin's *On the Origin of Species*, but only a portion of them appear to be extant. Since McCrady offered his comments early in 1860, he obviously obtained a copy of the volume published in England in 1859 or a copy of the American edition soon after it appeared. Darwin's theory on the origin of species shows "great originality," said McCrady, and "it has opened anew the question as to the relation of species *inter se*." In a summary of the comments made by McCrady on February 15, Gibbes noted that McCrady praised Darwin for striking "a mine of wealth which had been neglected by naturalists for a long time past." Gibbes added, however, that McCrady expressed disagreement with Darwin "in almost every other particular," and that he believed Darwin was apparently unfamiliar with the law of development. In conclusion, wrote Gibbes, McCrady stated that he must complete his own work on the law of development before commenting further on Darwin's theory. In fact, McCrady did not wait, for he offered comments twice more before he completed his initial paper on the law of development.[27]

At the meeting of the ESNH on October 1, 1860, McCrady read his paper titled "The Law of Development by Specialization, A Sketch of Its Probable Universality." Claiming that he had devoted nearly seven years to thinking about the topic, McCrady reminded his readers that he had previously noted that species undergo "form-changes" or a kind of "progressive morphology." Meanwhile, he added, "Mr. Darwin has produced his Theory of the origin of species by Natural Selection, which . . . furnishes a most beautiful explanation of the *modus operandi* which probably characterizes the law of development in the production of specific forms and varieties." He maintained, however, that Darwin had failed to see "any analogy with the grand law of development." In fact, Darwin was familiar with the ideas of the developmentalists, but he knew that they were unsatisfactory. That McCrady thought Darwin's theory was a "beautiful explanation" makes clear, however, that he had not yet grasped its full import.[28]

Grounding his law in religion, McCrady borrowed an idea from his mentor Agassiz, namely that nature represents the Thoughts of the Creator. Thus, averred McCrady, all laws governing nature "are the consistent course of his [God's] actions." Those actions have always been in effect, McCrady stated, and are evident in the long succession of geological eras. While changes in a species are obvious in the geological periods, he added, no species has ever lost its individuality. McCrady would thus have it both ways: a sort of evolution without complete transmutation. Again he argued that the law applied to civilization as well. He admitted, however, that it was difficult to explain such a law, though, as he saw it, the difficulty resulted from the different varieties of "Creative Thought," or, apparently, from the complex nature of Divine actions. It would thus take a long time before he could apprehend the variations and synthesize them into an all-embracing statement. In the meantime, McCrady observed, only a provisional sketch was possible.[29]

As he viewed it, every organic phenomenon—plant, animal, or civilization—began as a molecule, egg, or bud that possesses an individuality, or an "original synthetic whole" that develops from its embryonic form into specialized "inter-dependent parts." Mathematically expressed, the process was "like an infinite series," and "the further the development . . . the more and more the parts become differentiate among themselves." Yet, the greater the differentiation the greater the interdependence of the parts, thus making them even more important to the whole organism. To illustrate the process, he referred to the hydrozoan animal, which begins as a "spherical or ovoid embryo" (that is, the blastula and planula larva), then develops into "the Hydra" (that is, apparently, the hydroid stage), and finally develops into a medusa. Thus, there is a "succession of various orders of individuality." It seems, then, that, to McCrady, as to other developmentalists, evolution was simply embryogenesis, and, though he hinted at organic transmutation, he never explained how that was possible. He had no doubt observed, as had Agassiz and others, the similarities of features in the embryonic forms of, say, mammals, and thus viewed evolution as a process of developing the special characters of the various species. In any case, by 1860 he was more concerned with formulating his law than he was with doing research on the hydrozoans, but he was convinced that he was making a major contribution to science by trying to determine how nature works. Within a few weeks after the publication of the sketch of his law, Fort Sumter was under bombardment, and he had no time thereafter to think about the law of development.[30]

On the evening of December 20, 1860, amidst the din of parades, fireworks, and pealing church bells, delegates of the Convention of the People of South Carolina assembled in a Charleston hall and soon proclaimed the state free of the Union. A week later, the state's governor ordered three companies of South Carolina troops to board the guardboat *Nina* and seize Shute's Folly, a small, marshy island in Charleston Harbor, situated less than a mile from the Queen City. The objective was to capture Fort Castle Pinckney, one of the four United States military posts established in the Charleston area long before as a defense against foreign enemies. On board the *Nina* as a stowaway was John McCrady, who joined the troops as they rushed ashore to capture the fort. To the surprise and dismay of the invaders, however, the Union commander, Major Robert Anderson, had evacuated all but two of his troops to Fort Sumter. At home McCrady had left his young wife Sarah Dismukes, whom he had married in late 1859, and his baby daughter Esther Lynch. Since the College of Charleston was on Christmas vacation when McCrady joined the invading troops, the trustees did not learn of his absence until classes resumed. They would not see their assistant professor of mathematics again until four years later.[31]

Hyenas and Hybrids

uring the late 1850s, when John McCrady was attempting to articulate a science of culture, he was
continuing an effort begun several years before in
the guise of ethnology, or, to many of its early proponents, the science
of racial and cultural origins and differences. Widely embraced by
American scientists, the developing science relied heavily upon cultural impressions and such crude measures as physiognomy, skin
color, and skull size to ascertain differences. In later years, American
ethnologists turned their attention mainly to native American cultures,
but during the formative years of their work, they focused on African
Americans. The new field of study appealed to scientists in the North
as well as in the South, but it came to have special meaning in the latter
because it helped to justify the institution of slavery. One member of
the Charleston circle of naturalists stood virtually alone, however, in
refuting the arguments of the ethnologists that each human race constituted a separate species. Yet he also sanctioned slavery as a bibli-
cally ordained institution. The man was the clergyman-naturalist John
Bachman.

Contrary to the argument of William Stanton that Bachman was "half theologian, half scientist," however, the evidence clearly shows that Bachman was fully a scientist—and an excellent one at that. He was also completely faithful to his theology—and no more confused, as Stanton would have it, by conflicting elements in his personality than anyone else of his time. To judge Bachman otherwise is to lift him from the context of his culture and place him in the present. The issue must, of course, be evaluated in part by the contemporary mix of theology, culture, politics, and science, but it must also be viewed in light of who was following scientific principles the most faithfully, that is, in this case, who understood the scientific criteria for classifying mammalian species, including humans.[1]

That John Bachman condoned slavery had become clear by 1837, when, in response to the declaration of a Lutheran minister in the North that slave owners were guilty of "cruelty and luxury" and should thus be excluded from communion, Bachman said, "As you have no slave-holders with you, it would appear that you have traveled out of the way to denounce the acts of your brethren." Noting that his wife possessed four slaves at the time of their marriage and that they continued to own them and use them as domestic servants, he asserted that he and his wife nevertheless treated them as part of their family. Quite clearly, the Charleston minister sanctioned the institution of slavery, but he also believed that blacks were children of God and descendants of Adam and Eve. In his judgment, the Creator had allowed them to become inferior to whites. Their minds and souls were nonetheless important, and they must be guided by educated leaders of their own race. It is not surprising, then, that, when the young black schoolmaster Daniel Alexander Payne came to Bachman to ask him to identify a caterpillar, the minister readily provided the information for his visitor. When Payne returned later, Bachman showed the free black his garden, herbarium, and insect collection. On the occasion of a third visit from Payne, Bachman invited him into the parlor of his home and introduced him to his family. After a conversation with Payne, the minister asked one of his daughters to play the piano for his guest. Later, when Payne decided to enter a seminary in the North, he carried with him a letter of recommendation from Bachman. Payne attended the Lutheran seminary in Gettysburg, Pennsylvania, and eventually rose to a position of leadership in the African Methodist Episcopal Church. The admirable generosity of Bachman did not, of course, indicate any alteration of his belief that the continuing enslavement of blacks in the South bore the stamp of divine sanction.[2]

Indeed, Bachman firmly believed in divinely appointed racial distinctions. For example, in 1840, in an address on the importance of agricultural labor to national prosperity, he declared that American Indians had refused to comply "with the just condition which the Deity has coupled with the gift of life." In fact, said Bachman, the original inhabitants of America would have destroyed themselves if Europeans had not brought civilization and Christianity to the New World. He believed the conquerors would have done better, however, by enslaving the natives, for "the process of slavery would have implied labor, and labor, in turn, implies morals, strength, improving intellect, and the true erectness of manhood." Although the address focused mainly on the importance of agriculture to South Carolina's economy, it served also as a statement of the Charleston naturalist's belief in human "varieties." Soon thereafter, however, Bachman encountered arguments that each human race actually constituted a separate species. Among the first to make the case was Samuel G. Morton, the Philadelphia physician and naturalist who was held in high esteem by his Charleston confrères. The author of *Crania Americana* (1839) and *Crania Aegyptiaca* (1844) and collector of nearly 1,000 human skulls, Morton had concluded by 1839 that cranial capacity varied according to race, with blacks having a significantly smaller brain case. As Bachman knew, the mammalian skull is important to taxonomists for identifying species, but the number and nature of the bones in the cranium, the dental formula, and the character of the teeth are the diagnostic features. He was aware that skull size can vary among individuals of a mammalian species, including humans, but that cranial capacity, even if greater or smaller, is not a criterion of species identification. Morton either ignored or had no knowledge of the proper use of the skull as a taxonomic character. His special studies were appealing to many of Bachman's contemporaries, however, because they bore the imprimatur of scientific inquiry. Indeed, no one, not even Bachman, questioned his procedures, which, although Morton seemed to have been unaware, were biased in favor of the white race. In any case, before Morton could state absolutely that race and species were synonymous, he needed a satisfactory way to argue that separate species could mate and produce fertile offspring. By 1846, he thought he had the answer: the offspring of such unions are an intermediate stage between two species and that, contrary to standard views, are more often than not fertile, as are their progeny and their progeny's descendants.[3]

In 1847, Morton published his full argument in a two-part article in the *American Journal of Science*. In his "Hybridity in Animals, considered in reference to the question of the Unity of the Human Species," he admitted that naturalists do not agree on the definition of the concept of species, but he was convinced that he knew the best one: a species is any living or once-living thing that has a "separate origin and distinctness of race, evinced by the constant transmission of some characteristic peculiarity of organization." Morton then proceeded in the first part of his essay to cite cases of interspecific crossings of mammals whose offspring were fertile as confirmation of his argument that different species of humans possessed the same capacity. Among his examples were crossings within the genus *Equus* (horses, asses, and zebras) and within the order Artiodactyla (cows, deer, and pigs), most of which stretched credulity excessively, especially the cases of intergeneric crossings. Freely relying upon questionable authorities, he noted, for example, the case of "the Indian buck" (*Cervus unicolor?*) that successfully crossed with a member of "the Porcine species" (genus *Sus*) and produced "the well known intermediate stock called the *Spotted Hog-deer*." The hog deer of India, *Axis porcinus* and *A. maculatus*, are, of course, true cervids and related to hogs only by sharing the characteristics of the order Artiodactyla. Such facts were apparently of no significance to the Philadelphia ethnologist. In Part I of his article, Morton devoted special attention to the interfertility of canid species. "The wolf, the dog, the jackal and the fox, all intermix with each other," declared Morton, and their progeny are indeed often quite fertile. Before ending Part I, Morton revealed not only his gullibility but also his determination to prove his point at any cost by citing "a most unusual instance of hybridity," namely, the union of *Felis domesticus*, the domestic cat, and a mustelid carnivore, the marten *Mustela martes*. He gave this impossible hybrid no common name, but perhaps "Calico Weasel" would have been appropriate. Unfortunately, Morton noted, the owner of the four hybrid progeny had no success in breeding them with other cats.[4]

In the second part of his paper, Morton cited instances of hybridity in birds, amphibians, mollusks, and insects. The cases of hybrid birds were generally not quite so farfetched as those of the mammals ad-

Samuel George Morton. Reproduced from *Dictionary of American Portraits*
(Toronto: Dover Publications, 1967).

duced by Morton, but his documentation was dubious in most cases.
The sole example of an amphibian crossing was, however, quite remarkable. Although he had "but a single authenticated example,"
Morton was certain of a successful union between a species in the
genus *Bufo* and one in the genus *Rana*. A successful union between a

toad and a *frog* seemed not to surprise the naturalist Morton at all. More believable were cases of interspecific crossings of freshwater mollusks in the genus *Unio*, but Morton's examples also included some absurd cases of intergeneric crossings of insects. Although he was on more solid ground in cases of hybridity in plants, Morton devoted only a small amount of space to such cases. In conclusion, he maintained that, although hybridity is contrary to natural law, it nevertheless occurs often, not only between species within a genus but also between those in different genera.[5]

Morton's ideas were already well developed when Agassiz met the Philadelphia naturalist in 1846. Shortly before he came to the United States, Agassiz had written to Morton to say that he looked forward to meeting the man who had published "the most important work" on "the human races." Of course, as previously noted, it was in Philadelphia in 1846 that Agassiz saw a black person for the first time in his life. By the end of the following year, in Charleston and on Holbrook's plantation, he had seen many black people. Impressed by Morton's argument, Agassiz viewed it as a complement to the idea of "natural provinces" that he himself had formulated in 1845. In that formulation Agassiz argued against the orthodox interpretation of the biblical account of the creation of animals, contending that all species of animals were not created at the same time and in the same place. Instead, he argued that God created animals in successive acts and in their own natural provinces, or zoological zones. Thus, with Morton's notion of separate species of humans bolstering his own view of successive creations of animals in special zones, Agassiz was primed for the talk he presented before the Literary Club in Charleston in December 1847. That the members of the Club were without exception defenders of the institution of slavery and believers in the power of scientific inquiry meant that Agassiz had a particularly receptive audience. At least two of the members would not be swayed by his argument, however. They were John Bachman and the Reverend Thomas Smyth, pastor of the First Presbyterian Church of Charleston. Both men quickly recognized the import of Agassiz's argument, namely, that it undermined the biblical account of creation. It was one thing to them to entertain a liberal interpretation of the six days of creation, but it was quite another to play freely with the idea of special and repeated creations and with the notion that all humans were not descendants of the original pair. For Bachman, however, it was equally significant that the argument went against science. When invited by the Club soon thereafter to give his views on the subject of the plurality of origins and species of the races, he determined that he must counter the argument on the basis of the concept of species. In his judgment, Agassiz was abusing the principles of science.[6]

Bachman had read Morton's two-part article in the *American Journal of Science* soon after it was published in 1847, but his labors on the *Quadrupeds* volumes and his belief that others would challenge the argument kept him quiet on the matter. By 1849, however, the issue was gaining too much attention for him to ignore it. During that year the views of Josiah Nott were being disseminated throughout the South. By then a respected physician in Mobile, Alabama, Nott had studied first at South Carolina College, where he imbibed of the free spirit of Thomas Cooper. Later, he had enrolled in the University of Pennsylvania's medical department and, after completing his work in 1829, returned to Columbia, South Carolina, to establish a medical practice. Further influenced by Cooper, especially concerning views on the inferiority of blacks and on the mythic nature of the biblical story of creation, Nott decided in 1835 to enhance his knowledge of scientific medicine by studying in Paris. Returning to the United States a year later, he settled in Mobile, then a rapidly growing city. Influenced by Morton's *Crania Americana*, Nott became increasingly interested in the subject of hybridity. He was especially concerned over the intermating of whites and blacks because, in his judgment, it would eventually result in the demise of both races. Thus, in 1843 he took up his pen to show the consequences of producing progeny of mixed races. A conflict with Bachman was almost certain, for Nott's deep-seated racial views intensified in proportion to the clergyman-naturalist's success in refuting the proponents of pluralism, or polygenism.[7]

Of more immediate concern to Bachman in 1849, however, were the articles by Morton, mainly because Morton was so highly regarded by the Charleston naturalists and more widely recognized than Nott. In a letter to his old acquaintance on October 15, 1849, Bachman expressed his desire for a friendly discussion of the subject of racial differences. He informed Morton that Charleston's Literary Club had first asked him to review Nott's essay, and later to present a talk on "the whole subject . . . as a matter of science." Continued Bachman, "I advocated the Unity of the races. [The Charleston physician James] Moultrie and others who advocate the opposite view thought it best to have still a third meeting, and Holbrook . . . will throw his weight on the same side [with the pluralists]." Furthermore, noted Bachman, the club had urged him to publish a book on his argument, and he had agreed to do so. To that point, Morton probably read the letter with no displeasure, but his hackles surely rose when Bachman said, "I considered your authorities not such as could be depended on, and I endeavored to show that in several instances the facts stated were disproved by more recent and better authorities." Just as he had done in dealing with Morton's colleague George Ord over a decade earlier, Bachman spoke bluntly, but he endeavored to keep the discussion on a high plane.[8]

"We are both in the search of truth," he told Morton, assuring him that "these scientific investigations" could be kept apart from any theological issues. Indeed, Bachman added, scientific truth was always in harmony with Scriptural truth, for "the author of revelation is also the author of nature." From that agreeable comment, Bachman turned quickly to criticism of Morton's main source of authority, the English soldier-naturalist Charles Hamilton Smith. Noting that he had not found a single one of the American species described by Smith to be valid, Bachman asserted that Smith was unreliable as an authority. Then he offered to provide information to support his argument for the monospecific nature of all humans. "Let us battle for truth, not for victory," he urged Morton. Again he expressed a hope that the debate would not damage their friendship, but he warned Morton that he could "scatter some of your facts to the wind."[9]

Word of Bachman's forthcoming book spread quickly. On January 21, 1850, Robert Gibbes wrote to Morton that both Bachman and Thomas Smyth were preparing books in defense of the unity of the races. As Gibbes viewed the matter, Nott's essay had "brought down . . . the wrath of the clergy," but he noted that Bachman's book was supposed to deal "chiefly with the scientific question." Gibbes added that he had heard that Holbrook, Moultrie, and Eli Geddings, a professor in Charleston's medical college, would publish a refutation of Bachman's argument, though they never did. Meanwhile, one of Bachman's own colleagues was privately speaking against him. James Warley Miles, the College of Charleston professor of philosophy and Greek literature and an ordained Anglican minister, told an acquaintance, "I would not give a fig for the benevolent Dr.'s recommendation of anything except, perhaps, a partridge or a rat. I don't trust his judgment respecting the *genus homo* [*sic*]." In fact, none of the Charleston naturalists or the professors in the medical college came to the support of Bachman. As an independent soul and as a champion of rigorous scientific inquiry, however, Bachman could manage without their support. His book, titled *The Doctrine of the Unity of the Human Race Examined on the Principles of Science*, came off the press shortly before the meeting of the American Association for the Advancement of Science in March 1850, and Bachman presented a copy to the Association. Present at the meeting, of course, was Agassiz, who took advantage of the occasion to offer some comments on race. Expressing regret that, in his view, the subject could not be discussed rationally and that it had excited religious opposition, the esteemed naturalist stated that the human races differ even more among one another than do other species of vertebrates. In fact, he said, the teeth differ among the races more than they do between monkeys in separate genera. Thus, he concluded, evidence points clearly to separate origins of the races, hence to repeated acts of creation. That Agassiz had chosen to use mammalian genera as examples and that he had made such an unfounded assertion about variations in human teeth probably did not surprise Bachman, given the disappointing discussion he had held with him in December 1847.[10]

Probably more irritating to Bachman was a paper sent by Josiah Nott to be read at the meeting. Although he titled his paper "An examination of the Physical History of the Jews, in its bearing on the Question of the Unity of the Races," Nott used it partly to demonstrate the diversity of races since antiquity, partly to show the purity of the Jewish race, and partly to tout ethnology as a science "of great practical importance in its bearing upon races." After the paper had been read, Agassiz offered comments, the chief of which was, in the words of Thomas Smyth, that "the different races of men were descended from different stocks." Agassiz also reiterated his idea of zoological provinces and thus rejected the unity argument. Bachman was present at the session, but he declined an opportunity to counter Agassiz's argument, saying instead that he disagreed completely with it and that he had already published his views on the topic. Contending that the subject was unsuited to a public debate and that it ought to be investigated through research by the pluralists, he added that, because of his stature, Agassiz "had expressed an opinion . . . which would be likely to make a deep impression" on those who shared his view. Indeed, the

Charleston naturalist was right on two major counts: Agassiz was relying upon opinion, not upon research, and his views carried great weight, not only with those who were already inclined to believe that blacks were of a different species from whites but also with many who were not fully convinced by the arguments of Morton and Nott. In any case, Robert Gibbes was delighted, and he wrote to Morton two weeks later to say that the AAAS meeting was very successful. After the meeting, Agassiz had spent eight days with Gibbes, who took his guest around to several plantations to see slaves descended from various African tribes. Agassiz "found enough," announced Gibbes, "to satisfy him[self] that they have differences from other races." Indeed, Agassiz was so certain of the differences that he and Gibbes arranged for a local photographer to make a number of daguerreotypes of slaves, ordering both males and females to strip to the waist, apparently for the purpose of revealing variations in bodily types by tribal origin. The "examinations," as Gibbes called them, were hardly necessary, for Agassiz already knew what he would find. Indeed, for the March issue of the *Christian Examiner*, he had written an article in which he declared that the races had separate centers of origin. Gibbes also told Morton that Bachman and Smyth had "pressed Agassiz in Charleston, and he came out with his views on the unity question, and stirred them up very much," but Gibbes's view of pressure on Agassiz runs counter to Bachman's objection to debating the subject at the AAAS meeting. Soon thereafter, Gibbes penned another letter to Morton, saying, "Dr. Bachman's book is making quite a stir here." He had recently read the book, he added, and found "it contains many interesting facts," though Bachman "is visible on every page, and his confidence in his own observations is too dictatorial." Gibbes nevertheless conceded that Bachman "deserves great credit for his researches." Obviously, Robert Gibbes, a good paleontologist, was torn: on the one hand, he did not want to hear a view that ran contrary to his cultural conceptions, especially when advanced by a man who had criticized him in the newspapers; on the other hand, he recognized good research when he saw it.[11]

Bachman's book did in fact make "quite a stir." Despite some flaws and redundancies, it was an excellent scientific treatise on the subject of the specific relationship among all humans. However, the minds of the pluralists, or "advocates of diversity" as they were frequently called, were made up, and they remained unified in their determina-

Louis Agassiz, 1857. Courtesy of the Harvard University Archives.

tion to find fault with his defense. Bachman completed the text of the volume in November 1849, but he added three appendices while it was in press, running the total number of pages to 312. Although he vowed to avoid any argument on religious grounds, he was unable to avoid some comments on God as Creator. Yet fewer than thirteen pages, or barely over four percent of the work, referred to religious precepts. *The Doctrine of the Unity of the Human Races Examined on the Principles of Science* was almost wholly what its title stated. The standing of John Bachman as one of the best naturalists in the United States in 1850 should have been solidified by that work, but the message he delivered was not widely welcomed. Nevertheless, the stir caused by the book was considerable.[12]

At the outset of his treatise, Bachman stated that each side should strive for objectivity in dealing with the issue and that it should base its arguments upon the "teachings of nature." That he considered himself qualified to address the topic he made clear by referring to his studies in botany, ornithology, and mammalogy. In addition, he noted that he had joined with the leading European naturalists "in patient, minute, and varied examinations and comparisons" of plants and animals. It was no idle claim—and certainly one of which Morton could not boast. More specifically, Bachman mentioned particular mammals, birds, and hybrids he had studied in Europe, most of which Morton had used as examples in his paper on hybridity. After a brief discourse on the dangers of adopting a theory and then selecting evidence to support it, especially evidence from questionable authorities, Bachman promised to cite "original experiments and examinations pursued without any regard to theory."[13]

Since he viewed living forms as consisting of three types, Bachman carefully explained the characteristics of each. The basic form was the *species*, which he defined as "individuals resembling each other in dentition and general structure." Although his definition is far from modern conceptions of reproductive isolation and genetic associations, it was good in its own time because it included the diagnostic characters of mammalian species, which Morton did not understand. Of course, since Bachman was keenly aware of individual variations within a species, he thought of *varieties* as groups within a species that differed slightly in such characters as color of pelage or skin or texture of fur or hair. If the group remained sexually isolated from other members of the species over a long period of time, then, in Bachman's view, it became a *permanent variety*. That notion did not, however, constitute transmutation, for Bachman argued that a permanent variety could still breed successfully with another variety or with other individuals of the same species. Indeed, he roundly rejected the Lamarckian theory of evolution and called it "absurd," though later, unknowingly, he came close to accepting it. As biologists later established, of course, hybrids are not necessarily sterile, and, as Ernst Mayr notes, "high, if not complete, fertility is known for many species crosses . . . throughout the animal kingdom." Morton and the pluralists were not wrong in their conceptual argument. In their eagerness to prove their preconceptions, however, they readily accepted fanciful crossings. Bachman, on the other hand, was inclined a bit too strongly toward a belief in the extreme rarity of hybrid fecundity, but he clearly adhered more closely to the model of scientific inquiry.[14]

Although Bachman defined species as a group of individuals alike in external characters, he was aware that internal organs could be used as well. For example, he noted that "the American swan [probably the whistling swan, *Cygnus columbianus*] . . . and the trumpeter swan [*Cygnus buccinator*] . . . resemble each other so closely in form and colour, that until recently they were regarded as one species," but a recent study had indicated differences in the number of their ribs and in their tracheae. That illustration served as his invitation to Morton and other pluralists to compare the races of humans, by which method they would find that *all* members of the genus *Homo* have the same total number of bones, an identical number of bones in each section of the body, and the same number and kind of teeth. The pluralists, he charged, had also consistently ignored physiological factors—such as the respiratory, circulatory, and muscular systems, and mean body temperature—all of which, he observed, are alike in whites, blacks, and other races. All humans, moreover, he noted, have "the same power of speech, the same power of song, the same love of music," and their longevity is similar. The skulls of all races differ no more than they do among dogs, horses, cows, or sheep, he added. Certainly, admitted Bachman, skull shapes may vary among the races, but, Morton's studies notwithstanding, "merely having skulls . . . of races is insufficient." The pluralists, Bachman astutely observed, must acquire knowledge of "the structure of animals" before they can make judgments about races.[15]

Bachman then turned to Morton's cases of hybridity, and shredded them one by one. For example, Morton had cited a report of a caribou that copulated with a cow in Newfoundland and said he had no reason to doubt the authenticity of the account. Morton identified the caribou as *Cervus wapiti* (= *Cervus elaphus*), which, as Bachman noted, is not the caribou but the elk, the eastern range of which is hundreds of miles from Newfoundland. Bachman observed further that the animal to which Morton referred was likely a moose (*Alces alces*). In any case, Bachman saw good reason to reject the story of a male of the family Cervidae that found a sexually receptive female of the family Bovidae.

Morton's limited knowledge of mammals should have caused the Philadelphian to think twice before writing anything about Bachman's specialty. Morton had also referred to a supposed instance of the union of a bull and a sheep. Bachman had heard such stories before, and he noted that he had once gone ten miles out of his way "to see a similar prodigy," only to find that it was "a large ram with a hairy fleece, and rather straight horns." As for Morton's acceptance of Charles Hamilton Smith's notion that the hyena may have been the primitive stock from which the dogs had derived, Bachman showed that Morton knew nothing of the differences between those carnivores. In fact, the hyena is closer to the cats than to the dogs. Bachman spelled out some of the differences, stating that, unlike *Canis familiaris*, the hyena has four teats, fewer teeth, two anal gland sacs, and only four toes on each foot. It is obvious, however, that Bachman was familiar with only the Brown Hyena, *Hyaena brunnea*, for it is the only one of the three species of that carnivore possessing four teats. In any case, after piling up his telling criticisms, Bachman observed that he had "learned from experience to be very cautious in receiving all the strange productions of hybrids" among the mammals, and in a message aimed directly at Morton, he added, "This is at least a credulous world."[16]

In addition to refuting some of Morton's cases of hybridization among the birds by citing evidence from numerous experiments he had done in his own backyard in Charleston, Bachman made some comments on plants. He noted, for example, that he had experimented extensively in crossing roses in his yard, and that of the 200 varieties he had successfully raised, only three had developed fertile seeds. He noted further that in Edinburgh he had seen 106 varieties of wheat, but all of them were of the same species. The main point of those illustrations was to reinforce his argument for the monospecific nature of the human races, no matter how many varieties had developed. Morton had also asserted that a natural repugnance existed between species of animals but that long and close proximity tended to lessen the repugnance. Thus, maintained Morton, the "moral degradation consequent to the state of slavery" had resulted in numerous cases of unions between members of the black "species" and the Caucasoid "species." Quite the contrary was the case, replied Bachman, and he opined that many white males who had come to Charleston from the North and from Europe had displayed no repugnance toward black women in Charleston, or, if they did, the repugnance had vanished within a few days. In London, Edinburgh, and Paris, moreover, he had seen young whites, male and female, "walking arm in arm with Negroes," and in Stratford, England, he had witnessed "a rather pretty white woman leaning on the shoulder of her husband, a full blood African," with their offspring standing nearby. Such mixing, though successful in producing progeny, was unacceptable to Bachman, for he believed it generally led to "degradation and crime." Still, it served as evidence that the races were indeed of a common species.[17]

Why, wondered Bachman, had Morton previously ignored the fact that the Mongolian, the African, the Malay, the American Indian, and the Caucasoid can cross and "produce prolific offspring"? Why, he asked, had the pluralists formerly argued that other hybrid animals cannot produce fertile progeny while humans can? It was because they had recognized the weakness of their argument, and, in order to make their case, their spokesman Morton had now "ransacked" ancient tales as evidence. Bachman reminded Morton that he could cite from an ancient and reliable source that God had "made of one blood all nations of men," but that was not his intention in this work. Why, then, was Morton "dragging from the dust of antiquity every obscure and doubtful record, searching among rude and barbarous nations for ancient traditions, and striving to interpret in favour of their theory the hieroglyphics and sculptured heads on the moldering monuments of antiquity"? It was because the desire to prove his theory had over-whelmed his commitment to scientific inquiry, Bachman declared.[18]

Differences in physiognomy and in skin color were, of course, a favorite argument of the pluralists as indications of specific differences, but they were distinctions they had to handle with considerable dex-terity. Bachman scored heavily in his criticisms of their use of physiog-nomy and skin color as diagnostic characters of specific designation. Among Caucasoids alone, he observed, can be found "many interme-diate grades of . . . colour," and great variation in physical appearance. If the pluralists were honest, especially since they were now using animals to strengthen their position, they would try to deal with the fact, for example, that the common cow can vary greatly in form and color: some possess small horns, but others have "immense horns"; some have a shoulder hump, but others, none; some are as large as "a small elephant," but others no bigger than the largest dog; and some are of one color, and others of another. Thus, he insisted, if humans can be designated as separate species on the basis of variations in skin color and physical features, then the pluralists must divide *Bos taurus* into dozens of species. The same applied to hogs, sheep, dogs, domestic fowl, and the common pigeon, among the last of which were "an endless variety . . . of every form, size, and colour." Or let them make species of the "twenty-nine distinct varieties" of the canary, or of each of the widely varying types of goldfish. Morton and his supporters, added Bachman, conveniently ignore such cases in their effort to equate human varieties with species. After they had failed to prove that a hybrid from the union of different races was sterile, Bachman reminded them, they resorted to claiming that the hybrid's descendants remained "healthy and fertile." The pluralists had "suddenly shifted their sails on the other side," and now, he charged, they "would carry their sinking bark to a port of safety under the false colours of fertile hybrids in the lower races." Bachman would not let Morton escape from his contradiction, and, later, in reference to some Mexicans whose skin contained bluish spots, he opined that the pluralists might wish to designate them as a new species, "Homo maculatus."[19]

Bachman had, of course, given a great deal of thought to the causes of variation in the color of feathers, pelage, and skin, but he confessed that he had not found a satisfactory explanation. He believed that variation in color was related to "the laminae lying under the outer cuticle or scarf of skin," which was probably affected by climate. In a comment no doubt distressing even to his supporters, Bachman argued that the earliest humans were not as "light-coloured as are the Europeans and their descendants; . . . our ancient ancestors were not white, but they were not black either." Even among the Africans great variation exists in skin color, but the pluralists, he noted, have focused upon the American slaves, who are descended from peoples of the darkest skin. While Bachman argued that adaptations of humans to their climate result in "a succession of strikingly marked varieties" of skin color, he insisted that the pluralists went too far in asserting that climate resulted in distinctive changes. For example, in the United States, Bachman continued, Africans had shown no change in color as a consequence of climate, and that was because they had become a "permanent race whose characteristics have become organic." Such permanent races or varieties had their counterparts among the lower animals, said Bachman, a point he adduced by noting the perpetuation of such characteristics as a short tail in a dog or a cat and the absence of eyes in a species of fish and a species of crayfish found in Kentucky caves, the latter two of which "have descended from species that were not originally blind, but whose sight gradually became extinguished in the darkness of the cavern from want of use." The Charleston naturalist seemed to be unaware that he was advocating a view that coincided with what he called Lamarck's "absurd notions on the . . . progressive transmutation of species."[20]

More than once in his book, Bachman revealed his belief that even though blacks were of the same species as all other humans, they had been "stamped with inferiority" by the Creator. Yet, he maintained, they could be elevated morally and intellectually. In his judgment, however, northerners were taking the wrong approach, for they did not understand "the peculiarities in the African character." They were endeavoring, Bachman said, to raise the morals of blacks by philanthropic motives because of their sense of guilt, but, in fact, he argued, blacks are more debased in the North. The clergyman-naturalist genuinely believed that the South was seeking to convert blacks to Christianity. Viewing slaves as "a portion of our household," he said, "[they] have been the nurses of our mothers and wives, and are the playmates of our children." To think of them as another species, asserted Bachman, is to deny that they are equally God's children. The pluralists, Morton in particular, he added, were thus trying to deny a status ordained upon blacks by God. Morton would use his comparison of the volume of skulls to make his case, and that, mused Bachman, was tantamount to using phrenology. Morton's argument, charged Bachman, "is so mystical and transcendental that [I] scarcely feel warranted in attacking him as an opponent."[21]

After his manuscript went to press, Bachman secured a copy of Charles Hamilton Smith's *The Natural History of the Human Species*, published in 1848, and he added an appendix to *The Doctrine of the Unity* in order to offer a criticism of Smith's volume. Born in Flanders in 1776, Smith had been educated in England and then served in the British army for nearly thirty years. During his years of military service and later, he traveled to many places, including Africa, the West Indies, and the United States. A skilled artist and a prolific writer, he wrote many accounts, some of which included some far-fetched tales. In his work on separate species of humans, Smith cited some of his previous accounts of North American mammals, but Bachman noted that Smith referred mainly to the larger ones and ignored "the small rodentia where he might have been more successful" in making his case for varieties and species. In fact, said Bachman, none of the animals Smith had described was a new species. Moreover, Bachman observed that Smith considered the "flat-head Indians of South America" to be "naturally flat-headed." He also noted that Smith believed "the brain of man assumes first that of an adult fish, then of a reptile, and then of a bird," after which "it becomes the brain of mammalia, and then of man." In man, Smith had added, the brain passes through "the form of Negro, then . . . the Malay, the American [Indian], Mongolian, and finally attains that of the Caucasian." Bachman related other extraordinary statements by Smith, including "the ridiculous assertion that he has often seen a man's knuckles cut by striking the [coarse] hair of the Negro." Unfortunately, Bachman aptly concluded, Smith tended to "confound fable with facts." Smith was the authority most often referred to by Morton, and his Charleston critic would have more to say about that matter in due course.[22]

Having received no response to his missive to Morton in October 1849, Bachman decided to submit a paper to the *Charleston Medical Journal* in January 1850, probably as a means of drawing more attention to his argument. The editors, D. J. Cain and Francis Peyre Porcher, were professors in the medical college in Charleston, and the latter was an excellent botanist. Although they were supporters of Morton and Agassiz and later clearly revealed their dislike of Bachman's position on race, they were fair-minded, and thus agreed to publish Bachman's paper. Much of the text was taken from *The Doctrine of the Unity*, but Bachman directed it specifically to Morton. The title, "An Investigation of the Cases of Hybridity in Animals on Record, considered in reference to the Unity of the Human Species," clearly indicated that Bachman intended to rely upon substantiated cases and to discount those based on dubious accounts. Bachman stated his position without equivocation: "In nineteen out of twenty cases—and we might add, ninety-nine out of an hundred—where hybrids have been produced between two species, they have proved sterile." Cases of fertile hybrids are on record, he admitted, but he asserted that in most of those instances their progeny were weak. Even in instances of fertile hybrids that propagated for "a few generations," Bachman added, the ultimate result was sterility. All of the human races, or varieties, however, are able to propagate. It is "in this way, new races have been formed and perpetuated," said Bachman.[23]

Relying upon his extensive reading on the subject of hybridization,

Bachman went directly to the first group of animals used as examples by Morton—the horses, asses, and zebras (genus *Equus*). He stated that "by no process of domestication has a new intermediate race ever been, or ever can be, produced between the horse and either of the other species of asses and zebras," and he noted the difference in the dentition of those species. That is, of course, an important diagnostic character, and Bachman also informed Morton of the difference of dentition among species in the family Suidae (pigs). Another identifying character is the number of vertebrae, but, as the number occasionally varies in individuals of the same species, Bachman pointed out that Morton's reference to a single anomaly in one black person as an indication of specific variation from whites missed the mark. Variation in the number of toes among genera in the order Carnivora can also be diagnostic, but, since Morton apparently did not know that fact, he had, of course, accepted the word of Smith that dogs are descended from hyenas. Bachman repeated his statement about significant differences between the canids and the hyenas, the latter of which are more closely allied to the feliform carnivores, especially those in the family Viverridae (civets, genets, and mongooses).[24]

Morton's argument was shaky on other grounds as well, namely, his charge that the advocates of the unity of the races were following Lamarck, or what Bachman again called the "absurd notions on the progressive transmutation of species." To that charge Bachman responded that he was referring only to varieties. Thus, "by our theory," said Bachman "the dog begets a dog." It was true that Bachman believed that permanent types did occur after long periods of interbreeding within a variety, especially after long isolation from the original species, but he never argued that any of those constituted a new species. Morton, on the other hand, had maintained, for example, that "the d'hole, the jackal, the wolf, and the hyaena were required to make a dog," and a mixture of four species of wild ass to make " 'an improved austral horse.' " Indeed, as Bachman insisted, "La Mark [*sic*] would reject us as heterodox" and embrace Morton and his allies "as true believers." Bachman also pointed out Morton's readiness to construe any abnormal individual as a hybrid. In fact, Morton was willing to accept almost any unusual instance that he could use to support his theory, and he seemed bent on persuading his readers by piling up cases, no matter how dubious.[25]

The perceptive Bachman could not be gulled, however; he was too familiar with the older literature to let Morton escape by cleverly citing ancient writers as authorities. Bachman knew that most of the authorities cited by Morton were from the same work, *The Animal Kingdom Arranged in Conformity with Its Organization, by the Baron Cuvier, with Supplementary Additions in Each Order*, by Edward Griffith and others, published in fifteen volumes between 1827 and 1832. As the subtitle notes, the work contains additional descriptions, some of which were written by Charles Hamilton Smith, who, as Bachman pointed out, had divided the genus *Equus* into "three sub-genera . . . fifteen species, and four hybrids." The imaginative Smith "might . . . have increased his list to an hundred species," said Bachman. He then took Morton to task for accepting the dubious cases cited by Smith, providing extensive evidence to refute many of the claims of the authority he drew from so heavily. "We respectfully submit," said Bachman, "whether an author . . . so unscientific, so heedless in his writings, deserves to be quoted as having established any fact on so frail a foundation." He concluded that Morton had, in fact, offered not a shred of evidence to demonstrate "an admixture of any two or more species."[26]

Morton completed a reply on March 30, 1850, and the editors of the *Charleston Medical Journal* included it in the next issue. Opening with an expression of surprise that Bachman had rejected all of his authorities, Morton charged that the clergyman-naturalist had done so because they ran counter to his own opinions. In defense of Smith, Morton praised him as "a man who has grown gray in the pursuit of science" and who was in fact "a century in advance of most of his contemporaries." Trying to turn the matter around by noting a reference Bachman had made to the authority of the great British anatomist and naturalist Richard Owen, Morton declared that while Owen was good in some fields, he had not studied the subject of hybridity as "deeply" as others. Moreover, Morton said, "some questions in science . . . must always remain a matter of opinion." He admitted he had erred on one matter, but insisted that his authorities were reliable. In addition, Morton charged that Bachman had confined his criticisms to cases of mammalian hybrids. Oblivious to the proportion of space he himself had devoted to the mammals, Morton was seeking to show that Bachman's knowledge was limited. "Why," he asked, "were not the birds examined also?" It was a question that revealed his ignorance of Bachman's work in ornithology and apparently his failure to read all of

The Doctrine of the Unity. Remarkably, then, Morton danced deftly around his detractor's major criticisms.[27]

Morton also sought to demonstrate that Bachman was ignorant of ethnology and thus had failed to grasp the important features of hybridity. Resorting once more to such unscientific expressions as "the ancients averred," Morton declared that every astute observer knows that some wild species cross with other species, even species not generically related, and he gleefully reminded Bachman that the Charleston naturalist had himself collected two hybrid rabbits from the wild that were the progeny of "the American gray rabbit, *Lepus sylvaticus* [= *Sylvilagus floridanus*, the eastern cottontail], and the marsh hare, *Lepus palustris* [= *Sylvilagus palustris*]." Of course, Bachman had indeed made the statement, but it was entirely consistent with his view that interspecific crossings did occur occasionally in the wild. Morton plainly ignored Bachman's emphasis on the rarity of such instances, and he failed once again to comprehend both that the species were closely allied in the same genus and that their progeny had produced no offspring. He was unwilling, however, to retract his statement that these were "some of the first fully authenticated examples of hybridity in the *wild state* that have been noticed on our continent," for it supported his theory of "*a latent power of hybridity . . . in many animals*" that allows "*the several races of Mankind [to] produce with each other a more or less fertile progeny.*" Bachman must understand, added Morton, that "hybridity is divisible into four degrees or grades," ranging from total infertility to prolific fertility.[28]

Although Bachman had made few references to the biblical account of human origins, Morton was aware that Thomas Smyth and many other clergymen were upset because the pluralists had readily dismissed the Bible. Morton therefore devoted considerable attention in his paper to showing that his view did not contradict "the record of the Pentateuch," but, of course, it represented a liberal reading of that record. First, he insisted that the Noachian flood, as geological research had shown, was localized, the implication of which was that many animals and human races had not been affected. Second, Morton noted that the number of species of animals entering Noah's ark ranged from a single pair to seven individuals and thus included "all *varieties,*" probably every one of which was a domestic animal. The idea of all humans descending from a single pair struck him as morally wrong because "this incestuous intercourse tends eventually to the deterioration and extinction of the races that are subjected to it." To assuage those who might view the statement as a denial of the biblical account, Morton said, "the Sacred Text . . . does not require the interpretation usually put upon it." Instead, he argued, as "the recent judicious observations of Prof. Agassiz" show, species of animals originated in "their allotted regions of the earth." Moreover, they began with many pairs, not a single one for each species. Morton did not criticize the Bible or Christianity, but he preferred to speak of God as the ordainer of the laws of nature and as the architect of a great plan for the earth. Reverence for God, he maintained, should not be lessened by evidence that animals and human races "have existed *chiliads of centuries* upon the earth."[29]

Early in May 1850, before he had seen Morton's reply in the April issue, Nott wrote to Morton to say that he had been "skimming" Bachman's *Unity* and reading Agassiz's *Christian Examiner* article. In his view, Bachman wrote like "a blackguard" and included only "what suits his purpose," whereas Agassiz wrote like "a gentleman" and went straight to the truth. The arguments of the Charleston clergyman could be readily refuted by Morton and Agassiz, he added, but if they failed to "kill of[f] Bachman," then he would have to enter the fray. The Mobile physician ended his missive with an expression of hope that Morton would "take time to skin Bachman, [for] he is so rude that most of the gentlemen would be glad to see him used up." Slightly more than three weeks later, on May 26, Nott wrote again to Morton, to say that he had read his "smash up of old Bachman" in the *Charleston Medical Journal*. Indeed, Morton's reply had made him "feel good all over." It was especially pleasing, he said, to see "old Bachman . . . cut into sausage meat." Jubilant, the ardent racist opined that all of the opposing clergymen were "certainly in the way of being well licked," and he optimistically predicted "a change in public opinion coming in every direction."[30]

However, Bachman was not done with his defense of humans as a single species. On June 12, he completed a reply to Morton and sent it to the editors of the *Charleston Medical Journal*. Within a few days, the article had been set in type, and the editors quietly sent a copy of the proof to Morton. On July 10, after Bachman's article appeared, they requested "a rejoinder" from Morton for the next issue of the *Journal*, explaining that they did not wish for "Dr. B. to keep his ground unmolested." Although they told Morton they could not express their own opinion in the journal, they nevertheless assured him that the issue "appears to be pretty well decided in the city—most if not nearly all of the medical men leaning towards the side occupied by Mr. Agassiz and yourself." Meanwhile, on June 17, 1850, Robert Gibbes had penned a letter to Morton to say that a number of Charlestonians were displeased over "the course which Dr. Bachman [h]as pursued towards you." They liked Morton's reply, added Gibbes, and they were praising him for his "gentlemanly propriety." Gibbes also noted that the *Charleston Courier* had published a portion of Morton's article, in response to which Bachman had penned a caustic rejoinder. Embarrassed over the supposed affront to Morton, Gibbes assured him that "no Carolina gentleman would have written such an article." It is likely that Gibbes was even more displeased with Bachman two or three weeks later when he read the clergyman's "A Reply to the Letter of Samuel George Morton, M.D., on the question of Hybridity in Animals, considered in reference to the Unity of the Human Species."[31]

Filling over forty pages of the *Charleston Medical Journal,* Bachman's second reply was even stronger than his first, both in language and content. Although Bachman resorted to use of humorous sarcasm in several instances, he maintained a professional attitude overall. It was obvious, however, that he was particularly frustrated over Morton's refusal to admit that he had relied upon flawed sources and to acknowledge the scientific evidence he had adduced. In reference to his criticisms of Smith as an authority, said Bachman, Morton had "not ventured, in a single instance, to show that my statements were erroneous." Morton's decision to ignore those criticisms, he observed, was particularly unfortunate because the Philadelphian enjoyed high standing as a naturalist. Reminding Morton that the naturalist has a special duty to correct his errors, Bachman asserted that science advances in proportion to the investigator's willingness to discard information from old authorities when new and better evidence comes along. "You have carried us back to old Aristotle," he complained. Then "you hie us to Africa" to hear preposterous tales of hybrid animals, and "you carry us to Paraguay" to hear farfetched stories of strange progenies. "I confess," he said, "it is a severe affliction to be obliged to follow you so far out of the track of civilization." Indeed, lamented Bachman, many of the books cited by Morton are "out of print, out of date, and ought long since to have been restricted to the toy-shop of the antiquarian." He found it especially frustrating that he could not find a copy of many of those books in Charleston, and he asked Morton to lend him those he could not obtain. As Bachman would discover, however, Morton intended to ignore his request, for he knew it was advantageous to keep his most effective critic disarmed. After all, Morton was by then fully aware of Bachman's incisive skill in depreciating dubious sources.[32]

Exasperated, Bachman marched again through the alleged cases of fertile mammalian hybrids, slashing each of them to pieces. Not pausing for rest, he moved through Morton's examples of hybrid birds and left none standing. Impatiently, he turned to Morton's idea of a "latent power of hybridity . . . in many animals in a wild state," which effected changes in the reproductive organs of those animals when they were domesticated. "This," exclaimed Bachman, is "a new and incomprehensible development in physiology." The frequency of hybridity in domesticated animals had nothing to do with any mysterious power of latency, he said, but merely with "the artificial means which are used by man to produce these unnatural associations." After referring to his own, numerous experiments in crossing birds, Bachman admitted that, while crossings between species of birds do occur in the wild, they are quite rare. The problem with Morton, noted Bachman as he neared the end of his reply, was that he was determined to keep his theory "in spite of facts and arguments to the contrary," and now, despite Bachman's efforts to debate only on scientific grounds, Morton had introduced religion. He had done so, charged Bachman, because, "after making such a poor show of defense," he had to "hide" behind the Scriptures. Expressing displeasure over Morton's choice of biblical verses to support the plurality argument, Bachman called his opponent "a perverter of . . . [the] simple and expressive language" of the Scriptures.[33]

Nott had, of course, been following the debate, and on July 25, 1850, he told Morton, "Old Bachman is a pretty hard customer and has hit you some hard digs under the ribs." He expressed concern that "Morton could not make out anything satisfactory on the fertility of hybrids." The most effective attack on Bachman's argument, he suggested, would be to emphasize the vast variety of dogs, as that phenomenon was the most analogous to the variety of races. Reflecting his general distaste for the clergy and Christianity, Nott said that "Bachman *knows* that the Book of Genesis is unhistorical, and clearly abandons it in making as he does many centuries of creation in all other animals than man." The skeptic Nott was wrong if he thought Bachman secretly doubted the veracity of Genesis, but he was justified in criticizing the inconsistencies in the arguments of Bachman and others who embraced a strict biblical chronology for humans but possibly a vast span of time for animals. In any case, a month later Nott was still fretting over Morton's inability to demolish Bachman's arguments, and, in a letter on August 26, he told Morton he had been rereading Bachman's book and had decided to write a review of it. "I have never seen a more vulnerable Book," he said. By implication, then, Nott had told his Philadelphia friend a more effective attack was needed. Morton believed, however, that he was doing the job well and went on to write "Additional Observations on Hybridity in Animals" for the next issue of the *Charleston Medical Journal*.[34]

Calling himself "a pioneer" in the study of hybridity in animals, Morton said he had devoted virtually all his time to the subject since 1846. He seemed to be oblivious to the fact that Bachman could easily claim a record of attention to the subject that exceeded his by at least two decades. As before, Morton defended his sources. Indeed, while Bachman had said that his last essay "abound[ed] in errors in every paragraph," noted Morton, he had in fact commented on only thirty of the ninety-seven paragraphs in the paper. Moreover, Bachman's criticism of his authorities as old was irrelevant—"as if age, or antiquity itself, could invalidate truth." Morton denied that he had refused to lend books to Bachman. One of them he said he never owned but had seen a copy of it when "a gentleman from a distance" had shown it to him during a visit. Another he would have lent to Bachman if he had only known he wanted it. Continuing his disingenuous statement, he said he would "cheerfully" send it now if Bachman would indicate "the means by which it was to be conveyed a distance of five hundred miles."[35]

In his lengthy reply, Morton offered a new definition of species: "*a primordial organic form*." It was a rather useless definition, but it served Morton better than the one he had given in 1846 because he was realizing more and more that his case would be stronger if he emphasized primitive stocks of animals. In addition, he enhanced his theory by identifying three basic types of species: remote, allied, and proximate. Coupled with his idea of gradations of hybrids and Agassiz's zoological zones of creation, the new definition made it possible for Morton to get around some of the difficulties he had encountered.

It bothered him not at all that he was using the ancient bed of Procrustes to make his evidence fit his theory. Whether he was following the advice of Nott or acting on his own, Morton devoted the greatest space in this reply to the enormous variety of dogs, making twelve basic species of them and again referring to their primitive stocks, which he identified as "lupine," "vulpine," and "various species of wild dogs." He also repeated cases of intergeneric hybridity among other mammals and among birds. With respect to some of his authorities on birds, he said, "They are certainly better judges than either Dr. Bachman or myself." Morton simply would not accept that Bachman was an authority on birds—or on mammals, or on hybridity. Of course, he could not afford to do so. At the end of his long paper, Morton repeated his belief in a limited Deluge, but he asserted more clearly his belief in "the existence of man upon the earth for a period vastly longer than has generally been supposed."[36]

Bachman did not respond to that paper, nor to the next one, published in January 1851 in the same journal. Already in camp with Morton, whom they regarded as "the greatest champion" of the theory of the plurality of racial origins, the editors, Cain and Porcher, were very pleased. "We think these *notes* contain a complete refutation of several of Dr. Bachman's positions, in reference to the impossibility of fertile hybrids resulting from crosses, and which he supposed impregnable," they wrote to Morton. In their judgment, "should Dr. Bachman and those who embrace similar views ever succeed (which is extremely doubtful) in establishing their positions, the question might still be asked, Has this subject a very important bearing upon the Unity of the Human Races?" They were sure it would not, for "the natural history of the brute creation . . . is dissimilar in a multitude of respects." The fact that their great champion Morton was pushing the analogy of "the brute creation" to apply to races of humans apparently escaped them. In any case, in the first of two supplementary "Notes on Hybridity," Morton briefly reiterated his main argument and offered a long quotation from the Swedish naturalist Carl N. Hellenius's "Account of a prolific progeny from a Deer coupled with a Ram," published in 1794.[37]

By January 31, 1851, Morton had completed additional "Notes on Hybridity" for the journal. In that second supplementary paper, he offered more cases of fertile intergeneric and interspecific crossings, but his main point was to cite Peter A. Browne's recent studies of the similarities between sheep's wool and the hair of blacks. Browne, wrote Morton, had used "the microscope, micrometer and trichometer (an instrument invented by myself to determine the docility, elasticity and tenacity of filaments) in regard to *pile*," or wool and hair. Morton eagerly seized upon the conclusion of his fellow Philadelphian Browne that "there is no difference between the wool found upon the head of a pure Negro, and that found upon the back of a wooly sheep, except in the *degree* of the felting power." Although he offered no comment, Morton apparently believed he had found additional confirmation of his argument that blacks were not of the same species as whites. Thus, he neither questioned the assumptions made by Browne nor his knowledge of physiology. Had he lived to read the response of a British critic of Browne's monograph on the topic, Morton would have been chagrined. Said Richard Cull of Browne's treatise: "The author's entire want of physiological knowledge . . . renders his [study] but of little value." Cull told the members of The Ethnological Society of London that he would not discuss "a work of such unsatisfactory a character." Morton was content, of course, to accept Browne's spurious study. After all, it resembled his own studies of skulls.[38]

Near the end of April 1851, Morton published still another paper in the *Charleston Medical Journal*. Bachman seemed not to be bothering him now, leaving him free to develop further the argument he hoped to make into a book. In his latest paper, Morton dealt almost exclusively with races of humans, focusing upon the black race. He denied that white skin could become dominant in a remote descendant of a mulatto except by "a *diseased condition of the rete mucosum*," and when it did, it was a "dirty-pale, albinoid tint" that reflected "a morbid condition." Indeed, such was one of the results "when the primordial form is once broken by the crossing of dissimilar species." Whether in animals or in human races, the price of corrupting the original stock was "grotesque varieties." Growing bolder in the silence of Bachman and determined to clinch his argument, Morton declared that when a female was initially impregnated by a different species, all of her progeny thereafter bore the characteristics of that male, even the progeny from a later union with one of her own species. Thus, if a female wolf were first impregnated by a dog, she "would ever after implant the lineaments of that dog on her progeny." The same was true of human species: "a negro woman whose first conception was by a white man, would ever afterwards continue to produce mulattoes, even with a husband of her own race." At last Morton had come to a clear *conception* of the meaning of race as synonymous with species: each race or species had its primordial origin in a zoological province but, through breeding with other races, created new species, each growing weaker as it got further from its primordial stock. The immediate message was that "the mulatto race tends rapidly to extinction, unless strengthened by new crosses from the parent stock." Morton noted that Nott had been saying such, and he could now add that the institution of slavery in the South was a good thing: "the Negro race, in a genial climate, and well cared for, continues to multiply." The problem with slavery, as he now saw it, was not with the institution but with the failure to keep the black stock pure. The "mixed and motley colored population . . . is absolutely dying out," he concluded.[39]

Defenders of the peculiar institution could not have a better friend than Morton, and they did not want to hear what Bachman had to say. Just before his last paper was published, Morton heard from Nott, who, on April 6, wrote, "I have just received and read your last two pummelings of the *Old Hyena* with great gusto and must congratulate you on his annihilation." So thrilled was the devoted defender of white supremacy that he added: "I really feel as if a viper had been killed in the fair garden of science, and I hope his death will be a warning to all such blasphemies against God's laws." Mixing his metaphors, the passionate physician said that he found it "disgusting to see science polluted by passion," adding that whenever he had "fought back at these *skunks*," he had done so only to keep "the ball of truth rolling." Not so pleasing to Morton, however, was another letter from Cain and Porcher, on April 10. The editors informed him that Bachman had sent them a reply to his last three papers, and they warned their favored writer that his chief critic had not spared "personalities." Morton may have feared another onslaught by Bachman, but he could bask at least momentarily in the favorable comments published in the April issue of the *Southern Quarterly Review*. In a review of two of Morton's replies to Bachman, Agassiz's article in the *Christian Examiner*, and a book on the human races by the Scottish anatomist Robert Knox, the critic "L. S. M."—the initials of Louisa S. McCord, daughter of the prominent Charleston banker and statesman Langdon Cheves and a gifted writer in her own right—said, "We are not ashamed of our 'peculiar institution,' nor do we need any sugared epithets to cloak an iniquity of which we are entirely unconscious." Defenders of slavery and the notion of the inferiority of the black race need to entertain neither any "dogmatic assumption" about the unity of races nor "the cry of 'infidelity' " uttered against Morton, McCord stated. Of course, Bachman had never charged Morton with infidelity, but that Morton said he had was good enough for the reviewer. McCord praised Bachman and Thomas Smyth for their "zeal and sincerity" but added, "[they] will scarcely, we presume, feel themselves aggrieved if we rank their names, as *naturalists*, below the world-known ones of Agassiz and Morton." Bachman "has been answered," she concluded, and anyone who had read Morton's observations "will need further proof to convince them that the reverend, learned, and most estimable gentleman has not, in this discussion, got a little beyond his depth."[40]

Unyielding in defense of both science and religion, Bachman had not conceded, however, and in April 1851 he sent his "Additional Observations on Hybridity in Animals" to the editors of the *Charleston Medical Journal*, who published it in the next issue. His irritation over Morton's resistance to facts and the true scientific spirit was obvious in the caustic nature of this response, as Cain and Porcher had told Morton it would be. Nott may have thought the Old Hyena had been annihilated, but he was very much alive. Correctly accusing Morton of thwarting his request for the books he needed, Bachman noted that his opponent could have copied the information verbatim and sent it to him even if Morton could not send the books. "How long," he asked, "would the public suppose that, in these days of rapid traveling, it would require, to receive an answer from Philadelphia." Morton had taken "four months and nine days," and "a turtle could have accomplished the journey in half the time, and taken an occasional nap of a week on the passage." It was hard to dismiss the thought, said Bachman, that Morton "was averse to that fair and liberal examination and discussion characteristic of those who search for truth rather than victory." Morton's claim that his own arguments were "ingenious and elaborate" meant nothing, noted Bachman, in light of his erroneous views. Indeed, he added, *The Vestiges of the Natural History of Creation*, an argument espousing evolution, published in Britain in 1844, possessed the same qualities but was also wrong. He could not keep quiet, even in the face of charges that he was mistreating the esteemed naturalist, exclaimed Bachman. Repeating the apt phrase he had used before, he charged that Morton was continuing to support his own theory by dragging "from the dust of antiquity the exploded errors of bygone ages."[41]

Bachman had intended to continue the paper in the next issue, but Morton died suddenly on May 11, 1851. Out of respect for the memory of Morton, Bachman wrote to Cain and Porcher, he would not submit another paper until "a suitable time." His notice was published in the *Journal*, following a glowing tribute to Morton written by Robert Gibbes. "The death of this eminent man," lamented Gibbes, "leaves a large void in the world of science." In Europe, added Gibbes, Morton "was considered as the first in Ethnology," and in our own country, "we mourn his loss as a National bereavement." While Gibbes overestimated Morton's standing in Europe, he undoubtedly spoke for many, if not the majority, of his fellow scientists in the United States, both North and South. The idea of the inferiority of the black race, even if not as a separate species, was not the peculiar reserve of southerners.[42]

The Jawbone of an Ass

P erhaps Bachman thought he had once more helped to "stay the torrent of ignorance," but he held no illusions that he had halted the pluralist movement. The floodgate opened again early in 1854 with the publication of a compilation of essays advocating the plurality of human species. The first three words of its extremely long title, *Types of Mankind*, reveal the nature of its content. Conceived by its major contributors, Josiah Nott and George R. Gliddon, the massive volume included essays by the late Samuel George Morton, Louis Agassiz, and two relatively unknown physicians, William Usher and Henry S. Patterson. Gliddon, a native of England, who had resided for many years in Egypt and elsewhere in the Middle East, presented many lectures in the United States during the 1840s and eventually linked up with Morton and Nott, with whom he shared a dislike of clergymen and disdain for the biblical account of creation. Nearly a year after the death of Morton, Gliddon and Nott decided to pick up where their late friend had left off. Morton's widow gave them a manuscript on the pluralist view that her husband had nearly finished, and they worked it into shape for publication in *Types of Mankind*. Nott added chapters on hybridity and on the comparative anatomy of "racial types," while Gliddon drafted chapters on the biblical account of creation. Later, they persuaded Agassiz to write an essay, which he titled "Sketch of the Natural Provinces of the Animal World and their Relation to the Different Types of Man." Agassiz's essay was accompanied by a tableau of the so-called natural provinces. Usher contributed a paper on paleontology, designed to show that humans originated long before the time accepted by biblical literalists, and Patterson wrote a lengthy, hagiographic account of the life of Morton. In addition, the book contained more than 300 illustrations of "human types," or, in the view of the writers, the separate species of humans. The depictions of the lower types of men were grossly distorted, some even grotesque, while the Caucasoid types were drawn as fine-featured men. Indeed, the essays were hardly necessary since the illustrations clearly revealed the authors' perception of the differences between the higher and lower types.[1]

Soon after the 738-page volume appeared, John Bachman obtained a copy. That Nott and Gliddon would produce such a work surely did not surprise him, but that Louis Agassiz had contributed an essay to the book likely struck him as uncharacteristic of a man devoted to excellence in scientific publication. Like many of his fellow scientists, however, Agassiz believed that Nott was dedicated to promoting the relatively new science of ethnology, though he may have entertained second thoughts after seeing the other essays in the volume. That others shared Agassiz's faith in Nott is evident from the book's list of around 700 subscribers, including such northern naturalists as Timothy Conrad, John Eatton LeConte, and Joseph Leidy, and such southern naturalists as Robert Gibbes, Holmes, Holbrook, Ravenel, and Tuomey. Bachman could readily see that the application of scientific principles could elude even the best of naturalists when it came to the question of race. He would once more have to call forth his best efforts to combat another corruption of science, and he soon began to draft the first of a four-part critique of the work. His task would be similar to that of Samson, except that he would be slaying legions of misguided naturalists with the jawbone of an ass. In his case, however, Bachman knew precisely how many teeth were in the mandible of *Equus asinus*. He was also aware that the surly and dogmatic physician Nott had already resorted to a crude means of associating him with *E. asinus*, for, in a commentary on his *Unity*, published in *DeBow's Review* in February 1851, Nott had cleverly changed the last item in the list of honors following Bachman's name from "American Assoc." to "Amer. Ass." Nott plainly lied when he later claimed that it was a typographical error. The Charleston champion of the concept of humans as a single species knew that he faced a formidable foe who would readily engage in invective, ridicule, and mendacity in order to reach his goal.[2]

Bachman reviewed *Types of Mankind* in the *Charleston Medical Review*. In the first part of the review, he relied more on a biblical defense of unity than he had in any of his responses to Morton or than he would in the remaining parts of his criticism of *Types*. He did so because he believed that Gliddon had a "grand design of discrediting the Christian religion, and heaping on the Holy Scriptures all manner of epithets of derision and contempt." Indeed, Gliddon did have such a design, and Nott had already privately complained that his friend's criticisms of the Bible and Christianity were so strong they ran the risk of alienating some of their supporters. For Bachman, of course, his role as Christian minister came before his role as a naturalist. Thus, he must first counter Gliddon's depreciation of the Book of Genesis as a source of "crude and juvenile hypotheses about human creation," and he must reject Usher's attempt to place humans on the earth long before the time of Adam. Bachman could tolerate the idea of an ancient earth and even ancient animals, but he could not accept the idea of humans existing before the pair created in the Garden of Eden.[3]

Irritated over errors in Patterson's account of his debate with Morton, Bachman determined to set the record straight. He therefore gave an accurate and detailed chronology of the debate with Morton, but, because Patterson had died before *Types* came off the press, Bachman decided to offer no other criticisms of the sketch of Morton's life. Instead, he turned to the work of Nott and Gliddon. Noting that *Types* was merely an amalgam, he said, "compilations are in general evidences of poverty of mind." A book can be only as good as "the capacity of its judges," he added, and since the judges in this case had resorted to invective, exaggeration, and blatant misstatements, this book had no merit. Furthermore, asserted the indignant minister, Nott and Gliddon had suggested that he had conspired to undermine Morton, when in fact the conspiracy lay at the door of Josiah Nott. It was Nott, he observed, who had succeeded in persuading DeBow not to accept a rejoinder from him to Nott's article in *DeBow's Review*, and it was Nott who had omitted phrases or changed words in order to make Bachman look ridiculous. Unable to squelch his desire to employ a bit of sarcasm, the ruffled naturalist said, in reference to errors Nott had made about species of wolves, "We . . . exonerate Dr. Nott from any intentional design to misrepresent us . . . , as he seems thoroughly unacquainted with every branch of natural history, and we doubt whether he can distinguish a wolf from a hyena." A pointed jab was ineffectual against the self-appointed ethnologist, however, for he simply contrived a way to divert attention from his wound.[4]

Bachman realized that Nott held the advantage in the debate, for scientific truth could shine no light into the darkness of political reality. Southern politicians, he noted, eagerly embrace Nott's message because they think the strongest defense of slavery lies in proving that blacks are a separate species. Bachman did not believe that the plurality argument was wrong solely because it was unscientific; he believed it was also wrong as a defense of slavery. Since sectional hostility was increasing, however, Bachman had to assure his readers that he was a champion of southern ways. No one should doubt, he said, that he supported his adopted state and its institutions, nor should they construe his defense of the common origin and specific relationship among all races as a plea for the equality of blacks. "The Negro," he stated, "is a striking and now permanent variety, like the numerous varieties in domesticated animals." As with domesticated animals, insisted Bachman, the black race had developed special characters that kept it from ever reverting to the form of the original species. He did, of course, believe that blacks could improve themselves by marriage to whites, but he viewed such unions as repugnant. The best course of action, in his judgment, was to continue to hold blacks in benevolent bondage. Bachman contended that whites underestimated the intellectual capacity of blacks, but he nevertheless believed that Africans were decidedly inferior in mental ability. The inconsistencies of his arguments represented a conflict between his religious perceptions and his scientific knowledge, of course, and they indicate that he believed the principles of animal domestication were applicable to the acculturation of blacks. Nevertheless, during the mid-nineteenth century, John Bachman was one of the few naturalists in the United States to speak openly, and the only one to speak forcefully, for applying common taxonomic principles to the classification of all living beings. His fellow researchers could have been expected to follow the rules. Bachman understood that *Types of Mankind* was "filled with all manner of absurdities," but most of his peers accepted its flawed arguments and ignored the scientific principles that otherwise guided their studies of animals and plants.[5]

To Bachman, even a scientist of the stature of Louis Agassiz must be criticized whenever he constructed a theory without considering the evidence that failed to fit it. In fact, the idea of zoological zones, or natural provinces, was a potentially useful hypothesis, but Agassiz was too eager to use it to justify his personal belief about the races. The notion of natural provinces also conformed nicely with his idea of repeated creations. Since Bachman rejected the latter, he might have been expected to reject the notion of natural provinces out of hand. Instead, he found it unacceptable because he saw it as mainly a scheme for justifying a predilection and because it omitted important facts. Thus, he devoted the other three parts of his review to the essay by Agassiz. His critique was cogent, which may be the reason Agassiz chose not to respond, though he privately maintained that it was better to ignore what he viewed as the contemptible criticisms of a man who misunderstood science.[6]

In fact, as Mary P. Winsor has shown in her insightful essay, "Louis Agassiz and the Species Question," Agassiz possessed supreme confidence in himself as a scientist and "was rather prone to the fallacy that because he was a scientist, his opinion on every issue was a scientific one." That he considered himself to be "sophisticated in his awareness of the demands of scientific method" was sufficient for him to hold Bachman's views in disdain. Moreover, Agassiz believed that he had "a special knack for recognizing species." Motivation also played another role in shaping the argument by Agassiz, for he understood that the position held by Bachman and other defenders of the idea of the common ancestry of humans "would be to demonstrate evolution." Of course, Bachman also rejected the possibility of evolutionary descent, though, inadvertently, he himself came very close to advocating it. As with Bachman, Agassiz understood that the issue revolved around the definition of species, and he rejected the notion that "morphological detail and sexual preference" were sufficient criteria. As he would endeavor to show in 1857 in his "Essay on Classification," observes Winsor, Agassiz insisted that a species had to be defined also in terms of "its characteristic mode of reproduction and growth, its geographic distribution and fossil history, and the manifold relations that the organism bears to the world around it." In practice, Agassiz was "an extreme 'splitter,'" which, of course, made it easier for him to employ the additional characteristics he established for identifying species—and, quite obviously, to split *Homo sapiens* into several species, all of which happened to possess differing racial features. His theory of "natural provinces" bolstered his notion of the specific origin of the races.[7]

In his essay on "Natural Provinces," Agassiz declared he would "show that the differences observed between the races of men are of the same kind as the differences observed between the different families, genera, and species of monkeys." In fact, he argued that the chimpanzee and the gorilla differ less from one another than do the descendants of the Mandingo and the Guinea tribes of Africa. He was sure he knew, for he had personally observed descendants of those tribes in 1850 when Robert Gibbes had taken him around to several plantations in South Carolina. Agassiz did not mention any details of his examination of the slaves, which had been cursory at best. Nor did he admit to a limited knowledge of the primates. In his response to Agassiz's essay, Bachman observed that Agassiz had failed to compare the chimpanzee and the gorilla with the orangutan or with any of the numerous species of monkeys. Furthermore, he asserted that Agassiz could offer no valid comparisons with the gorilla since little was known about an animal so recently discovered and since only four skulls of it had been sent to the United States. In fact, Bachman was wrong about the number of gorilla skulls in the country, but he was correct about the limited information on them. Quite naturally, Nott would pick up on Bachman's minor mistake and use it against the clergyman. In any case, Bachman went on to note that Agassiz had made an inadequate comparison of the Mandingo and the Guinea. A thorough comparison between the two groups would have to include examinations of the skulls of each, and, had Agassiz done so, noted Bachman, he would have discovered that they vary no more than do the skulls of two tribes of American Indians. Indeed, added the expert mammalogist, Agassiz should "not fly to conjectures," for the facts are "abundant and within our reach." In a statement that surely irritated Agassiz, Bachman suggested that the esteemed naturalist would do well to compare his conjectures with the conclusion of the renowned British anatomist Richard Owen that the human is the only species in the genus *Homo*. Finally, Bachman offered, in tabular form, details on the number of vertebrae in nine species of apes and monkeys and in three races of *Homo sapiens*, the latter based upon a study by none other than Cuvier.[8]

Bachman turned away from *Types of Mankind* momentarily in order to deal with another analogy that had appeared in a review signed by "A. L." Bachman believed that Agassiz had penned the review and reversed his initials. His guess was almost certainly wrong, for the comments of the reviewer suggest that he lived in the South and was better grounded in theology and philology than was Agassiz. In addition, it is unlikely that Agassiz would have reviewed a book in which he had published an essay, and it would have been uncharacteristic of him to publish a review anonymously. The reviewer may, in fact, have been Aaron Leland, a longtime professor of theology at the Columbia Theological Seminary, a Presbyterian institution located in the capital of South Carolina. In any case, A. L. declared that while blackbirds exist in both the United States and in Europe, they are separate species. Neither species taught the other to sing, he added, "and yet their note is similar, because their throats are alike." Aside from the fallacy of associating two species of birds solely on the basis of color, A. L. demonstrated poor knowledge of avian anatomy. Bachman, an authority on the subject, noted that although the larynx differs in the two species of birds, it is identical in all human races. For the benefit of the incompetent ornithologist, Bachman noted that the blackbird of Europe is in fact *Turdus merula*, a thrush in the family Muscicapidae that is related to the American robin, *Turdus migratorius*. It is, as Bachman said, "in no wise" akin to American blackbirds in the family Emberizidae. Indeed, as he indicated, the notes not only differ considerably between species in the family Muscicapidae and those in the family Emberizidae but also appreciably among the species within the former. After demolishing the argument by citing numerous examples of specific bird calls, Bachman asked: "If the tribes of men are of different species, why do they not exhibit by their organs of sound those differences that pertain to the species in all other animals?" Though their languages vary, all humans possess an identical capacity in speech and song, he observed.[9]

Despite his intention, Bachman never directly criticized Agassiz's idea of natural provinces in the second part of his critique. In the third part, titled "An examination of the Characteristics of Genera and Species as applicable to the doctrine of the Unity of the Human Race," Bachman fired a few opening shots at Gliddon and Nott and then aimed directly at Agassiz. His intention was to clarify what the concepts of genus and species mean to most naturalists and to show thereby that

Agassiz's theory of natural provinces failed because it violated those concepts. Moreover, Bachman intended to hit Agassiz where he was weakest in his knowledge of the vertebrates—the mammals.[10]

Bachman emphasized the diagnostic characters used for placing mammals in a particular genus and those used to identify species in order to show that the pluralists, including Agassiz, were ignoring those characters in their argument for separate species of humans. He noted that mammalogists assign species to a genus on the basis of "the number and distribution of the teeth, number of toes, the possession or absence of cheek pouches, character of the nails, etc." As an example, he noted that the horse, the zebra, the ass, and the quagga are assigned to the genus *Equus* because all of them are identical in the number and character of their teeth, possess a single toe or hoof on each foot, and have two teats. Yet, he added, each is a separate species, which, he reminded the pluralists, is evident from its distinctive morphological characters and its ability to produce fertile offspring. Certainly, interspecific crossings occur within the genus *Equus*, he observed, but, unlike in the crossings of human races, the progeny are sterile. Moreover, said Bachman, crossings of horses and asses in the wild are uncommon because the two species do not normally associate, whereas human races mingle as a common rule. Of course, Bachman believed that each species represented a divine creation and that every species was created in the beginning. No species was "the production of blind chance." If each species had, as Morton argued, "a primordial existence," it could only be true in the sense of the original creation, Bachman added. There could be no such thing as evolution; all humans stemmed from the union of the first pair. Varieties of humans had developed, and some, by isolation, had developed into a permanent variety. However, a permanent variety was permanent only in the sense of having a tendency toward reproductive isolation from the original species and toward developing certain unchangeable characters, for the variety could still produce fertile offspring in a union with the original species. Bachman contended that, by following his definition of species, he had "never found any difficulty" in identifying the species to which any variety of domesticated animal belonged, but he did not point out, as he had done in *The Quadrupeds*, the great difficulties he encountered with some of the wild mammals, especially some of the squirrels. On the whole, however, he was right in saying that the naturalist could use "characters of primary importance" to guide him in determining species, and he reminded Agassiz that he had himself previously said that all humans are of one species. Of course, Agassiz had made that statement before he had seen blacks.[11]

After several delays, Bachman finally completed the fourth part of his review of Agassiz's essay on natural provinces. It was the longest and most devastating of his four-part critique. Agassiz had established eight natural provinces or zoological zones and had listed some of the animals and the variety of humans in each zone. For the latter, he presented a figure of the type of head and one of the skull to depict the variety. Bachman immediately stated three objections to the scheme: (1) it would be difficult to determine the natural boundaries of each province; (2) it was impossible to ascertain that human types and animals are confined naturally to the zones; and (3) it was impossible to determine the connection of the human type to the animal types in the zone. As he noted, naturalists had never agreed on the boundaries of such zones, and animals, like humans, were not bound by arbitrary lines. In fact, he noted, he and Audubon had faced a problem in trying to decide upon the limitations of the mammals to be included in *The Quadrupeds*, and, as he pointed out, the animals could not be confined to "artificial boundaries of States." Thus, he and Audubon had decided to include all mammals that inhabited some portion of the United States and Canada but to exclude any in Mexico unless their range extended northward beyond the arbitrary line between that country and the United States. They had also chosen to give the distribution of each animal within the United States, insofar as they possessed knowledge of it, not solely by official state boundaries but by natural limits. He cited examples of mammals that ranged over several states or over only portions of them. The problem of assigning birds to a natural province was even more difficult, especially for the migratory species, and it was hardly easier to determine the boundaries of the insects, the fishes, and the marine mammals. Zoogeography would later become an important field of study, but it would depend upon a multitude of precise records of animal distribution. Bachman recognized that Agassiz's zones were based upon inadequate data and arbitrarily selected species. Agassiz had drawn the lines to satisfy his conceptions of racial types, not according to reliable species-distribution records. As Bachman noted in reference to Agassiz's provinces of birds, "whilst they may apply to a few species, [they] are arbitrary in themselves, and positively inaccurate in regard to a very large number of [migratory] species."[12]

It was Agassiz's "arctic realm" as a natural province that Bachman found to be especially flawed, and he discussed its shortcomings at length. Agassiz drew the lines of that supposed zone around what he deemed as "peculiar species of animals" and "a peculiar race of men known in America under the name of Esquimo." Bachman quickly spotted the flaws in the argument. If Agassiz had simply listed all of the animals within his arctic zone and given their range of distribution, said Bachman, he would have provided no grounds for argument. The champion of zoological provinces had fudged, however, and Bachman pounced upon the arbitrary designations. The rodents that Agassiz included in the zone, observed Bachman, are in fact "restricted within comparatively narrow boundaries," while the polar bear, *Ursus maritimus*, and the reindeer, *Rangifer tarandus*, often range outside the zone. Moreover, Bachman continued, Agassiz did not include in the zone some of its quite common mammalian inhabitants: the gray wolf, *Canis lupus*; the ermine, *Mustela erminea*; the beaver, *Castor canadensis*; and the wolverine, *Gulo gulo*. He also listed numerous birds omitted by Agassiz. "Why," asked Bachman, "did he not select the raven (*Corvus corax*)" since it inhabits "the most northern parts of the arctic regions" but is "also a permanent resident of Canada, and in all the States of the Union, except on the sea-board." Or, why, he asked, did Agassiz select the right whale "as a representative of his arctic fauna" when it can also be found in the Southern Pacific Ocean, and indeed "navigate[s] all the seas of the world." In fact, three species of whales have been called "the right whale" at various times. Only one of the three, *Balaena mysticetus*, is circumpolar; of the other two, *Eubalaena glacialis* is found in the North Atlantic and the North Pacific, and *Eubalaena australis* inhabits the oceans of the southern hemisphere. It is likely that Bachman was referring to *E. glacialis* and, like others of his time, did not recognize *E. australis* as a separate species. In any case, the former is not confined to "the arctic realm," for it ranges in the Atlantic as far south as Florida and in the Pacific as far south as Baja California. Bachman also questioned why Agassiz selected reindeer moss (presumably the lichen *Cladonia rangiferina*) as characteristic of the zone when it flourishes in northern Asia and northern Europe. In fact, added the sometime botanist, he had collected it from the mountains of Virginia.[13]

Equally perturbing to Bachman were Agassiz's notions about the Eskimo. He objected strongly to Agassiz's decision to place the "typical figure of his esquimo . . . on a line with the mongol, evidently to impress his readers with an idea of the wide difference between these two forms of men." In addition, Bachman protested the illustration of the typical Eskimo because it "must have been selected from the ugliest specimen of Eskimo humanity that could have been chosen by the painter to degrade and bestialize the tribe." To be fair, he said, Agassiz should have selected a "less hairy and less hideous" representation. Bachman knew that other portraits of Eskimos were available, for he had seen some of them, including two done by none other than Charles Hamilton Smith. The paintings by Smith, he observed, depict a male Eskimo who "is by no means ill looking" and a woman whose "features . . . are mild, thoughtful, and on the whole, rather pretty." Bachman also objected to Agassiz's conjecture that the Eskimo probably "originated in the Arctic." Bachman rejected that hypothesis and said a special act of creation was not needed to explain the presence of the Eskimo in such a harsh environment, for he migrated to the region. Worse, added Bachman, Agassiz had wrongly assumed that the Eskimo was a solely carnivorous species of human. Obviously, he said, the great naturalist had never examined the teeth of the Eskimo, for, if he had, he would have found that they varied "not a shade" from those of other humans. The Eskimo, Bachman asserted, differs not a whit from other human beings in terms of basic physical structure, mental abilities, and moral capacity.[14]

Bachman went on to criticize Agassiz's notions about the other natural provinces, especially Australia and Africa, in every case raising pertinent objections. But he reserved his most vehement strictures for the figures of heads and skulls given in the tableau. The former, he declared "are not characteristic of the varieties to which they belong," while many of the latter, both of animals and of men, were "so very inaccurate that they are calculated to mislead." The purpose of the distortions, he correctly charged, was to support the theory of pluralism. He added that he had no desire to "inflict a wound either on science or on Prof. Agassiz," but he had doubtless wounded Agassiz by presenting his catalogue of telling objections. Unrelenting, he ended his cri-

Josiah Clark Nott. Courtesy of Reynolds Historical Library, the University of Alabama at Birmingham.

tique by correctly accusing both Agassiz and Morton of "discarding the rules by which they had until now been governed in describing all their new species."[15]

Josiah Nott commenced a new chorus soon after Bachman had published the first part of his review of *Types of Mankind*. In a letter to the editor of the *Charleston Medical Journal*, Nott claimed that he had intended to ignore Bachman's "garbling and special pleading" and to refrain from entering into his "'warfare' of personalities," but the plea of his friend Gliddon to defend the memories of Morton and Patterson had persuaded him to respond. His claim was typical of his clever and devious approach—to paint himself as a humble scientist who was driven from his reluctance toward controversy only by a sense of duty to proclaim truth. That his response contained more slander than substance should have caused the naturalists and physicians who read it to question his motives, but, since they were already in the pluralist camp, they were probably delighted over Nott's charge that Bachman's *Unity* displays "bigotry and fanaticism" and is filled with "bold assertions, palpable contradictions, [and] false reasoning." Only the most insensitive of them, however, could commend the vehement racist for attempting to depreciate Bachman's professional stature.[16]

Nott was especially incensed because Bachman had declared that "the world of Science has never admitted . . . [Nott and several other advocates of pluralism] into their ranks as Naturalists. Their names are utterly unknown among them—not one . . . has ever described a single animal." Of course, Bachman could not correctly include Agassiz in that group, but he was certainly right that Nott belonged to it. Now that *Types of Mankind* was off the press and selling well, Nott decided to reverse the charge: it was Bachman, he averred, who was little known among the true scientists, the ethnologists. Nott told his readers to go back to his review of Bachman's book and see his "exposure of some of the absurdities of the Doctor, which has not served much to improve his Christian spirit." There they would also see that Bachman had precipitated the controversy. Be aware also, he added, that the private letters containing criticisms of Bachman were published in *Types* after the manuscript left his own hands. Patterson made that decision, he declared, for "I should not have had the egotism" to print them. Nott wore humility poorly, but he never ceased to try it on for size. Later in his response, however, he did not even feign humility: "When my offence against Dr. Bachman is narrowed down to the simple truth, it is this—*I have published scientific opinions, which conflict with his sectarian dogma.*" As Nott viewed the debate, the proud parson had no bona fide scientists on his side. He did not know that Bachman had received an encouraging letter from Asa Gray in January 1855, expressing "admiration for the ability with which you have defended the doctrine of the specific unity of the human kind," and he could not know that James Dwight Dana would soon publish an article advocating the unity of humans as a single species. It would have mattered little to Nott anyway, for Gray was opposed to the views of Agassiz and was one of the few contemporary American scientists who expressed approval of the idea championed by Bachman. Nott never responded to the article by Dana.[17]

In July 1855, soon after the last part of Bachman's review appeared, Nott penned his "Reply to Dr. Bachman's Review of Agassiz's Natural Provinces." Claiming that he had heard indirectly that Agassiz planned to ignore Bachman's criticisms, Nott declared that he was obliged to perform the task. Unquestionably, he would have preferred that the master do it himself, but Nott no doubt believed he was the next best person to write a response. And a response must be given, for Bachman had devastated Agassiz's argument. Nott began by explaining that the busy professor had written the article "in haste, to oblige the authors." Then, he noted that Agassiz had intended it "as a mere 'outline delineation,' and, necessarily, very incomplete." After conceding that Bachman was certainly correct about the distribution of species in the Arctic zone, Nott said the clergyman had failed utterly to disprove that natural provinces and special centers of origin exist. Nott's admission that Bachman had given facts about animal distribution might be construed as a major concession, but it was merely a way of saying that Bachman had dealt only with part of Agassiz's argument. "Does Dr. Bachman suppose that Prof. Agassiz *could* 'overlook' such patent facts as he charges? . . . Can Dr. Bachman imagine that Prof. Agassiz would be weak enough to suppress such common-place facts, or that he would attempt to determine the point of such animals?" Indeed, Bachman did imagine such suppressions, and Nott, realizing the weakness of Agassiz's argument, had to make it appear that Bachman was "quibbling." Moreover, Nott contended that Agassiz had merely intended to give representative examples. Why should Bachman expect the noted naturalist to "overload" his essay with examples "with which every one of [his] pupils is familiar," he wondered. Besides, the querulous parson had made his own natural provinces in his study of mammals, said Nott, and, if anyone "were inclined to dissect those provinces . . . in the same spirit," he would find Bachman guilty of "the same sins of commission and omission." It is obvious, of course, that Nott was trying to equate Bachman's information on the range and distribution of a mammalian species with Agassiz's idea of natural provinces and that he failed to see, or chose to ignore, the difference between determining the range only after accumulating the evidence and constructing provinces a priori.[18]

Nott also argued that Bachman erred in recognizing only one species of some mammals, whereas, Nott insisted, other naturalists regard their varieties as entirely separate. He listed only "wolves, etc." and named no naturalists who disagreed with Bachman. More important to Nott was Bachman's criticism of Agassiz's representative human types for the Arctic realm. Nott dismissed the objection to the portrait of the Eskimo by declaring "it is needless to say anything on this point." Indeed, Nott knew that the picture of the Eskimo as a low type of human was more valuable than a thousand words, and to respond to Bachman's criticism was to run the risk of diminishing the effect of the portrait. The part of Bachman's review that astonished him most, said Nott, was the argument against the origin of the Eskimo in the Arctic province. In response, Nott insisted that no one could show that a group had migrated to the area and that "without any stretch of imagination," one could believe that the Creator could have created the Eskimo right there, and, if necessary, supplied him with fishing gear and "roasted seals." As for the resemblance of the teeth of the Eskimo to those of other humans, Nott, oblivious to the importance of diagnostic characters, said that nothing could be made of such trivial information. The polar bear is distinct from other bears, just as the Eskimo is a distinct type of human, he concluded. But the self-styled ethnologist did not return to the example of the bears when he once again rejected Bachman's defense of unity on grounds of the fertility of the progeny of racial crossings. He had "neither time or space" to defend the question of whether "there be eight or more *natural provinces*, or whether these provinces can be, as Dr. Bachman would have them, encompassed by the exact lines of a land-surveyor." Besides, he added, a precise number would not alter the basic idea proposed by Agassiz. For the first time, Nott had largely eschewed personal attacks in a response to his opponent, but he could not resist the temptation to throw a barb in each of his final paragraphs. The reason that Agassiz chose not to respond, he opined, was probably because he regarded Bachman's reviews "as more popular than scientific in character." Moreover, he added, while "the Dr. uses his best endeavors" to refute Agassiz's natural provinces, "he has unfortunately so lost his balance in this controversy" that he has forgotten to deal with Agassiz in a gentlemanly way. Any disagreement with the savant was tantamount to a sin in the eyes of Agassiz's apostle, but Bachman thought otherwise.[19]

Nott continued to publish criticisms of Bachman's argument in favor of humans as a common species. One appeared in 1856 as a three-part appendix to an English translation of selected portions of Joseph Arthur de Gobineau's *Essai sur l'inégalité des races humaines*, edited by the Swiss immigrant Henry Hotz. Nott's criticisms were not new, but they kept the fire stoked. Then in mid-1856, Nott sent to the *Charleston Medical Journal* a lengthy letter to Gliddon from Luke Burke, former editor of the defunct *Ethnological Journal*. In his accompanying letter, which the editors also published, Nott referred to his previous response to Bachman, saying that even though his opponent had thrown "villainous brimstone" at him, he had come to "the rescue" of Agassiz. Now, he had support from the "distinguished" English ethnologist Burke. Again feigning humility, Nott said he had thought of omitting Burke's flattering references to him, but he had decided it would be better to let Burke "tell his story in his own way." In fact, Nott relished the praise, and he wanted to see the letter published because he thought it presented a better case than he had made. Totally misrepresenting Bachman's arguments, Burke declared that Agassiz had written his essay on natural provinces "as a scientific man, as a Geologist, as a Naturalist," while Bachman had written as a theologian: "faith, miracles, traditions, writings, internal inspirations, and such like are his only concern." Bachman, Burke continued, was trying "to serve two masters; he wishes to be wholly a scientific man, and yet wholly a theologian." Agassiz's tableau, he added, "has completely bewildered poor Dr. Bachman," for he "has not the least idea" of its purpose. *Types of Mankind*, he concluded, has deservedly received acclaim. Nott was delighted, but Bachman was no doubt depressed. The other Charleston naturalists remained silent, though young John McCrady had begun to form his views on racial types, following Agassiz as his guide. The torrent of ignorance indeed appeared to be sweeping science away.[20]

Nott had been eager to publish Burke's letter because it supported his anti-black and anti-biblical views and because it vilified Bachman. The self-styled British ethnologist was hardly representative of the best in his country, however, and Nott carefully avoided any reference to the short life of Burke's *Ethnological Journal*. Nor did he refer to the separate *Journal of the Ethnological Society of London*, which represented the rigorous inquiries of several serious students of past and present cultures. Although those scholars held a number of erroneous assumptions regarding non-Anglo-Saxon cultures, they endeavored to be objective. Thus, they viewed *Types of Mankind* with disfavor. At a meeting of the Society on May 26, 1854, for example, the Society's secretary, Richard Cull, charged that the authors of the book had written it not to report any new research but simply to depreciate the concept of the specific relationship among all races. Any expectation of a scholarly contribution in *Types* was "doomed to disappointment," he declared. The essay by Usher, he added, was worthless because it was based on ideas long since rejected by serious students of ethnology. The essay by Agassiz, said Cull, is sadly wanting. Indeed, he observed that Agassiz seemed to be totally unaware of earlier studies of so-called natural provinces and extensive studies by philologists. He pointed to the flaws in Agassiz's zones and stated that Agassiz's scheme satisfied neither zoologists nor ethnologists. Equally critical of Nott, Cull cited the Mobile physician's undocumented claims about the nature of the brain of the black man. Any anatomist would be astonished, said Cull, for "all is vague generality where precision of detail is emphatically demanded." Gliddon fared no better in Cull's estimation. Indeed, the self-appointed expert on the language of the Hebrew Scriptures was guilty, argued Cull, not only of distorting meanings in his translations but also of writing in a frivolous mood. Along with Bachman's cogent critique of *Types of Mankind*, Cull's remarks effectively discredited the book. Message triumphed over merit, however, and *Types of Mankind* served the purpose for which Nott and Gliddon had thrown it together—helping to fix racial stereotypes in the minds of its readers.[21]

Meanwhile, during the debate with the pluralists from 1849 to 1856, Bachman had been involved in another controversy. Because it was mainly confined to Charleston and because it pitted his sectarian views against those of another, the controversy was construed by many Charlestonians as merely a quarrel between a Lutheran clergyman and a Roman Catholic layman, but it also indicated a larger issue. Bachman had never been tolerant toward Catholic clergy who criticized Luther and the doctrines of the Lutheran church. Early in 1838, he had been involved in a controversy over church doctrine with Bishop John England, and both men had sharpened their epistolary skills on each other. It had begun in November 1837, when Bachman delivered a sermon on errors in the doctrine of "the Romish Church," and Charleston's German Friendly Society decided to publish it. Bachman claimed that he was "willing to unite with every lover of the Gospel of Christ in producing the downfall of Sectarianism, though not the obliteration of sects." Certainly, he was sincere, but deep in his soul he could not believe that the Catholic clergy were true lovers of the Gospel he viewed as correct. In mid-January 1838, Bishop England arranged for the *United States Catholic Miscellany*, located in Charleston and founded by England, to publish Bachman's sermon, after which he published a series of letters to refute the principal arguments in the sermon. A formidable thinker and master of the English language, the transplanted Irishman wrote some twenty-one lengthy letters in the *Catholic Miscellany* between February 1 and August 30, 1838. The bishop began with an accusation that Bachman had gone out of his way "to assail those who gave you no provocation," but he expressed his high regard for his fellow clergyman because of his candor and total lack of hypocrisy. The result of Bachman's sermon, said England, was to give sustenance to those who hate the Catholic religion. Such criticisms, the bishop added, were especially unfortunate because Bachman had been "disciplined by science." Bishop England challenged Bachman's right to claim that his interpretations were superior to those of the Catholic Church, and he proceeded to argue for the validity of Catholic doctrine. The *Catholic Miscellany* published Bachman's replies, which ended in June 1838, when Bachman left for Europe, though England continued his letters for two months afterward.[22]

The quarrel—petty, sectarian, and clearly precipitated by Bachman—indicated not only the views of a Lutheran pastor but also a more general disregard for Catholic beliefs in the Queen City of the South. Still, Charlestonians were tolerant of Catholics, and they held Bishop England in high regard, especially since he sanctioned the institution of slavery. A more serious quarrel between Charleston Catholics and Protestants broke out in March 1852, when an itinerant preacher entered the city and announced his intention to denounce the Catholic religion. On March 13, the preacher, styling himself as the "Rev. E. Leahey, a late monk of La Trappe," placed an advertisement in the *Charleston Courier*, indicating that he would present a lecture in the Masonic Hall on the evening of March 15, in which he would reveal "the unchristian treatment of Females in the confessional by Popish Priests." Leahey added that women and children were "positively prohibited from coming to this Lecture, as some awful disclosures will be made." The admission charge, as it had been for those wishing to take a peek at the mermaid in 1843, was fifty cents, "the proceeds of [which] will be used to propagate the Gospel among the Catholics." As it turned out, Monk Leahey had previously done the same in Baltimore, but the admission charge there was only twenty-five cents. As with the mermaid exhibitor, the polemical monk set a higher price for his Charleston audience.[23]

The *Charleston Mercury* had refused to run the advertisement, contending that it was too offensive. The *Charleston Courier*, however, argued that to do otherwise was to violate Leahey's right to freedom of speech. Catholic leaders in Charleston, The Right Reverend J. A. Reynolds and Bishop P. N. Lynch, did not agree, insisting that the *Courier* had violated the journalistic principle of impartiality and calling Leahey "an obscene lecturer." In addition, the *Catholic Miscellany* charged that the *Courier*'s editor was following the example of the northern presses. Charleston, it declared, is "not like other cities, and especially her northern neighbors. Our standard of public morality has been higher." The matter became an issue for the City Council when the owner of the American Hotel, where Leahey was lodging, asked whether the city would be responsible for damages to his hotel in the event of a riot over Leahey's lecture. Leahey attended a meeting of the Council on the afternoon of the 13th. Two aldermen expressed doubt that a riot would ensue and argued that, in any case, Leahey had a right to speak. The mayor voiced concern, however, and he was joined by a third alderman. Then the alderman of the Fourth Ward, John Bellinger, a native South Carolinian and professor of surgery in the Medical College, stood to announce his disagreement with the argument of free speech. Had he confined his remarks to that issue, Bellinger might have carried with him the two who feared a riot. Instead, Bellinger, a recent convert to Catholicism, used the occasion not only to express fear over a riot but also to refer to "the immoral practices" of Martin Luther. The council decided that it had no authority to prevent Leahey from presenting his lecture. No riot occurred, but Bellinger had jumped into a boiling pot by castigating Bachman's hero.[24]

Fresh off the first round of debates over the race-species issue, Bachman fired off a letter to Bellinger through the *Courier*, charging that he had assailed the name of a man revered by Protestants. He promised to refute the attack in subsequent letters. Bellinger was not easily cowed, however, and stood his ground. As the sparring increased, the *Catholic Miscellany* entered the fray. Bachman spoke of "priestly profligacy," and Bellinger replied with charges that Luther abused women, while the *Catholic Miscellany* continued to condemn the elusive Monk Leahey. Before long, other Protestant ministers joined with Bachman in castigating Catholic critics of Leahey. By the end of July, all of the opponents had grown weary, and in early August they dropped the subject. Bachman was not finished, however, for he drew together all of the exchanges and enlarged upon his views in a large tome titled *A Defence of Luther and the Reformation Against the Charges of John Bellinger, M.D., and Others, to Which Are Appended Various Communications of Other Protestant and Roman Catholic Writers Who Engaged in the Controversy*. Although Bachman surely did not think so, the ugly affair had ended in a draw. As a naturalist, he scrupulously followed the principles of scientific inquiry. As a defender of Luther and as a champion of the right of South Carolinians and other southerners to own slaves, however, he allowed passion to be his guide.[25]

Despite his defense of all races as a common species and despite his sectarian propensities, Bachman continued to enjoy the esteem of many people in South Carolina. Certainly, he received some praise from those who were offended by the potential threat of the pluralists' argument to the sanctity of the biblical account of the creation of the original pair of humans. Perhaps others did not understand his scientific argument or merely thought it did not matter, and perhaps some of them considered him too sensitive about Luther, but all honored his sterling integrity. Until the Compromise of 1850, Bachman had tried to remain aloof from politics, but as sectional hostility increased he found it more difficult to maintain his silence. In late August 1851, several prominent South Carolinians evidently persuaded the clergyman to travel to Washington to see whether he could get some of America's leaders to hear the sentiment prevailing in their state. Bachman reported that he was "sought out" and received "undeserved attention." The northerners with whom he talked "smile[d] at the [prospect of the] secession of a single state, but look[ed] grim" when he "spoke of a southern confederacy." Somehow Bachman received an audience with President Millard Fillmore, who asked Bachman what could be done "to satisfy the South." In response, Bachman said that the nation must follow "the letter of the constitution." As Bachman reported the President's reaction, Fillmore indicated that he thought that had been done, but Bachman cited "the case of Texas, where a state had been threatened with a Federal army." Fillmore contended, however, that he had done so because he feared the state was on the brink of a civil war. Bachman believed that the president was embarrassed over the matter, and he said that Fillmore promised him he would confer with Congress. After the conversation, Bachman opined that "our course can best be promoted by battling for our rights." Clearly, he hoped the issues would be settled in South Carolina's favor, and he was unwilling to settle for less.[26]

A few days later, in a letter to Victor Audubon, Bachman expressed concern that the secessionists were in the majority, but he believed "the evil day may be staved off another year." He admitted, however, that he was "growing every day less attached to the Union as it now exists, and if South Carolina declares for secession, I will, for weal or woe, go with her." In addition, he complained that New York and Massachusetts seemed bent on electing senators who went to Washington "to read abolitionist petitions and abuse and insult the institutions, the morals and the religion of the South." The southerner-by-adoption expressed fear, however, that South Carolina would experience "long years of poverty and misery" if she elected to secede. The Union would be foolish, he boasted, to send troops if South Carolina seceded, for the other slave states would come to her aid with legions of troops. Victor replied soon thereafter and told his former father-in-law that the "deep-rooted angry feeling" in South Carolina was much greater than he had ever imagined. He said that Bachman could persuade others "of the unutterable horrors" that would result from secession. Audubon did not doubt the South would fight, but he believed the consequences would be disastrous—the country "would be ruined" and a generation of young men would be "cut down." He assured Bachman that he loved his country, that he knew "no North, no South." Victor Audubon lacked the knowledge to be a good mammalogist, but he was a wise and compassionate man.[27]

Six years later, early in 1857, just short of his sixty-seventh birthday, Bachman had not mellowed on the issues of slavery and states' rights. In a letter to Victor Audubon, he complained that "the black republicans are fast rising into power, [and] when they do, the South will walk out of the Union, peaceably if she can, forcibly if she must." If secession occurred, he apprised Victor, he would "sink or swim in the Southern ship." Bothered by what he viewed as "the outrageous lies of the northern papers," he had come to believe that the two sections of the country were no longer brothers but separate nations geared to war. Bachman lamented the actions of the underground railroads and wondered how "stealing our property" could be "considered an act of heroism." As tensions rose, Bachman and the Audubons corresponded less frequently, but their attachments remained until the outbreak of the Civil War. By 1860, Bachman and the firebrand Edmund Ruffin were sharing their lamentations and fulminations with each other. In a letter to Ruffin on January 18, Bachman said that "frenzy and madness rule the hour." The only course, he continued, is secession, for the abolitionists will "rob and plunder and bully us in the Union until they [have] their feet on our necks and their daggers in our throats." He was torn, he told Ruffin, for "My religion bids me forgive, [but] God help me I would rather have them hanged first and forgive them afterwards." On May 23, Bachman told Ruffin he was concerned over the forthcoming presidential election and feared the Republican candidate would win. "We will then see who will swallow black republicanism."[28]

Meanwhile, Bachman was worrying over the declining health of his wife. In a letter dated March 30, 1856, he told Lucy Audubon that Maria could no longer use her right hand. Sixteen years earlier she had fallen and injured her elbow, and now the joint had become enlarged to twice its normal size. Bachman said he was filled with uneasiness, for Maria has "been all to me—a mother to all my children, my adviser, my companion, my help in all things." A year later, Bachman wrote to Victor Audubon that Maria had been severely ill with pneumonia, but indications pointed to tuberculosis. In mid-August he reported that her health was continuing to decline, and in December he indicated that she weighed only seventy-eight pounds. Consumption was once again wasting away the body of another of Bachman's loved ones. Maria hung on, but early in May 1858 the Grim Reaper suddenly snatched up Bachman's thirty-five-year-old daughter Harriet Eva Haskell and the baby to whom she had just given birth. Of Bachman's fourteen children, only five were now living, and his second wife was in the throes of a deadly disease. The disease was also beginning to bother Bachman again, but he continued to work as actively as ever and to support the secessionist movement with the zeal of a new convert.[29]

In a letter dated July 16, 1860, Edward Harris, a friend since the days of working on *The Quadrupeds* with Audubon, told Bachman, that he could not "suppose that you subscribe to such an ultra doctrine" as secession, but it is apparent that he feared otherwise. Harris declared he would travel any distance to see Bachman, for he considered him to be "my good old friend, whom I still, and always have held in the highest regard and esteem." Bachman could not be swayed; his mind was set. Toward the end of the year, he was violating his long-held promise to avoid politics in the pulpit, preaching to his congregation on "the duty of the Christian to his country" and telling them that "truth and justice [are] on our side." The importance of the scientific case for the unity of the races had long since ceased to occupy the concerns of Bachman. He and his old nemesis Nott were now linked by a common cause.[30]

The Broken Circle

From piazzas, rooftops, wharves, and the battery, Charlestonians cheered as Confederate General P. G. T. Beauregard's troops bombarded Fort Sumter on April 12, 1861. News of the assault likely caused pangs of anxiety in young Sallie McCrady, however, for, although her husband seemed to be safe at Fort Pickens, on Battery Island near the mouth of the Stono River, he had told her in a letter six days earlier that his position could be attacked. She had seen little of John McCrady since he had left to stow away on the guardboat *Nina* just before Christmas, for immediately afterward he accepted responsibility for helping construct a line of defenses to protect Charleston should war ensue. In a letter to Sallie back on January 18, the thirty-year-old husband, father, and patriot had complained that his feet were "swollen with long standing and much walking, night and day," and he had expressed concern that he might be unable to complete the battery at Fort Morris before the commander at Fort Sumter, Major Robert Anderson, opened fire upon the engineers. Soon thereafter, the Confederate legislature had established a corps of military engineers and appointed McCrady as one of its four lieutenants. By April 11, he had completed work on the battery at Fort Pickens and told his sister that he hoped God would bless his efforts.[1]

Meanwhile, the College of Charleston had assigned McCrady's duties to Lewis Gibbes, but McCrady did not relinquish his faculty position until July, when it was clear that he would not be able to return any time soon. By late September, the college trustees had decided to leave McCrady's position vacant, for, with the loss of many of their students to the army and an increasingly uncertain economy, they were compelled to reduce expenses. They also suspended the museum's taxidermist and reduced appropriations for chemical and museum supplies. Nevertheless, at the end of October, Francis Holmes was able to report that the museum had received "two large contributions of great value." The first came from James Hamilton Couper, a prominent planter of coastal Georgia, who donated his collections of fossils, minerals, and mollusk shells. The collection of mollusk shells was so large, stated Holmes, that it filled two-thirds of a room in the museum. The other donation, "the gift of Miss A. M. Annelly of this city," was "a very handsome cabinet of recent shells" including both native and foreign specimens. That collection filled the other one-third of the room set aside for Couper's collection of shells. Holmes estimated that it would require several months of sustained effort to label and arrange the specimens. The trustees believed the museum's shell collection might now be "the amplest and richest in the Western world." Their estimate was too high, but the collection was indeed superb. The new specimens would have to wait for labels, however, for Holmes was engaged in helping with the war effort as often as he could. Moreover, the trustees were worried that the museum might be damaged or destroyed by Union bombardments. Thus, they asked Holmes to devise a plan for moving the specimens to a safe location. Since he had leased a plantation in Edgefield, located in the western section of South Carolina, mainly for the safety of his family, Holmes proposed moving the collections there. The trustees agreed, and Holmes arranged for crating most of the fossils and minerals, all of the shells and specimens in alcohol, the rarer birds, mammal skins donated by Audubon, and Elliott's herbarium. The crates were eventually transported to Aiken by rail and on to Edgefield by wagon.[2]

On November 14, the college freed Holmes from his duties so that he could direct the Wayside House and Hospital in Charleston. The recovery house was the idea of Holmes's brother-in-law George A. Trenholm, a successful Charleston merchant. Holmes began at once to raise money and solicit supplies for soldiers passing through the Wayside House. By late December he was providing for 3,000 soldiers, and, as a friend indicated, he was "nearly broken down by the press of work." Worse was to come a month later, however, when Holmes added to his burden the task of providing aid for the victims of a monstrous fire that had broken out between 8:30 and 9:00 in the evening of December 12 and raged unchecked until daybreak. As reported by the *Charleston Mercury*, "heavy gusts of wind . . . swept the dust and smoke and sparks hither and thither in blinding clouds." The fire began in the southern part of the city among closely built wooden houses, and the "fierce and roaring march" of the flames over 540 acres of the city consumed residences, factories, and churches. The Friend's Meeting House, St. Peter's Episcopal Church, the Circular Church, the Cathedral of St. Finbar, St. Andrew's Hall, Institute Hall—all fell. So too did 575 homes, including the residence of John Edwards Holbrook.[3] As the fateful year of 1861 drew to a close, inhabitants of the city no doubt prayed the new year would be better.

The prospects were not bright for John Bachman, however, for his wife Maria's condition had worsened. Maria herself could no longer deny the severity of her situation, and, in a letter to Catherine (Kate), the youngest of the Bachman children, she said, "I am now the subject of chronic consumption, which is incurable and will end my conflict here long before peace is restored." Expressing fear that Federal troops would sooner or later attack Charleston, she added, "all is dark and gloomy to me." Bachman was worried as well. He knew the ravages of consumption, and he told a friend that Maria would likely be experiencing lung hemorrhages again. Unlike his wife, however, he had "not the slightest apprehension of Lincoln's hoard attacking Charleston." Yet, early in May 1862, he moved Maria and Kate to the home of his son William, in Columbia, where he thought they would be safer. William himself was serving in the Confederate Army. Dedicated to doing everything he could to help the South, John Bachman was spending three hours of every day at the hospital in Charleston, tending to his flock, and raising money to aid sick and wounded soldiers. "How terrible an evil is war," he told a friend, "yet we must go through with it." He was perturbed, however, by the greed of some southern manufacturers and complained that some of them were robbing the people by setting exorbitant prices on scarce goods. On May 10, the *Charleston Courier* published his "open letter" to J. G. Gibbes & Company of Columbia. Verbally flailing the firm for reaping excessive profits on cotton and woolen goods, he declared that its owners were unwilling to share in the public burden and were thus further injuring the already bleeding Confederacy. Considering Christian charity inappropriate in this case, the irate patriot wrote, "God grant our country soon a safe deliverance from the enormities of all such beings—devoid of conscience, enemies to their country's welfare, [and] friend to none but themselves." The seventy-two-year-old Bachman had lost none of his fire or courage. Even one of his old opponents in the race-species controversy, Dr. Frances Peyre Porcher, was now praising him for displaying "the character of the true patriot during our present struggles."[4]

During the summer of 1862, Bachman wrote to his friend Edmund Ruffin to console him upon the death of a young grandson killed in battle and on the loss of valuables to plundering Union troops. Proudly he told Ruffin that his own sixteen-year-old grandson John Bachman Haskell had joined a military unit in Charleston to perform guard duty during the summer while the college was in recess, and he praised his son William, who had been severely wounded in the abdomen during a battle in Virginia. He told Ruffin not to worry about an invasion of Charleston. "If the Devils come," he opined, "they will receive a Devil's welcome . . . [and] never go back out of this harbour without the benefit of drowning." The depth of John Bachman's commitment to the cause could not be doubted. The cause was in fact now a part of the ministry of the aging pastor, and he pushed himself to the limits of his strength. He assisted in publishing the new *Southern Lutheran*, in which, in late 1861, he penned "a particularly savage attack on a northern newspaper editor," explaining that his purpose was "to show to our southern people the impossibility of any reunion with our northern church." He also helped to establish the Southern General Synod and aided ill and wounded soldiers, even traveling as far as the Potomac River in Virginia to deliver clothing and other necessities to military hospitals. Meanwhile, at home, he visited the sick, buried the dead, and comforted the bereaved. He pushed himself so hard that he reached the point of exhaustion and was found unconscious in his buggy one day in August 1863. Accompanied by his daughter Kate, Maria, who was extremely weak herself, came for him and carried him to Columbia. Within a few weeks, the old patriot was back at his work in Charleston. "I trudge on foot," he told a friend, "[and] return home, mourning over our scattered people." He bore the burden, he added, because it was his lot "by a Wise Providence." Complain he would not. He was laboring "for the liberties of my country." Only the rapidly failing health of his wife drew him away from his work. Maria died on December 26, 1863. Bachman knew death intimately, but once more, after grieving over his great loss, he pressed on.[5]

Meanwhile, by the spring of 1862, concern over an invasion of Charleston and the surrounding area had intensified, and the South Carolina Executive Council appointed Holmes the task of procuring slaves to strengthen fortifications around the district. The Council authorized Holmes to pay $10 per month per slave, or to impress slaves in neighboring districts if necessary. Soon thereafter, he was also placed in charge of a district of the Confederate States Nitre and Mining Bureau. He had closed the Wayside House earlier, since he needed to be on the plantation near Edgefield in order to plant spring crops. But his duties in obtaining slaves kept him occupied in and around Charleston, while the superintendency of the nitre district demanded his services in the western and central sections of the state. Simultaneously, Lewis Gibbes endeavored to carry on with his duties at the college. The task was difficult, for only a handful of students were now enrolled. The trustees elected to keep the institution open, however. The crisis in Charleston worsened on August 22, 1863, when Union forces began to bombard the city, forcing most residents in the southernmost section to abandon their homes. Gibbes began to inquire about a place to move his family for safety. In November, he wrote to the South Carolina College physics professor John LeConte to ask if he could find the Gibbes family a place in Columbia, but, after a diligent search, LeConte replied that he could find no house for rent. Eventually, Gibbes found a place in Anderson and sent his family there. As the bombardment of Charleston increased, and as shells exploded all around the college, the devoted professor endeavored to keep his beloved institution in operation.[6]

The war had not been going as McCrady had expected either, and in February 1862 he wrote to Sallie that "hard times are upon us and we must submit to hard separations." He advised her to take their two baby daughters to a place of safety. In March 1862, he was appointed to the rank of captain in the Corps of Engineers of the Provisional

Army of the Confederate States of America and on June 7 was ordered to Savannah, Georgia, to strengthen fortifications around that city. Shortly after arriving in the old city, McCrady issued a circular calling for a thousand slaves to help in erecting defensive works around Savannah, and soon he was at work in constructing a ring of protective embankments. When General Beauregard came to inspect the works on September 20, 1862, McCrady proudly told his wife, the commanding officer invited the young captain to meet with the general so highly esteemed by Charlestonians. Three days later he received a letter from his brother Edward, telling him that their brother Thomas had been wounded. Edward himself had been wounded earlier, but both he and Thomas returned to duty even before their wounds had healed. The McCradys were gladly paying the price for the independence of their beloved land. By November, Beauregard had appointed McCrady as chief engineer for Georgia. "I must say I was quite surprised," McCrady wrote to his father. His constant concern for his wife and two infant daughters, whom he had seen only infrequently since 1861, and the increasing possibility of a Union invasion of Charleston worried him greatly, however, and in late April 1863 he complained in a letter to Sallie, "I am broken up. My mind is so much worn out that I can neither work nor sleep well." Word that President Abraham Lincoln had issued an Emancipation Proclamation on January 1, 1863, had already set his mind to dwelling upon "the fearful train of evils" that would surely follow a victory by Union troops, one of which, he believed, would be "an organized abolition agency in our very midst."[7] The prospect of freed slaves was one of McCrady's worst nightmares.

In August 1864, McCrady received word that his two-year-old daughter had been severely ill and that one of her arms was paralyzed. Indications point to poliomyelitis, but neither the McCradys nor the physician had any idea of the cause. By September, McCrady was nearing a nervous breakdown, and a Savannah physician recommended that he take leave and rest himself completely. He received a furlough of a few weeks, but was back in Savannah in late November. There he "found everyone in a considerable state of excitement over Sherman's last movements." Atlanta had fallen, and none of the southern military leaders knew where the Union general would head next. McCrady feared he would aim for Savannah and then move on to Charleston. If Sherman attacked Savannah, McCrady told his father, the city would certainly fall. He held out hope, however, that the Confederacy could muster 20,000 troops and stop Sherman in the middle of the state. In reality, he understood not only the improbability of mustering such a force but also the impossibility of getting it there in time even if the troops were available. Thus, McCrady told his father that he expected to be captured. The ever-scrupulous man fretted, however, over whether he was obligated to pay the remainder of the rental money due on the house he had leased in November for six months. A few days later, he penned a morose letter to his wife. "Should we never meet again," he said, she must remember that she had his "whole heart's love." He instructed her to rear their children as Christians and vowed he would meet her again "in a better place." Still thinking of the possibility of resuming his scientific career and fearing plunder in Charleston, he later directed that his books, specimens, and microscope be taken to Columbia.[8]

By mid-December, the enemy, outnumbering the defenders by three to one, was flanking Savannah, and minié balls were flying all around McCrady as he inspected the lines. The situation was hopeless, and four days before Christmas the Confederate defenders evacuated the city. On December 30, McCrady wrote to Sallie from Augusta, telling her that Savannah had held out for eleven days because he had done such a good job of constructing a defensive line. Deep in his soul, however, he sensed that the cause was lost and that the Confederacy would have to capitulate within a matter of months. Over in Edgefield, Holmes knew that the prospect of stopping the relentless march of Sherman was dim at best, and soon he watched helplessly as enemy troops in search of plunder ripped off the tops of the crates holding the specimens of the Charleston Museum. To his great delight, however, the soldiers quickly abandoned the task as they discovered the contents of the boxes. McCrady moved on with his unit, barely staying out of the path of the enemy. He had heard of Sherman's capture of Columbia and the fire that devastated the heart of the city, and he feared the worst for his precious scientific items. Indeed, his fear was well founded, for, like John Bachman, he lost all the books and manuscripts he had sent to the city. On March 17, 1865, he wrote to Sallie from Charlotte, North Carolina, that he was "no longer fit for duty" and that he had been in such pain for the past six days that he was compelled to lie down often.[9]

Things had been going no better in Charleston. Back in mid-November 1864, Bachman had written to Ruffin to console him over the loss of a son in battle. The death had reminded him of his own recent loss, he said, and he was keeping busy to avoid thinking about it. "I wander everywhere collecting for the hospitals, [and I am] preaching every Sunday." His son Wilson had escaped the enemy as they came through his plantation near Atlanta, but he had lost his slaves and all his furniture and was unable to find his wife and children. "Are not the Yankees barbarians," Bachman cried. His son William had been in seventeen battles but was still safe, he proudly added. The situation was growing worse in Charleston, however, and by mid-February 1865, it had become apparent that Union forces would soon be in the city. Lewis Gibbes fled at the last moment, apparently headed for Anderson, where he had moved his family. Like others, he had sent some items to Columbia, but he was fortunate in losing few of them. His friend John LeConte had placed most of the items in a building of South Carolina College, which escaped destruction. Holbrook, who had lost his wife in 1863, apparently stayed on in Charleston until mid-March, when he caught a train from the city. A young girl who sat next to him reported that the passengers had heard rifle fire, but Holbrook was "not perturbed in the slightest" and alternated between reading "a yellow paper backed French novel" and napping. After being awakened several times by the conductor, he finally "lost patience and began to damn the conductor. Damn it, he couldn't help it if they were to meet the Yankees; a little sleep before then would help." The aging naturalist made it through without harm.[10]

Such was not the good fortune of Bachman. After insisting that he must stay in the city because he had "so many duties to perform," he finally gave in to the pleas of his daughter Kate, his son William, and several of his parishioners. On the extremely cold day of February 13, he and Kate caught a train from the city, only four days before the evacuation of Charleston commenced. They got off at "Cash's Station," near Cheraw, located in the northeastern corner of South Carolina, and trudged to the home of a Mrs. Ellerbe, where several young women had come for protection. On the third day after their arrival, the frightened refugees spied Union troops approaching. Fearing the soldiers might set fire to a nearby railway car filled with ammunition, the resourceful old clergyman quickly rounded up a number of blacks and ordered them to push the car down an incline. It came to a halt near the troops, who then set it on fire. Soon the captain and his company rode up to the house and promised no harm would come to the occupants. Ere long, however, some of the soldiers entered the house and broke the locks on drawers, in which they found nothing but calico cloth. They soon rode off for "orgies at the negro quarters" on a nearby plantation, according to Kate in a later recollection of the event.[11]

On the following day, the troops returned and began to query Bachman about buried valuables. After Bachman repeatedly denied knowledge of any such treasure, some of them led him behind the stable and threatened to shoot him. Eventually, "they cocked their pistols and held them over my head," Bachman reported. One of the soldiers then began to kick the old man until he fell. He rose but was knocked down repeatedly. Finally, Bachman asked them to shoot him if they wished but to stop battering "a defenceless old man." The captain then struck him on both arms with his sheathed sword. The pain, said Bachman was "excruciating," and he thought his left arm had been broken. When Kate learned from an elderly black woman what they were doing, she ran to her father, but the brutal men had moved on to search for other things. Kate helped her "faint and tottering" father to the piazza. Soon, however, one of the soldiers came up and pointed a pistol to the chest of the old man. Kate threw her arms around her father to protect him, and the soldier began to search Bachman's pockets and found a small comb given to him by his eldest daughter. So dear to Bachman was the item that, despite his pain, he grabbed it, but Kate persuaded him to let go. The soldiers finally left, but returned on the next day and set fire to Mrs. Ellerbe's smoke house. Fearing the flames would spread to the house, the injured minister mustered strength to order the women to fight the blaze. Suddenly, a number of the soldiers took charge and extinguished the flames, and soon thereafter the cruel victors departed for good. Several days later, after a responsible Union officer learned of the incident, the culprits were brought before Bachman, but he said he did not recognize any of them. A spirit of Christian charity had returned to the heart of the minister who had taught it to his Charleston congregation for fifty years. Four months later, he returned to his pulpit, his left arm still useless.[12]

Francis Holmes considered himself fortunate to have escaped the wrath of the Union soldiers who had come through his plantation, and the museum collections were safe. Soon after the war ended in April, he began to prepare for returning to Charleston during the fall. His luck suddenly ended before dawn on September 2, 1865, when arsonists, probably former slaves, set fire to a small building housing his library of 2,800 volumes, some of his best fossils, his microscope, his table silver, and his manuscripts and notes. The building burned to the ground. Discouraged over his great loss, Holmes returned to Charleston in November, where he soon learned that the College of Charleston trustees were endeavoring to open the institution by the first of the new year. Equally eager to return to their faculty posts were Gibbes and McCrady, both of whom were virtually in a state of poverty. The college was also lacking funds to resume as it had been operating four years earlier, and the trustees were compelled to eliminate the professorship in classical languages. Unbeknownst to Holmes, they had even considered eliminating the professorship in geology and paleontology and replacing it with one in French. They had no intention of abolishing the museum curator's post, however, even though it might be considered a luxury for so small a college. Gibbes was so poor that the trustees had to advance $250 of his salary so he could move back to Charleston.[13]

The college opened its doors on February 1, 1866, but by late May, McCrady was suffering from severe depression. In a note to Gibbes, he said, "The shattered condition of my constitution will not permit me to resume my duties at the College for four or five months at least, and perhaps never again." He asked Gibbes to assume his duties, for which he would pay him $100 per month from his salary. Gibbes readily agreed to assist his young colleague but declined to accept any money. Upon the advice of a physician, McCrady decided to go abroad for three months. His wife Sallie, in her fifth month of pregnancy with their fourth child, was left to manage affairs. As she had already shown during the four years they were separated because of war and as she would later demonstrate, Sarah Dismukes McCrady was entirely capable of making the best of adversity. John McCrady departed for England on July 27. Suffering various pains throughout the journey, he dosed himself frequently with "stimulants" and suffered "bad dreams." In England, he spent some time in visiting historic sites, but he was usually too ill to enjoy them. When he returned to Charleston in mid-October, McCrady was still quite unwell, and he asked Gibbes to continue to carry out his duties until January 1867.[14]

The war had exacted a heavy toll upon the scientific enterprise in the South and even a greater one upon perhaps the most brilliant of its naturalists. But some of the costs had yet to be calculated.

Edmund Ravenel had been blind for several years and unable even to try to straighten out his huge shell collection, which his son and a daughter had hastily packed just before the war began. Holbrook, who had retired, was unable to arrange for the return of his personal library, which he had sent away during the war to an acquaintance in northwestern South Carolina. Moreover, he was spending about half of each year with relatives in Massachusetts. Bachman was in poor health, fast approaching four-score years of life, and dedicated to spending his final years in ministering to the needs of his beloved congregation. Only Gibbes, Holmes, and McCrady were still active, but all were determined to restore the study of natural history to its once-prominent place in Charleston. The efforts of Holmes, however, soon foundered upon the jagged rocks of a devastated economy. In fact, the college trustees could not pay for transporting the 108 crates of museum collections back to Charleston, and Holmes bore the expense from his own meager resources. Aside from his family, Holmes considered the museum to be the most important thing in his life. Thus, he began at once to set it in order again. The most demanding task he faced was to identify and reconstruct collection data for a number of the specimens, as many labels had been lost during the hasty packing of the items four years earlier.[15] As a good naturalist and curator, Holmes understood the virtual uselessness of a specimen without such vital information as the location from which it was taken, the date it was collected, and the name of the collector. No one could know, for example, whether it was a type specimen or use it to determine the distribution and range of a species. By relying on scattered records and personal memory, he could provide a correct label for many specimens, but he obviously encountered difficulty with a number of items. Time and assistance would be necessary to complete the task. The dedicated curator soon learned, however, that he would have neither.

In March 1868, Holmes reported that many people were visiting the museum each week, but the trustees had appropriated nothing for its maintenance. The college was facing a serious shortage of money, and during the next month the trustees began to discuss the need for reducing faculty salaries. On their scale of priorities, the professorship in geology and paleontology fell last, and the museum curatorship just above it. Holmes's joint positions thus became the focus of attention, though the man who had lifted the museum to high standing seemed to believe the trustees could not be serious. Moreover, he apparently thought he would have the support of his faculty colleagues. Mellowed by time and tribulation, he had rejoined the Elliott Society, which, under the leadership of Lewis Gibbes, had reorganized itself as the Elliott Society of Science and Arts, in an effort to broaden its interests and its membership. Holmes had not changed his views, however, about the importance of the museum and about his role in promoting its interests. Thus, in May he took it upon himself to inform the college's Chrestomathic Society that it must give up its meeting room to the museum. Incensed when he heard the news, the College of Charleston president stated that only he and the faculty jointly possessed authority to make decisions about the assignment of rooms. At a meeting of the faculty, Holmes insisted that he had brought up the matter with the faculty a year earlier and had been authorized at that time to confer with the Chrestomathic Society. It was only after he received approval from the Society that he had proceeded to appropriate the space, he added. Then he offered a motion to limit authority for assignment of rooms to the faculty, and it passed. Acting as faculty secretary, John McCrady recorded in the minutes that Holmes "acknowledged the blunder of not communicating with the Faculty." The choice of wording was probably not unintentional, for McCrady believed that Holmes was pushing the museum at the expense of the college. Moreover, it soon became clear that McCrady wanted to be the curator. Stubbornly refusing to alter his phrase at the next faculty meeting, McCrady harshly criticized his colleague. Later, he apologized for his "excited and savage manner," but it was now obvious that Holmes's position was even more vulnerable.[16]

Still, Holmes did not recognize the gravity of the situation, and soon after the incident, he boasted to the president that he would soon be "independent" of the decisions of the faculty, for he intended to bypass the trustees and go directly to the city council for "nearly the whole of an appropriation for the benefit of the Museum." His confident attitude was misguided, however, for the city council, which made appropriations to the college, was unsympathetic and in August 1868 introduced a bill "to repeal the Ordinance providing the payment of the salary of the Curator of the Museum of the College of Charleston." On August 31, the trustees voted to abolish the curator's position as of October 1 and thereafter to add $200 per year to Holmes's professorship for maintaining the museum. Of course, Holmes would lose one-half of his salary by that arrangement. Shocked and hurt, he penned an irate letter to the trustees. He understood, he said, the necessity for reducing expenses and would not have objected if the trustees had cut all of the faculty salaries equally. It was "invidious," however, "to select *one* out of *seven* [positions] for decapitation." Holmes continued: "I have given the flower of my youth, the prime of my manhood to this institution for nearly twenty years . . . [and] cherished [it] as one of my children."[17] Indeed, he was right, but he quickly decided that he must figure a way to save his position with the museum.

Holmes soon took another approach, informing the trustees that the city ordinance did not eliminate his position, that it only forbade an appropriation for the salary of the curator. In fact, he added, since the curator's salary of $1,200 was only a part of his salary of $2,000 as a professor, the trustees could retain him as curator and cut his salary only slightly. The suggestion backfired, however, for, after uttering many words of praise for Holmes as curator, the trustees voted to abolish his professorship and keep him as curator. Since his salary as curator was, by his own admission, $1,200 per year, the trustees would pay him as much of that amount as they could, namely $1,000, which would come from the city and from a portion of the salaries of the president and the faculty. The trustees no doubt believed they were acting in a generous manner, and in a sense they were. Yet Holmes's salary would fall by a whopping 50 percent. Fearing that Holmes might reject the offer and concerned that he might remove from the museum a number of specimens that he considered as his personal property, the trustees directed Holmes to present "a minute and detailed statement" of items in the museum and of those he claimed as his own possessions.[18]

Perhaps the trustees might have endeavored to find other ways to keep from reducing the salary of Holmes so drastically had they dismissed from mind his other sources of income. Experience in business had triggered the enterprising imagination of Holmes once he returned to Charleston, however, and he had not only opened a bookstore but also launched a phosphate-producing industry. Perhaps some of his critics believed he was devoting too much time to those successful ventures, but almost certainly some believed he needed less income from the college. Holmes resented their decision, and he objected to their call for a detailed inventory of the museum collections. He could not provide a full list on such short notice, he informed them: "It would require an Agassiz many months to give a *minute and detailed* statement in writing of the property . . . in this extensive collection." He agreed, however, to provide a general list, and he declared that it would take only a few minutes to remove his own specimens from the museum. After they received the general list from Holmes, the trustees appointed a committee to inspect the museum holdings, and although they reported that the specimens claimed by Holmes were "not numerous," they expressed concern over the "blending of personal and official acts." Some of the items claimed by Holmes, they added, had been collected and prepared while Holmes was carrying out his official duties as curator. Since no rules had been drafted, however, the trustees chose not to press the matter. Instead, they simply reaffirmed their decision to abolish his professorship and to pay him $1,000 per year as curator.[19]

Unhappy but unwilling to sever the connection with his beloved museum, Holmes decided to remain. The sordid story had not ended, however, for in December the trustees adopted a set of rules that placed direct responsibility for operating the museum under a supervisory committee that would make an inspection of the museum every quarter, "or oftener if need be." In addition, the rules specified that a catalogue of the collections must be maintained and kept in the president's office; that "no one connected with the Museum shall work for himself in the Museum during the working hours"; and that any specimens collected by the curator in his official capacity were to be considered the property of the Museum. The rules were not unreasonable, but when the committee made its first visit on January 19, 1869, Holmes used the occasion to scold them, and he "censured with severity . . . the new Rules." Said the chairman of the committee soon afterwards, "The meeting was a short one. There could be no compromise under the state of sentiment and feeling manifested by the Curator." In fact, Holmes already knew what he planned to do, and he handed the chairman a previously prepared letter of resignation. The trustees quickly accepted his offer. Humiliated and angry, Holmes was finished as a contributing naturalist. In 1871, he left Charleston for a plantation near Goose Creek, about fifteen miles north of the city where he had risen to fame. There he again became a successful farmer. Controversy followed him, however, as a quarrel developed over whether he, Charles Upham Shepard, or St. Julien Ravenel was the true discoverer of the value of the abundant beds of phosphate in the region.[20]

More significant for the study of natural history, however, was the loss of Holmes as an active scientist. With his removal, the circle of Charleston naturalists was further broken, and only two links of it remained: Lewis Gibbes and John McCrady. Holmes also made another decision that represented a loss to natural history research in Charleston and the South. Stung by his treatment, he decided not to offer to sell his own specimen collections to the Charleston Museum. Of course, the museum had no funds for such a purchase anyway. Moreover, although he had intended to write a third volume on South Carolina fossils, he gave up that idea too some time after his resignation from the College of Charleston faculty. As early as April 1869, he notified Joseph Leidy that his collection of fossils was for sale and asked whether the ANSP might be interested in purchasing it, saying he preferred to place the items in Philadelphia. He thought his collection was worth $5,000. Having no luck with the ANSP, Holmes offered the Smithsonian Institution an opportunity to purchase the collection, but Joseph Henry informed him that the Smithsonian had no funds to do so. Holmes's luck changed in October 1870, however, when the recently established American Museum of Natural History in New York expressed an interest. By June 1872, Holmes had begun to negotiate with the AMNH. He had finally succeeded in getting Leidy to return some of the specimens he had borrowed, but he struck out with Louis Agassiz, as he had done before. Agassiz died in late 1873, and Holmes did not try again to get his specimens back until 1876, when he wrote to Agassiz's son Alexander. Even without some of the specimens borrowed by Louis Agassiz, Holmes had a very valuable collection, which he offered to the AMNH for the price of $3,000. Either the AMNH did not want the entire collection or it persuaded Holmes to sell the items at a lower price. Most valuable among them were the type specimens described and figured in the Tuomey and Holmes volume and those described and figured in Holmes's *Post-Pleiocene Fossils of South Carolina*. The AMNH reported soon thereafter that it had obtained its "first valuable series of fossils." New York's gain was Charleston's loss.[21]

Meanwhile, less than a week after Holmes's resignation as curator, the college trustees had appointed John McCrady to the post on a half-time basis. McCrady had wanted the position before Holmes relinquished it, and he had already secured a promise from some of the trustees that they would support him for it if Holmes resigned. The curator's job was important to McCrady for several reasons. First, although his students praised him, McCrady seemed to be unable to derive the pleasure from teaching that had characterized his interest before the war. In large measure, the problem emanated from his penurious situation. By 1868, he had a family of four children (another had died in infancy), and he could not make ends meet on the salary he received. Moreover, he was still shaken by the defeat of the South and the loss of his manuscripts and books. In addition, he had come to believe that his scientific work had been forgotten. Instead of returning to his work as a hydrozoan zoologist, however, he fixed his mind upon working out the problems of his law of development and, as a corollary, refuting Darwin's theory of evolution by natural selection. Indeed, he became obsessed with those ideas, much as he had with the idea of southern civilization during the late 1850s. Adding to his difficulties were his frequent, debilitating illnesses. Related to his brilliance was an intensity that drove him mercilessly, and, as a consequence, placed enormous strain upon his psychological composure. His physical ailments were nonetheless real, however, and the physicians who treated him prescribed some very potent drugs that likely damaged his liver or his kidneys, or both. By 1868, he had begun to contemplate finding another way to increase his income, but he wanted to maintain his association with the College of Charleston. Thus, if Holmes resigned, McCrady could take his place as curator of the museum, give up his teaching position, and find a more lucrative job. After consulting with his father, whose approval he frequently felt he must have, he made an arrangement with James Adger, a prominent Charleston entrepreneur, to launch an oyster-farming business. This proved to be a mistake, as McCrady was more interested in doing research on oysters than in the commercial enterprise, while Adger expected a good return on his investment. That McCrady stuck with the business from the time he resigned his faculty position in the spring of 1869 until February 1872 was due to his desire not to fail his father and to his father's strong influence upon Adger.[22]

McCrady clung to his faculty position until he got the oyster business under way, much to the dismay of the college trustees. Meanwhile, in late January, he began to act as the museum curator. At once he found great difficulty in labeling specimens, a job left unfinished by Holmes. He called upon Gibbes for assistance, especially with the Crustacea, and eventually, at the insistence of Gibbes, he employed Alexander Hume, an elderly physician and former professor at The Citadel. The old man had no income, and Gibbes prevailed upon McCrady to pay Hume the meager sum of $200 per annum set aside for an assistant. McCrady had wanted to hire young Gabriel Manigault, who was well qualified for the position and who eventually succeeded McCrady when he resigned as curator, but since he needed the help of Gibbes, he decided to give in to his request. The mammals proved to be especially vexing for McCrady, as he knew little more about them than did Agassiz.[23] In all likelihood, he could have received help from one of the best mammalogists in the country if he had called upon Bachman, but McCrady probably could not allow himself to think of calling for the services of the man who had dared to criticize his mentor Agassiz.

In keeping with his prewar vision of a South with great institutions of research, McCrady apparently thought of modeling the Charleston Museum after the one established by Agassiz, that is, the Museum of Comparative Zoology (MCZ) at Harvard. It did not take long for him to realize that he would be lucky if he could just keep the museum alive. Nevertheless, unlike Holmes, he viewed the museum as primarily an institution for research and only secondarily as an educational institution. Thus, he complained when he had to open it to the public on Saturdays, because he had to watch children during their visits and answer questions when adults made inquiries. Actually, he was not himself collecting for the museum or using its collections for research. Instead, he was working on his law of development. He did persuade the Elliott Society to turn its collections over to the museum, and he made a concerted effort to label specimens, or at least to see that Gibbes and Hume labeled specimens. Like Holmes, McCrady understood the importance of collection data, and he complained that some of them could not be recovered without the aid of Holmes. In fact, he finally swallowed his pride and asked for help from Holmes, who, despite the mistreatment he had received, readily complied. Of course, Holmes was not aware that McCrady had several times suggested that the former curator had taken items that belonged to the museum, nor did he know that McCrady felt insulted when he read in the *Charleston*

Courier in August 1869 that he had been invited to give an address at "the Humboldt Festival" only after the organizing committee had first invited Holmes to do so. The *Courier* had made a trivial mistake, but it was significant to McCrady, who insisted that the editor print a correction.[24]

Like Gibbes, McCrady wanted to revive the Elliott Society, and he did his part in trying to put it on steady feet again. After locating a few of the Society's papers delivered before the war, he prepared them for the second volume of the Society's proceedings. In addition, he secured a promise from Holbrook to write descriptions of some fishes for the second volume of the Society's journal. Holbrook never fulfilled his promise, but even if he had, the Society had no funds for publishing the journal. Indeed, two decades passed before the Society was able to publish volume 2 of its proceedings. Meanwhile, McCrady presented several papers before meetings of the Society, often with only a half dozen, or usually fewer, members in attendance. Few of his presentations dealt with original research, however. Although Mc-Crady did use a microscope to study specimens during the period from 1866 to 1872, he was much more interested in speculative subjects, one of which was an intriguing hypothesis that there is "a sort of atmospheric Gulf Stream" that flows constantly between the equator and the North pole and by meeting with its returning current creates "an endless succession of cyclones," some of which become storms on the surface of the earth. After developing the idea into a paper, he sent it to Joseph Henry for a Smithsonian Institution publication, but Henry declined it as being only hypothetical. McCrady knew that the idea needed confirmation, but he sent the paper on to the British journal *Nature*, which published it on November 10, 1870. A year later, as a member of a committee to study the causes of yellow fever, McCrady wrote a stimulating paper in which he attributed the cause to a germ and advocated disinfectants and freezing as ways to kill it. A leading Charleston physician, Eli Geddings, strongly objected, saying that no one "*out of the profession*" of medicine was "capable of solving the mystery."[25]

Also important to McCrady was the question of racial origins and differences, and he drafted a number of comments on that subject, though he published none of them. Of course, the emancipation of blacks presented new problems for him after the Civil War, and although he drafted a few speculative commentaries on the origins of races, he was much more concerned during the first six years after the war over miscegenation and over keeping blacks under control. In May 1869, for example, he complained in his diary that "a low white Radical miscegenationist . . . brought me a negro applicant for the [oyster] carrying trade, who attempted to take a seat on my sofa, [but] I ordered him up and made him stand for the rest of the interview." Later, he noted that the sexton of St. Philip's Church had been criticized for forbidding black pallbearers to sit in a pew during a funeral service for a black person. The critics, opined McCrady, are endeavoring "to drive us to miscegenation." Blacks, he added, are incapable of ever becoming equal to whites in intelligence and morality. More pressing in 1870, in McCrady's view, was the issue of allowing Chinese immigrants to enter South Carolina to work on farms. At a meeting of the State's agricultural convention in early May 1870, delegates debated the issue of encouraging the immigration of Chinese laborers. McCrady voted against a proposition in favor of immigration, later recording in his diary that "I clearly perceived how dangerous such a movement is likely to be to us by . . . bringing in another inferior race." Soon thereafter, he wrote a letter to the *Charleston Courier*, expressing opposition to Chinese immigrants.[26] Some people in South Carolina did not share his view about forbidding Chinese immigrants to settle in their state, for they believed Asians would serve as more reliable laborers than the freed blacks.

McCrady continued to labor diligently to work out his law of development, and he endeavored to get some scientists to read his manuscript and to offer comments. Among them was the noted English philologist Friedrich Max Müller, who said it should be published. In fact, Müller tried to get a British publisher to do so, but he had no success. McCrady asked Gibbes to note any flaws in his argument, but the latter showed no interest in the subject. After trying again to get his colleague to review his paper, he gave up, as Gibbes "intimated that my frequent coming was something of a bore." Offended by his colleague's attitude, McCrady then contacted Charles Venable, a professor of mathematics at the University of Virginia, who merely said McCrady's essay was "ingeniously presented." Undaunted, McCrady sent the manuscript to the editors of the *American Journal of Science*, but they declined it. "Perhaps they think me suffering under some kind of aberration of mind," recorded McCrady in his diary on September 20, 1870. He believed that no one had yet grasped his argument, but he would not give up. His law of development, he firmly believed, would show that there is a type of evolution, and it would replace the materialistic scheme devised by Darwin. At least Agassiz shared his concern about the implications of Darwin's theory, and McCrady thought for awhile that his old teacher would eventually subscribe to his law of development. In fact, when Agassiz visited Charleston in late April and early May of 1869, he asked McCrady to write out some of his thoughts as they pertained to race. Agassiz was no doubt pleased to read McCrady's views that "we find the races distributed according to a law [the law of development]" and that "the highest and most developed races occupy the center" of "an irregularly oval tract of the most valuable portion of the Earth's surface," which, as he saw it, extended from the Euphrates River westward to the Mediterranean Sea, while the "aberrant and degraded forms" lived outside that special oval. The superior, he added, always triumph in "the struggle for existence," apparently oblivious to his use of a Darwinian term. Later, after hearing that Agassiz planned to write a book in opposition to Darwin, McCrady noted in his diary that he suspected Agassiz planned to steal "my thunder, unacknowledged." He need not have worried, however, for Agassiz never wrote the book. Actually, Agassiz held McCrady in very high regard as a hydrozoan zoologist, and he was thinking about trying to find a position for the Charleston naturalist in the MCZ.[27]

On March 29, 1873, only a month after McCrady gave up the oyster-farming business, he received a letter from Agassiz, saying he expected soon to be able to appoint him to the MCZ staff. McCrady was delighted, though he reminded Agassiz that for the last thirteen years he had not worked actively in hydrozoan zoology and thus was "entirely behind the world in information." He also noted that "my opinions, my religious convictions, and my opposition to Darwinism are all alike unpopular." On April 7, however, McCrady notified Agassiz that he would accept the position, even though it had not been clearly defined. "I may yet be able to do something for the advancement of genuine Science, and at the same time strike a sturdy blow for religion," he said. A week later, Agassiz replied, expressing pleasure over the acceptance and specifying that McCrady would instruct students in zoology and serve as an assistant curator in the MCZ. He expressed concern, however, over McCrady's statement about religion, saying, "I hardly think these topics belong in the lecture room beyond a general

discussion of principles, [for] . . . tenet teaching is no part of science."
McCrady hastened to assure him that he had no intention of introducing religion in the classroom, that he meant only that the position would give him "the opportunity . . . of combating the current erroneous views of (so-called) 'Evolution' . . . on scientific grounds." He added, "the idea of making tenet-teaching a part of instruction . . . never entered my brain." Satisfied, Agassiz urged him to come on in time for the summer term, but McCrady deferred his departure until September.[28] Of the circle of Charleston naturalists, then, only Lewis Gibbes was left, but, through McCrady, the prospect of continuing its legacy seemed promising.

Last Links

The stylish attire of Cambridge pedestrians likely contrasted sharply with the threadbare suit worn by the impoverished southerner John McCrady as he rode into the city in the fall of 1873. A student there some twenty years before, he would now be an associate of the master naturalist who had influenced him so strongly. McCrady was proud of the recognition symbolized by the appointment, but he was filled with apprehension because he had been out of touch with developments in his field. Never prone to waste time or even to enjoy leisure, McCrady was sure he could soon catch up. He learned quickly, however, that a weary and ill Agassiz intended to assign to his new colleague every duty he felt unable to perform in the Museum of Comparative Zoology. In addition, McCrady must prepare for teaching two courses in zoology. "I have had suddenly thrown upon me without preparation all the lectures delivered to undergraduates at Harvard in Zoology," he wrote to Gabriel Manigault, his successor as curator of the Charleston Museum, "and my whole time has been taken up in informing myself what has been done [in zoology] since 1860." Within three months after his arrival, moreover, the situation took an unexpected and worrisome turn, for on December 14, 1873, Agassiz died, shortly after suffering a stroke.[1]

Someone, probably Elizabeth Cary Agassiz, who had asked Mc-Crady a few days earlier to present lectures for her ill husband, invited McCrady to give an address at a memorial service for the deceased naturalist. Honored, McCrady gladly obliged. In a lengthy tribute to his old master in science, McCrady said, "Yesterday we laid in the grave one of the greatest naturalists, not only of his own day, but of all history, a philosophical inquirer into the Order of Nature, whose name will go down with those of Aristotle, Linnaeus, and Cuvier." Of Mc-Crady's esteem for Agassiz there could be no doubt, and as a devoted disciple, he heaped paragraphs of praise upon the man who had helped him rise to recognition. But he also used the address to pronounce his dislike of Darwin's theory; to criticize, indirectly, Agassiz's critic John Fiske, an outspoken advocate of evolution; and to promote his own idea of the law of development, using Agassiz as his vehicle. Defending Agassiz against his pro-Darwin critics, McCrady asked, "Has the so-called 'Evolution Theory' become a very standard of Scientific Orthodoxy, so that it is heresy to review the facts as they are and to demand whether those facts warrant a belief in transmutation?" Neither he nor Agassiz rejected the whole of Darwin's idea, he added, and both recognized the possibility of integrating it with the "true doctrine of Development." As it stood now, however, Darwin's theory was an "utterly unsatisfactory" explanation. Indeed, he said both he and Agassiz recognized it as a hypothesis, not as a theory. Thus, as only a partial explanation, he argued, it left unanswered "the great question what is the Universal Law of Development." Until scientists had found a satisfactory answer to that question, concluded McCrady, any theory is pure speculation. In the address, McCrady had endeavored to distinguish between his views and those of Agassiz, and Agassiz's widow later told him he had done so, though she admitted she did not understand all of his comments. In fact, it is likely that most of his audience understood neither his law of development nor Darwin's theory and that they simply viewed his comments as esoteric praise of Agassiz.[2]

"No one can see," said McCrady in a letter to his father soon after the death of Agassiz, "how this unexpected event will affect my future." He was moderately optimistic, however, for a few days later

Alexander Agassiz told him that, after a brief discussion with Harvard's president, Charles W. Eliot, the name of Count Louis de Pourtalès stood "first on the list of [Museum] Assistants and mine second," for appointment as director of the MCZ. McCrady believed that Alexander Agassiz desired the appointment of Pourtalès as the chief curator of the MCZ and McCrady as professor of zoology. As the winter of 1874 progressed, however, McCrady began to grow more apprehensive about his future at Harvard. His demanding expectations created tension with the students in his senior section of zoology. Eliot initially supported him, but changed his mind after members of a supervisory committee reported that McCrady demanded too much from his students. Eliot suggested that McCrady ease up a bit, but McCrady did not follow his advice. At some point, the dean of the faculty, Ephraim Whitman Gurney, who was sympathetic toward McCrady, urged the southerner to relax his expectation that students in an introductory zoology course should know as much as experts in the subject, but he found McCrady "deaf to all arguments." In fact, said Gurney, "I talked as to one to whom the word of the Lord had come recently." The problem stemmed in part from McCrady's intensity, but it was also due to his perception of the "Yankee" students as possessing "intensely all the faults of their people," in whom "treachery is . . . a constitutional trait." Although he confined his remarks to his diary, McCrady likely conveyed something of his feelings through his actions. The opposite was also likely, that is, that his students did not approve of his southernness, a feeling that seemed evident as well in the way McCrady was treated by the Harvard librarian, who, on January 31, told him that the bag he was carrying was "suspicious and that if they did not know me . . . he would suspect me of intending to steal books." As McCrady correctly noted, "nothing could have been more insulting." The librarian continued his subtle form of harassing the unfortunate fellow. McCrady might have avoided going to the library often had Alexander Agassiz continued to allow him to borrow books from his personal library at the Museum. After a report that McCrady had left one of Agassiz's books overnight in the laboratory, however, Agassiz demanded that he return all of the books he had borrowed. The report was erroneous, but McCrady felt that he was "subjected to mortification through continued dependence on others." Later, Agassiz removed to his home some of the books McCrady needed most. As

McCrady viewed the matter, and not without some justification, the decision "looks like a settled design . . . to keep me in a subordinate position."[3]

From January to July, McCrady was not sure where he stood in the scheme of things. If he could believe Agassiz, his future at Harvard was bright. In late January, for example, McCrady's friend St. Julien Ravenel told the McCrady family in Charleston that he had received a letter from Agassiz, indicating that Agassiz and other MCZ and Harvard colleagues were pleased with McCrady and that he was to be the "successor" of Louis Agassiz. In fact, Agassiz was encouraged that McCrady had spoken of publishing a textbook in zoology, but what he meant by "successor" was not as McCrady and his friend and relatives imagined. They apparently believed that McCrady would be appointed as director of the MCZ, but Agassiz had in mind that he would become the professor of zoology. In fact, in late February, Ravenel reported that Agassiz, who had been in Charleston only days before, "thought there was no doubt that Johnnie would be elected in April to the Professorship of Zoology," but, according to Ravenel, Agassiz also said that McCrady "would have control of the Museum in some way." Ravenel noted, however, that Agassiz had not said that McCrady would be the curator. In mid-April, Agassiz told McCrady directly that Eliot had said he would likely be appointed as professor, but his salary would not be increased. McCrady confided to his diary that "this is impossible," for "books, etc. alone would cost $600–700 a year." Since no word of his appointment had come by June 2, McCrady went directly to Eliot and showed him Louis Agassiz's "first letter" to him, but Eliot said that Louis Agassiz had never mentioned McCrady as his successor. Eliot also informed McCrady that his salary would be the same, that is, $2,500 per annum. Although he did not tell Eliot so, McCrady believed that he and Nathaniel Southgate Shaler, who was appointed to take Louis Agassiz's place in geology, should each receive one-half of Agassiz's salary of $7,000. Six days later, Alexander Agassiz told McCrady that the Harvard Corporation would appoint him at its meeting on that very day. He also informed him that Eliot would add $500 a year to the salary for three years, as he had received a contribution for that purpose from "a friend of the College." McCrady noted in his diary that he asked Agassiz if he planned to succeed his father, in which case he would not seek the appointment. Agassiz, according to McCrady, said he was not seeking the position.[4]

A month passed before McCrady heard from Eliot, by which time McCrady was in Charleston for the summer. "You have been duly appointed and confirmed Professor of Zoology in this University," wrote Eliot. The salary was to be $2,500 per year for three years. Eliot stressed that the appointment was for three years, although the Corporation could make the position permanent at the end of that period if it chose to do so. Eliot added that funds from a private source made it possible to supplement McCrady's salary by $500 a year for each of the three years. McCrady sent off a letter of acceptance. A few days later, however, he told Eliot that the arrangement gave him "food for much anxious thought." The income, he noted, was barely enough for supporting his family and left him nothing to spend on research. He expressed particular concern over the three-year term of the appointment and said he accepted the offer under that condition only because "a Higher and Wiser than myself guides and governs my concerns." In fact, Eliot was already thinking that he probably would not reappoint McCrady at the end of the three-year term unless he received more favorable reports about McCrady's teaching and unless McCrady renewed his research in zoology. McCrady did, in fact, make an effort to resume his original research during August, when he began to study some species of Scyphomedusae. Once he was back in Cambridge, however, he discontinued the work.[5]

During the fall of 1874, McCrady found his students to be more receptive to his instruction. He was especially effective with his "special students," or those pursuing an advanced degree. Some even sought his views on evolution, and McCrady began to "strike a sturdy blow for religion" by telling one of them that in a quarter of a century no one would accept the Darwinian theory. He told another of them that "zoology could no more do without metaphysics than metaphysics without zoology . . . [and] that the process of research could never invalidate any part of religious truth handed down to man." For McCrady, revelation in religion and research in science could never conflict, a view that would have made John Bachman proud of him. His conservative views were not confined to religion, however, and in late December 1874, when he learned that Eliot intended to drop studies in the humanities and knowledge of Greek as requirements for the Ph.D. degree, he tried to persuade Eliot of their value. Eliot was unbending, and in his diary McCrady wrote, "I am more than ever impressed with the idea that my own schemes for the advancement of real study are so much higher than those entertained here, that I shall be hated for holding them."[6] Whether he was right or wrong, John McCrady was in the wrong place.

By early January 1875, the unfortunate southerner had begun to suffer bouts of illness again. He complained that the cold weather in Cambridge was the cause, but it was really only a factor in exacerbating his misery. On at least two occasions he had to sit in order to continue his lectures. Headaches, weakness in his limbs, drowsiness, and pain in the abdominal area were the typical complaints until the end of March, when his legs began to swell. The swelling spread upward and eventually reached his arms. His skin became so sensitive that the mere touch of his clothes caused pain. Two physicians diagnosed the cause as enlargement and malfunction of his liver and could offer no remedy. McCrady struggled through his lectures, but he could do no work in the museum during April and May. He found some consolation through the support of two of the younger Harvard faculty—Shaler and William James, the latter an instructor in physiology. Although McCrady had originally hesitated to develop a friendship with Shaler, because that Kentucky native had served in the Union army, he later confided in him. Both were "restive," McCrady reported in mid-February, because Alexander Agassiz had listed them in an MCZ publication as "Assistants in the Museum," and each believed he should be listed as a member of the museum faculty. McCrady maintained that Agassiz wished "to make us appear secondary" to him and to some of his longtime curators. In fact, following the pattern set by his father, Agassiz intended to keep his subordinates in place. A few weeks later, in early March, Shaler cautioned McCrady against opposing Eliot on the issue of dropping Greek as a requirement for the Ph.D. degree, noting that the Harvard president viewed disagreement as "personal opposition." McCrady ignored the advice, however, thinking he could yet persuade Eliot against the proposal. On March 6, he went to see Eliot about the matter and argued that the omission of Greek was the design of such men as Thomas Huxley "to cut modern culture loose from its moorings in the past, especially the Christian past." The argument carried no weight with Eliot, who dismissed McCrady's charge against Huxley. William James liked McCrady and offered him advice—both about his research and his teaching. James suggested that McCrady could help himself by returning to new research in zoology. Moreover, he hinted that McCrady should lighten his expectations for students in his zoology classes.[7]

Meanwhile, despite his illness, McCrady impressed his graduate students with his knowledge of the lower invertebrates, especially the hydrozoans and the scyphozoans. Among the ablest of them were William Keith Brooks and Jesse Walter Fewkes. On several occasions, McCrady invited his advanced students to his home for evening discussions on, as he viewed it, philosophical matters that pertained to zoology, including the theory of evolution. He did not always believe that he was succeeding in inculcating his own views, however, and on one occasion noted in his diary that one of his students "is in love with the Darwinian theory." In fact, nearly all of those students held a favorable view toward Darwin but avoided any mention of it in McCrady's presence. Brooks and another student believed they could profit from doing field and laboratory work with McCrady during the summer of 1875, and they offered to come to Charleston for that purpose. The experience would have been profitable for both McCrady and the students, but for unknown reasons the students did not make the trip. It was the last opportunity for Brooks to work in the field with McCrady, for he would receive the Ph.D. degree at the end of the next spring. The summer of 1875 passed without McCrady doing any scientific research. Instead, he drafted comments on speculative subjects.[8]

Back in Cambridge in late September, McCrady found that only a handful of undergraduates had registered for his courses Natural History 4 and Natural History 5. He complained that students were flocking to the easier elective courses. Whether or not he realized it, he would be the subject of special scrutiny during this year at Harvard, concerning both his teaching and his research. By November, McCrady was spending more time than ever since coming to Harvard in making microscopic studies, but they came to naught as another bout with illness began in December and as McCrady could not let go of his efforts to work out the problems of his law of development. Meanwhile, reports on his teaching by members of the visiting committee did nothing to help his case. James sat in his class several times and, wishing to help McCrady, frankly, and correctly, told him that his lectures were much too detailed for undergraduates. Soon thereafter, Agassiz put it more bluntly. He and two other members of the visiting committee had reviewed McCrady's examinations and concluded that they contained questions about things "which nobody in America beyond two or three knew anything about." Agassiz also noted that McCrady used illustrations of animals unknown in the United States, to which McCrady replied that it was not the specific animal but its structure that counted. In his diary, he observed that Agassiz did not like his independence, but he made no comment about the criticism regarding esoteric knowledge. Two days later, he recorded his view that Agassiz had "completely reversed his course with regard to my plan of teaching . . . [which] is precisely the same as that with which I started and which he admitted was the only possible plan." In his judgment, he had "made enemies" by refusing to compromise his standards and, as a result, Eliot would try to get him to resign. On the next day, he discussed the matter with James, who repeated that his courses were too difficult and that he should be aware that Eliot ran the university "as a Superintendent works a factory."[9]

Also among those in whom McCrady confided was Francis Wharton, a professor in the Episcopal Theological School in Cambridge. On December 5, McCrady told Wharton that he sensed increasing opposition to him because of his high standards of instruction and believed that Eliot would probably ask him to resign. Prospects for McCrady did not look bright, and on top of the criticisms of his teaching came a return of the swelling in his legs. In late December, his MCZ colleague Theodore Lyman visited one of his classes and afterwards suggested that it might be good if McCrady taught only graduate students. It was a reasonable suggestion, for McCrady apparently did an excellent job with those students. Lyman had also observed that the penniless professor did not own an overcoat, and on December 24, he brought him a good secondhand one. It was a genuine expression of regard for McCrady, but, quite naturally, the proud southerner believed that it only enhanced his dependence. On the same day, Elizabeth Agassiz sent Christmas gifts to the McCrady children. Although McCrady believed he and his family were the objects of pity, the truth is that several people in Cambridge liked him and hoped to save him from injuring his future at Harvard. Quite ill in late December, McCrady persuaded his physician to tell Agassiz that he could not survive the winter in Cambridge. Within hours, Agassiz was at the bedside of McCrady, "show[ing] some of the same warm feeling to me, which he showed when I first came here." Agassiz insisted that McCrady go to Charleston, and he promised to find a way to pay his salary for the entire term. When McCrady replied that he did not even have enough money for transportation, Agassiz said he would personally provide a $500 advance on McCrady's salary. McCrady replied that he would accept only if he could sign a note. Agassiz insisted, however, that he did not want a note and added that "it would make no difference" if McCrady were never able to repay the loan. Struck by the genuine altruism of Agassiz and too ill to remain in Cambridge, McCrady accepted.[10]

On New Year's Day 1876, Agassiz called on McCrady and reported that Eliot and the Corporation had rejected his request for leave as it would only encourage other faculty to make similar requests. The flimsy explanation justifiably reaffirmed McCrady's belief that Eliot wanted to push him out. Agassiz then told McCrady that he thought it best for him to seek another position and promised to recommend him to Johns Hopkins University. On January 12, Agassiz wrote a letter to Daniel C. Gilman, the president of Johns Hopkins, recommending McCrady, especially for work with graduate students, with whom "he has taken endless pains" at Harvard. Aside from European zoologists who "occupy the summit," he added, "I consider McCrady as a most desirable man." Moreover, Agassiz continued, "I am utterly at a loss how to replace him." He stressed that McCrady had to leave only because of the cold climate of Cambridge. Finally, on January 13, McCrady felt well enough to depart, but one of his students had to lift him into the carriage that would take him to a ship bound for Charleston. To McCrady's great delight, Pourtalès, James, and two other colleagues came to bid farewell "very warmly." McCrady arrived in Charleston on January 18, but he was too ill to record much in his diary until June 27, when he noted, "I have not done as much as I might perhaps in the way of Nat[ural] Hist[ory] research." During the next two months he was well enough to make an effort, collecting and studying sea jellies. One of the specimens he collected was actually in the class Cubozoa (commonly called box jellies or sea wasps). He had collected a specimen of the same species in 1858 and made notes about it in his "Journal of Observations," which was burned in Columbia in 1865. Had he published a description in 1858, he would have earned priority, but when he described it in 1876 as *Cheiropsalmus carolinensis*, he had been preceded by Fritz Müller, who had described it in 1859 as *Chiropsalmus quadrumanus*. McCrady also collected another species of Cubozoa during the summer of 1876, which he described in a manuscript as *Cheiropsalmus vitreus*. His comment that the animal had a "brilliantly glassy appearance" suggests that it was *Tamoya haplonema*, the only other box jelly normally found in temperate western North Atlantic waters. It had also been described by Müller in 1859.[11] Of course, McCrady had been unable to keep up with the literature on the cnidarians, but had he persisted in doing original research, there seems little doubt that he would have made further contributions.

Upon returning to Harvard in late September 1876, McCrady immediately went to the MCZ to see how many students were enrolled in his courses. The result was depressing—two in the first course and one in the second. Two more students signed up for the latter soon afterwards, but such small numbers did not bode well for his future at Harvard. Agassiz had already informed him that another lengthy illness would certainly result in his termination. McCrady did not help his case a few weeks later when, at Wharton's invitation, he began to present a series of lectures on "Relations between Science and Religion" to the students of the Episcopal Theological School. William James cautioned him "not to talk beyond what they could understand," and McCrady took the warning as an indication that James was telling him "this is my fault as a teacher." Still, McCrady's lectures were well received at the seminary, and several of the faculty contributed to the purchase of "a very costly fur overcoat" for him "as an expression of their warm regard and their sense of my value to this community as a Charleston scientist." Unlike the instance of the secondhand coat given to him a year earlier by Lyman, this gift struck McCrady as an honor, not as charity. Five days later, he called Lyman to his office at the museum and asked him to take back the first overcoat, saying he had "never worn it" because he viewed it as a poor way for the university to compensate him for an inadequate salary. He expressed to Lyman a hope that he would not be offended, but it is likely that Lyman viewed McCrady as ungrateful.[12] Certainly, McCrady had not cemented a bond where he needed it most.

Attracting a number of followers at the seminary, McCrady continued his free lectures on into January and February of 1877, but some of the more liberal theologians on the faculty were dissatisfied with his views, and a few considered him to be a bigot. The latter found further confirmation of their belief in late February when McCrady "refused in the most positive manner to meet . . . the negro priest of our [Episcopal] Church." He said that blacks were "unfit" to serve in any administrative functions of the Church. Unfortunately, McCrady also went out of his way to defend the customs of the South, and in late February he left with Eliot an article he had drafted on the subject. Eliot later read the article and sent McCrady a four-page note, urging him not to publish the article and telling him it could only create bad feelings. Such, McCrady told Shaler, was what he had expected of "men who are considered thoughtful, cultured, and conservative in Massachusetts." Unable to allow Eliot's comments to pass without a response, McCrady composed a letter to the Harvard president and asked Shaler to review it. Shaler agreed that Eliot had misread McCrady's paper, but he told his colleague that the letter was "too spicy" and should be "toned down." McCrady took the advice and rewrote the letter. What had incensed McCrady most was Eliot's charge that McCrady was condoning the continuing desire of southerners "to break up the Union." McCrady told Eliot that he had "said nothing of the kind," that he had simply referred to "the *ideal* of the Southern people, . . . [for] *honest democratic government.*" That ideal, he added, "will infallibly survive all others in the struggle for existence, . . . [for] it is the only ideal consistent with morality, good sense, and true patriotism." He accused Eliot of misrepresenting his views, and said that "the Northern people are . . . not yet prepared to look the whole truth in the face." His goal, he said in conclusion, was to apprise readers in the North that the federal government was creating anarchy in South Carolina by keeping "the civilized white people, . . . surrounded by savages," from restoring law and order.[13]

Eliot did not respond, but it is certain that he viewed the matter as yet another mark against renewing McCrady's appointment. In less than a month, on March 26, 1877, McCrady received a note from Eliot, informing him that "the Corporation had determined to make no appropriation" for his post thereafter, mainly because his instruction was too specialized for undergraduates. Eliot added that the Corporation did not believe it was practical for McCrady to alter his approach. McCrady viewed the termination as a reversal of Eliot's promise that "my professorship was for life," and he vowed he would tell Eliot just that, though he was sure the president would "deny the fact." Of course, McCrady was still clinging to the belief that Louis Agassiz would not have brought him to Harvard on a trial or temporary arrangement, and he was overlooking the letter in which Eliot had specified that reappointment at the end of three years was not guaranteed. In a meeting on March 24, the president had told McCrady that one of the reasons for not renewing the agreement was that McCrady had "published no new researches" in two decades and was therefore "no better known" since then. To his father, however, McCrady averred that the real reason was that "after four years here, I [am] still absolutely and wholly a Southerner and that I have in no way concealed this." Indeed, he had not concealed his views, but, while they were doubtless unpalatable to some New Englanders, and perhaps even to Eliot, they were not the basic reason for the decision. Nor, as he wrote to his father a few days later, was the reason that he had provoked "the liberalism of Massachusetts" by championing religious "orthodoxy" in a region of "infidel circles." If his defense of traditional Christianity amounted to bigotry, he added, then he "thanked God for enabling me to be a bigot."[14]

McCrady quickly sought out Alexander Agassiz for his view, and Agassiz said that Eliot had deceived McCrady. Whether or not he was misinformed or only trying to assuage McCrady's feelings, Agassiz said that two members of the Corporation told him the decision was "merely a necessity of retrenchment and not any fault" of McCrady. Endeavoring to avoid any notion that he had been fired, McCrady submitted a notice of his resignation on March 26, "to take effect on and after the first day of September next." Also wanting to get on record his view of what led to his decision to resign, he added, "there is an irreconcilable difference of opinion between the University Authorities and myself as to the scientific standard to be maintained in the study of zoology, . . . and I am conscientiously unable to adopt a more popular standard." Shaler and James assured McCrady that Harvard would look bad because it was condoning low standards. Both talked with Eliot, but they reported conflicting views to McCrady. Shaler thought he had persuaded Eliot to reverse the decision, but James offered little hope. Wharton said the resignation "would be ruinous," for he considered McCrady to be "a man of genius." The only hope, in McCrady's view, lay with Agassiz and Lyman, for he believed they could get Eliot to change his mind. James talked with Agassiz, and he assured McCrady that Agassiz supported his reappointment and that he would urge Eliot to reconsider. Whether or not Lyman tried to assist is unknown, but any efforts by the once-sympathetic supporter would likely have been lacking in enthusiasm for his unbending colleague.[15]

Meanwhile, on April 5, 1876, McCrady presented one of the Harvard Natural History Society's "Popular Scientific Lectures." As headlined by the *Boston Daily Advertiser*, the address was on the topic of "What Jelly-Fishes Can Teach Us," but McCrady also used the subject to offer criticisms of Darwin's theory, which, he declared, "rests for verification absolutely and solely upon circumstantial evidence." Around 700 people attended the lecture, and McCrady, who spoke without notes, was "frequently applauded." William James was there, and he told McCrady that the lecture was "capital" and that his criticisms of Darwin were "perfectly true." Wharton was equally impressed and complimented McCrady as "one of the clearest extemporaneous lecturers on abstract points he had ever heard." McCrady was worried, however, that after he left Cambridge stories would circulate that he had been dismissed "for scientific incompetency," and he wrote to his brother Edward that "I must try to publish some researches before I leave here." No one worked harder than did James to get the Harvard Corporation to retain McCrady, as he was convinced that McCrady had been guaranteed a lifetime appointment. The members of the Corporation did not change their minds, however, and on the day after their vote, James went to see McCrady to give him the bad news. He was able to inform McCrady that the Corporation had voted in favor of paying McCrady his salary for an additional quarter. Then, in McCrady's presence, James tore up a $300, postdated check written by McCrady to repay a loan to him. James told McCrady to repay him only when he was able to do so. James also drafted a note for the *Boston Daily Advertiser* and for *Nature*, as a "true statement" of the reason for McCrady's resignation.[16]

After reading the statement in the *Advertiser*, the noted Boston naturalist Alpheus Hyatt told McCrady that he was astonished to learn of his resignation. Hyatt opined that Eliot would have to respond openly to McCrady's letter of resignation. It is likely that McCrady derived special comfort from Hyatt's praise of him for his religious conviction. A leading proponent of the neo-Lamarckian school of evolution, Hyatt told McCrady that he had done much to reinforce his belief in "a personal God" and that he was "much less of a Darwinian than he used to be." Several of the other professors, curators, and assistants in the museum likewise expressed regret over McCrady's resignation. Meanwhile, Shaler urged McCrady not to write a letter to the newspaper because it might dampen the possibility of his "recall."

In addition, Shaler said he believed the Episcopal Theological School would likely offer McCrady a position, but McCrady feared that he had "stirred up the Liberals" too much. Back in Charleston, Mc-Crady's father, his brother Edward, his relative William Henry Trescott, and St. Julien Ravenel thought McCrady had made a mistake by resigning. Trescott aptly ventured the opinion that McCrady had used the resignation to become a martyr. By early May, it was becoming apparent to McCrady that the situation was settled, but he still hoped he might be appointed to the faculty of the theology school. Wharton told him that he knew of no active effort and suggested he should pursue the position he had heard of at the University of the South in Sewanee, Tennessee. McCrady took his advice, and the Episcopal institution quickly invited him to come for an interview.[17]

On June 1, 1877, McCrady presented his final lecture at Harvard, and during the next few weeks prepared for his visit to Sewanee. When he returned the last of borrowed books to the Harvard library, he did not escape one final insult, namely an accusation that he still had one book out. He walked to an alcove, pulled the book from the shelf, and took it to the librarian, but he received no apology. On June 26, McCrady departed for Charleston. While in the city, he visited Lewis Gibbes, now nearing four decades of service to the College of Charleston. After hearing McCrady's story, Gibbes commended him for holding to his high standards. On July 4, McCrady departed for Sewanee. For the next five weeks he was "very handsomely treated" by the University of the South, and he wrote joyfully to his wife in Cambridge that "The people are all Southern people, the atmosphere is Southern, and the culture . . . [is] excellent." He could not refrain from telling Sallie that large crowds were attending his lectures and that the response to them was excellent. By the end of July, it was certain that he was in great favor, for the head of the university named him among those he considered as the great scientific defenders of Christianity. Meanwhile, Shaler had sent a letter to the University of the South, praising McCrady as "the ablest philosophical Naturalist now living."[18]

It came as no surprise to McCrady, then, that on August 3 the trustees appointed him as professor of biology and the relations of religion and science. Elated, McCrady wrote to Sallie that he had been "elected Professor here . . . with great enthusiasm" and that someone had even compared him favorably with General Robert E. Lee. The salary was low, only $1,500 per year, but he was expected to teach for only five months of each year and was free to do whatever he wished during the rest of the year. On top of his success in finding a suitable position, McCrady was uplifted by the comments of the Irish scientist George J. Allman in the May 10th issue of *Nature*. A noted authority on hydroids and professor at the University of Edinburgh, Allman referred to McCrady as "an eminent original worker in an important department of zoological research" and expressed regret that he had received no support in his efforts "to raise the standard of zoological education." The hand of William James is evident in the latter part of Allman's comment, but readers were unaware of that. Accounts in some American sources had attributed the resignation of McCrady to his criticisms of the Darwinian theory, and, when asked by his student Fewkes if that were true, McCrady simply replied that he had made no "open opposition."[19] In fact, his criticisms of Darwin's theory had nothing to do with the decision to terminate his position at Harvard.

By the end of August, McCrady and his family had settled in Sewanee, and a few days later McCrady was in Nashville for the annual meeting of the American Association for the Advancement of Science. He presented his paper on *Cheiropsalmus carolinensis* (= *Chiropsalmus quadrumanus*), but it is likely that no one there knew that the species had previously been described. At least he was making an effort to become professionally active again. Among the papers he was eager to hear was one by the renowned paleontologist and zoologist E. D. Cope, on the subject of intraspecific variation. At the end of the presentation, McCrady whispered to T. Sterry Hunt, a noted American geologist, that he had treated the topic even further than Cope. Hunt urged him to rise and comment on the matter, which McCrady did. After four minutes of talking, however, the session chairman, William H. Dall, a former student of Louis Agassiz and a leading authority on mollusks, said he had spoken overtime. Cope responded to McCrady's comments, after which McCrady rose again and briefly stated that he agreed with Cope's idea. When the session ended, A. R. Grote, director of the natural history museum in Buffalo, New York, told McCrady that Dall had stopped him solely because he was "not in harmony with modern scientific thought and had left Harvard on that account." Probably there was some truth in Grote's statement.[20]

Back in Sewanee, McCrady was astounded to find that at least a few southerners were inclined to make some changes in their relationship with blacks, and his reaction was ugly at best. In the inn where he and his family were lodging temporarily, a black servant sat at the table to take her meals with the guests and was served by a white woman. McCrady left in disgust, and immediately thereafter wrote a note to the proprietor, saying that he and his family would take no more meals in the inn if the black servant were allowed to sit at the table. John McCrady remained unrelenting in his desire to preserve southern customs. Apparently, no one in Sewanee openly opposed his views. McCrady had already endeared himself to the leaders of the university, and through a series of evening lectures for the public, he was gaining greater esteem every day. The favorable environment prompted him to resume work on his law of development, for which he began to read more widely—in works of metaphysics, linguistics, mythology, religion, and physics. It also incubated his increasing intention to refute Darwin's theory. One bold critic asserted that he failed to see what McCrady's lectures "have to do with Zoology." McCrady replied that he was "not lecturing on Zoology but upon the Law of Development and the Relations of Science and Religion." The numbers attending the lectures and the compliments he received convinced McCrady that he was doing the right thing.[21]

As the first weeks of 1878 passed, McCrady labored away at trying to refine his law of development, often working into the wee hours of the morning. In a moment of despair over his heavy teaching load, his financial debts, and his inability to formulate a clear conception of his law of development, McCrady penned a poem, titled "The Forge of Thought," in which he expressed his feelings about his efforts and his dreams: "Work! Work! Work! / From the crack of day to the close / and Half the Night / . . . Clang! Clang! Clang! / And a wonder shall be wrought / whose head sublime / Shall tower o'er time / The masterpiece of thought." He ended the verse, however, on a note of pessimism: "The work shall grow / Yet never shall perfect be." Indeed, he found the task of making the law apply universally to be insuperable. In his diary entry for January 25th, after working until 3:00 A.M., he noted: "It seems to me . . . that I am always producing yet never bringing any fruit to perfection. I cannot complete anything and shall probably die leaving the greater part of my researches and even their results unpublished." Only a few days earlier, he had read an editorial in *Popular Science Monthly* attacking the Reverend Joseph Cook for his recently published criticisms of some statements made by T. H.

Huxley. The editor of the magazine rightly contended that Cook had misrepresented Huxley's views, but McCrady was sympathetic toward Cook, and unquestionably he agreed entirely with a friend who wrote to him in January 1878 to say that "the [evolutionist] creed of the Huxley, Spencer, and Tyndall school is the dreariest, most hopeless, and soul-destroying belief ever attempted to be foisted on mankind." In March, McCrady completed a 10,000-word essay titled "Materialistic Systems" and sent it off to the *Literary World*, but the editor rejected it as being entirely too long. Perhaps, McCrady reflected to himself, he should collect his essays and publish them as a book. He sent a prospectus to Macmillan and Company, saying the possible title would be "Broken Words in a Biological Theory of the Universe" and the number of pages of manuscript would run to 600. Nothing came of the inquiry, but McCrady continued to write, ever convinced that he would master all of the diverse elements related to his law of development and also show the errors in Darwin's theory and the dangerous teachings of Darwin's disciples.[22]

Wrought up over the *Popular Science Monthly* criticism of Cook, McCrady penned an essay in defense of the theologian's argument and sent it off to the editor of the *Literary World*, who published it in November 1877. Although he noted that he did not "grant the validity of several of the arguments," McCrady contended that the Reverend Cook had succeeded very well in revealing "the actual state of scientific thought on biological questions." A group of scientists, he said, had gained the attention of the public by touting the theory of evolution as the truth, not merely as "a peculiar school of scientific thought." Long-established truths were thus under threat by "the gaudy flies of materialism . . . that swarm and buzz about" the new babe known as the law of development. The time would soon come, however, when that law would "dissipate nearly all the twilight fogs of the 'Evolutionists.' " The arguments by Cook constituted part of that fog-dissipating sunshine, said McCrady, but they were insufficient to complete the task, as they failed to show that science itself would reveal the errors of Darwin's theory. Thus, he concluded, Christianity needs more, not less, science, and it needs more scientific "speculation" governed by "business-like and systematic use . . . under well-ascertained and rigorous law." That process was, of course, precisely what McCrady thought he was following. In one of his unpublished lectures in Sewanee, he equated speculation with "imagination," which he viewed as a plausible hy-

pothesis, such as used by Newton in developing the law of gravity. The
inductive process was necessary but not sufficient. Deduction in science was also important in order to get beyond the mere accumulation
of facts. McCrady had come very close to a tenet of modern scientific
inquiry, but a dogged adherence to an orthodox assumption kept the
bright thinker he was from rending the veil.[23]

By mid-April 1878, McCrady was again suffering abdominal pains,
nausea, drowsiness, and depression, and a local physician advised him
to give up mercury (probably calomel) as a medicine, which, no doubt,
had severely exacerbated his illness. In September he became quite ill
once more, and the same physician prescribed morphine and calomel.
Nearly a year later, he was suffering from excruciating pains in his side
and back, and he took a medication containing opium. By that point,
the various remedies had no doubt damaged his liver even further.
Conditions at the University of the South also took a bad turn for
McCrady. In May 1878, he had found few students enrolled for his
courses, but he was less concerned over that matter than he was over
the rumor that the "nationalizing" of the university was necessary for
attracting students from the North. In August the trustees appointed
Dr. Telfair Hodgson to head the school, and Hodgson soon told
McCrady he would have to add English composition, rhetoric, and
literature to his instructional duties, with no additional pay. McCrady
consented to the new duties, and soon thereafter, when he learned that
financial difficulties would compel the university to reduce faculty
salaries, he volunteered to resign. His generous colleagues agreed to
further reduction of their own salaries and persuaded McCrady to stay.
The financial crisis persisted, however, and in May 1879 the trustees
decided they must eliminate a professorship. Once again, McCrady
volunteered to be the sacrificial lamb, for he believed that he must save
the university. As he had said in *The University Record* only two
months earlier, "On Sewanee Mountain, Southerners may . . . have
their sons educated under the sanctifying influence of that Church
which rocked the cradle of Southern civilization in Virginia and the
Carolinas." For the good of the university, then, he must give up his
position since he was the last hired, but once more his colleagues
requested that he remain. The trustees found a way to weather the
crisis, but McCrady had to assume even more duties. In August 1879,
then, he became "Acting Professor of Physics (including Astronomy),
Acting Professor of Engineering, and Acting Professor of English

Language and Literature" in addition to his regular appointment. His salary had been cut to $1,150 for the year, and McCrady began to explore the possibility of selling his beloved microscope. "Here, then," he said on October 9, "is the beginning of a new phase in my life." Soon thereafter, the local sheriff called to inform him that he would have to sell McCrady's possessions if he did not pay one of his creditors within three days. As before, McCrady's father and brother Edward came to his rescue. On the last day of the year, McCrady recorded in his diary, "The old year closes with me tonight in turmoil and annoyance."[24]

As early as 1878, officials of the University of the South were exploring ways to increase student enrollment, and they believed they should look beyond the region after which the institution had been named. This view did not sit well with McCrady, who considered it an effort to "nationalize" the university—and thereby to lose the school's special identity. He protested at length to John Kershaw, a prominent Episcopal spokesman in Camden, South Carolina. In a lengthy missive written on February 11, 1878, he told Kershaw that the university should not be "shedding tears over a 'Lost Cause' . . . [for] the cause is not lost at all: . . . it is only waiting its proper time in the manifold & ponderous workings of human development." The South had not, he added, been annihilated by the triumph of the Union, for the victory was merely political and did not include "the conquest of the mind of the conquered by the mind of the conqueror." He reminded Kershaw that "if any people have a marked idiosyncracy [*sic*] of tradition & principle it is the Southern People." Opposed to the Atlanta journalist Henry Grady's idea of a "New South," he objected to the "rippling whisper about 'nationalizing' the University of the South," which to him was the beginning of the abandonment of "Southern tradition, Southern ideas, & Southern social life as the basis of the University." McCrady grew increasingly annoyed over what he considered to be the liberal tendency of the university's presiding officer, and his annoyance reached a peak during the summer of 1880. By early June he was hearing rumors that a U.S. Army officer would be appointed to head the University Cadet Corps. "I dread his presence here," said McCrady. Worse, he noted that he expected the cadets' uniform would be "changed from Confederate Gray to that of the U.S. A[rmy] and that a flag staff supporting the Stars and Stripes will be set up on the drill grounds." The unreconstructed Confederate officer was right: the

U.S. flag was hoisted on June 23, and his son Edward, a cadet, was ordered to take part in the ceremony. Immediately, McCrady drafted a letter to Hodgson, saying the incident "would practically give to the . . . Cadet Corps *the character of a U.S. Garrison.*" Three days later, Hodgson replied. "Your letter startled me," he said. "I cannot see why . . . simply using the flag of our government could convert us into a garrison." McCrady then accused Hodgson of having the flag raised as "a temporary bid for Northern Capital and Students," and he warned him that people in the North would see through "such . . . gushes of 'patriotism' and 'loyalty' in Southerners." Hodgson persisted, however, and John McCrady lost yet another battle for the South he idealized. At the end of the year, he sent a letter to his old colleague Lewis Gibbes, saying that he could not "help wishing that I might join you . . . in an attempt to get our old [Elliott] Society at work again and once more engage in original zoological research."[25]

Although McCrady believed that the scientific world had forgotten him, it had not, and on January 12, 1881, he received "an affectionate letter" from his former student and then Johns Hopkins University professor William Keith Brooks, saying that he appreciated "the extreme accuracy" of McCrady's 1857 descriptions and illustrations of medusae, which he had recently used to identify species collected near Beaufort, North Carolina. Soon thereafter, Daniel C. Gilman, the president of Johns Hopkins, invited McCrady to present a series of six lectures at the university. McCrady readily accepted and told Gilman he would lecture on "the Theory of Development and Its Philosophical Significance." Brooks apparently assumed that his former instructor would lecture on the cnidarians, and it is certain that Gilman had expected something else. When McCrady arrived in Baltimore on March 28, he went immediately to see Gilman, who gave McCrady a copy of the program for his lectures. McCrady later wrote to his wife that neither Gilman nor Henry Newell Martin, a professor of biology at the institution and a former associate of Huxley, was able to "understand the statement I made of the subject of my lectures [and] simply left it out of their printed statement." That action, added McCrady, "at once showed me that I was regarded with suspicion." Brooks was apparently concerned also, but after talking with McCrady, he said he thought the audience would find the subject interesting. The first lecture went well, and Martin congratulated McCrady on keeping the attention of the audience on a difficult topic. The lecture hall "was crowded" for the second lecture, and McCrady proudly told his wife, "I am so popular that I can hardly get an hour to myself." Gilman soon told McCrady no other speaker had "attracted so much attention." William James came all the way from Harvard to hear the comments of his old friend. At the end of the series, McCrady said, "Everybody is urging me to publish" the lectures, and Gilman, he added, was "so enthusiastic" that he had asked the editor of the *Baltimore American* to obtain an abstract of them. Quite obviously, McCrady had succeeded with many people in his effort to cast doubt upon Darwin's theory. It is unlikely, however, that they fully followed the more abstruse aspects of his argument. Gilman was probably pleased mainly because McCrady's lectures had helped to allay some of the ill feeling toward Johns Hopkins as a den of Darwinism. Moreover, he found that McCrady had presented himself not as a rabble-rouser against evolution but as a thoughtful scholar who not only carefully analyzed the theory of evolution by natural selection but also offered an erudite theory that placed the Divine Being at its center. Gilman said he would arrange for McCrady to return a year later for another series. "I don't think I ever made so complete a conquest of a Yankee as I have of Gilman," he told his wife.[26]

Back in Sewanee, McCrady resumed his writing on topics related to his law of development and on Darwin's theory. By summer, however, illness had struck him again, and it continued intermittently into September. Meanwhile, Hodgson was mercilessly pressing him to pay overdue rent on the university-owned home in which he lived. McCrady, who had sacrificed so much for the university, claimed that he did not owe anything, that, in fact, the university owed him for repairs he had made to the house. The question became moot because, just before midnight on September 26, "flames burst . . . out around the kitchen chimney" and spread quickly throughout the house. McCrady was too ill to do anything, but students rushed to help and saved some items, including McCrady's books, manuscripts, and correspondence, just before the house burned to the ground. Still sick and in shock over his loss, McCrady received word a day later that his daughter Sabina, visiting a maternal uncle in Nashville, was seriously ill, and he and Sallie went immediately to be with her. Sabina recovered, but her father's condition worsened. McCrady died in Nashville on Sunday, October 16, 1881, one day after his fiftieth birthday.[27]

The circle of Charleston naturalists had long since been broken, and now death was claiming its former members one by one. Edmund Ravenel had died on July 27, 1871, two weeks after receiving injuries from a fall on the stairs of his home. John Edwards Holbrook, the only one of the group to be elected to the National Academy of Sciences, had passed away quietly at the home of a relative, in North Wrentham, Massachusetts, on September 8, 1871. John Bachman, who continued to minister to his congregation even after suffering a stroke, had died of a cerebral hemorrhage at his home on February 24, 1874, and was buried under the altar of the church he had pastored for six decades. Almost exactly one year after the passing of McCrady, Francis Holmes succumbed to death, on October 19, 1882. Only Lewis Gibbes remained.[28]

After the departure of McCrady from Charleston in 1873 and the death of Bachman in 1874, Gibbes had become virtually isolated as a naturalist. His principal contact with others interested in natural history in Charleston was through the curator of the Charleston Museum, Gabriel Manigault, and through meetings of the Elliott Society. Although Manigault was a superb osteologist, he was only modestly active in publishing in natural history. However, he attended meetings of the Elliott Society fairly regularly and presented seven brief papers at its sessions. William G. Mazÿck, who had briefly attended the Lawrence Scientific School, came to a number of the meetings of the Society. Mainly interested in malacology, Mazÿck published a few studies of land snails. The meetings of the Elliott Society were important to Gibbes, but by the early 1870s, he could usually count on only three to attend meetings: Manigault, Mazÿck, and J. F. M. Geddings, a local physician. At the meeting on November 5, 1875, Gibbes presented a paper titled "Synoptical Table of the Elements." He also sent the paper to Dana for publication in the *American Journal of Science*, but on December 4, 1875, Dana informed him that a chemist said Gibbes had "not seen the best works that have been written on the subject" and that others had developed a more complete periodic table of the elements. It was another misfortune for Gibbes, as he had been working on the idea for a number of years and had developed an excellent system of classifying the elements. He was unaware, however, of the periodic system developed in 1869. In fact, in his report to the College of Charleston president in March 1875, Gibbes complained of the difficulty of keeping up with progress in the fields he was teaching, namely, astronomy, physics, mathematics, and chemistry, and he said "it is indeed preposterous for any *one* teacher to 'profess' a single one of the four subjects, especially in our city, where we are debarred all access to the current literature of Science." The paucity of scientific books and journals was many times more limiting by then than it had been during the antebellum period. Moreover, until 1885 the Elliott Society had been too poor to publish the part of the second volume of its proceedings in which Gibbes's periodic table appeared.[29]

In his faculty report for 1876, Gibbes complained again of the problems of keeping abreast of new knowledge in science. He noted that, with a large family to support, he could no longer purchase books and journals, and he concluded that he "must quietly and patiently await the intellectual darkness that seems to be slowly enveloping us." Nor could he keep up with botany or with the study of crustacea. Responding in 1881 to a request from Richard Rathbun, of the Smithsonian Institution, Gibbes prepared a list of edible crustacea from Charleston waters, but he told Rathbun the scientific names were those with which he "was familiar 20 years ago . . . [and] I have no books for reference." Asa Gray continued into 1883 to ask Gibbes for information about southern flora, but Gibbes could offer nothing of value. He would not give up entirely, however, and in January 1885 he revived the Elliott Society again and served as its president through 1888. His efforts were more successful this time. Attendance at meetings fluctuated between six and eight during those years but fell to three or four in 1889. In 1885, Manigault presented a paper on the right whale, *Eubalaena glacialis*, taken in Charleston Harbor in January 1880. He had prepared the skeleton of that baleen whale and placed it in the Charleston Museum, where it remains on display today. Young Arthur Trezevant Wayne, who eventually became an authority on the avifauna of South Carolina, presented a paper on Swainson's warbler, *Limnothlypis swainsonii*, in August 1885, and on the prothonotary warbler, *Protonotaria citrea*, in February 1886. But it was Gibbes who presented the greatest number of papers during the period 1885–1888. They included a report on three comets; a superb account of the devastating earthquake that struck Charleston in 1886; a report on his invention of a portable heliotrope (a device used to reflect solar rays for signaling); a comment on calculating the area of a triangle; a note on a stalactite formed in an artificial structure; and a note on some butterflies he had observed in Charleston between 1870 and 1879.[30]

In 1888, Gibbes was seventy-eight years old, and his vision was failing. Thus, he declined reelection as president of the Elliott Society. He presented his last paper before the dying Society in 1889, on "The Annular Phase of Venus." The paper was reprinted a few weeks later in *Science*. Meanwhile, Gibbes had asked the trustees of the College of Charleston if they wished for him to retire, but the trustees expressed "unabated confidence" in the ability of the man who had given fifty years of service to the institution. He continued to serve until 1892, when he said he could go on no longer. The trustees generously praised the old professor, but they could do little to reward him financially. Gibbes was to receive a meager pension of $500 per year. His alma mater, the University of South Carolina, had conferred an honorary doctorate upon him in 1890, and the honor no doubt gave some pleasure to the penniless and virtually blind old scientist during the four remaining years of his life. He died on November 21, 1894.[31]

The struggles of Lewis Gibbes to keep the scientific enterprise going in the city that he and his fellow naturalists loved so dearly paralleled the struggles of Charleston to recover from the war that toppled it from its queenly throne. After twelve years of "reconstruction," white Democrats fully regained political and social control of the city. Signs of economic prosperity were soon evident, but, as Walter J. Fraser Jr. indicates in his comprehensive history of Charleston, "the prosperity of the early 1880s was illusionary." By and large, Charleston business leaders in 1880 were the same men who had controlled its economy before the war, and they were generally inflexible in their views toward changing things. Thus, while new urban centers were prospering in the South, Charleston had to cope with old problems. As Fraser notes, two-thirds of the city's nearly fifty-four miles of streets remained unpaved, and the drainage system in the lower part of the city was in disrepair. Moreover, by 1880, the growth of Charleston's population was nearly stagnant, and once again the number of blacks exceeded the number of whites. With the increase of the proportion of blacks came confinement of their residence to certain sections of the city—and soon thereafter the enactment of segregation laws. Much of the money that might have been spent for improvements in the city went for the support of the city's police force, which, on a per capita basis, remained larger in 1880 than that in the city of New York. Charleston's leaders would have the city return to its golden age, but the advancement of science was largely left to its own fortunes.[32]

Worse was to come for Charleston during the 1880s, however, when the city fell victim to natural forces. The first occurred on August 25, 1885, when a powerful hurricane swept over the city, killing twenty-one of its citizens, flooding houses on the Battery, and damaging or destroying nine-tenths of the city's homes. Barely over a year later, a major earthquake rumbled through the old city, causing the immediate death of twenty-seven of its residents and extensive damage to 2,000 buildings and homes. Many of the injured died during the weeks following the devastating jolts. In 1893, another destructive hurricane struck Charleston, resulting in four deaths and considerable damage to buildings. Charleston had experienced traumatic events before, however, and its resilient residents once again set about to restore their scarred city. Ever mindful of their cultural heritage as well, a group of Charlestonians sought to revive the Elliott Society in 1900, but the life of the old society was so feeble that it had succumbed by the following year. Meanwhile, the fortunes of the Charleston Museum had declined, and the College of Charleston eventually decided that it could no longer support it. Charlestonians would not concede that the venerable institution must fade away, and in 1907 the city assumed direct responsibility for managing and supporting it. Despite hard times in the years ahead, the museum survived and once again became a major showpiece of the city, the wealth of its collections adding great richness to the history of natural history.[33]

Epilogue

ong before the last of the Charleston naturalists had passed, natural history as a subject had disappeared from the curriculum of nearly every American college, being replaced by a field called biology. Still, taxonomy and systematics retained a place of great importance in the new subject, though they were no longer the main focus of research. Advances in the field could not be made, however, without a system of classifying living things and without scientific descriptions of them. Thus, the pioneering work of the much overlooked naturalists of antebellum Charleston must be counted among the significant contributions to science in America. The work of the six Charleston naturalists before the Civil War was comparable to that of naturalists elsewhere in the nation. Certainly, there were more productive naturalists in America, but the Charleston group faced greater obstacles overall. Lacking ready access to the pertinent scientific literature and more isolated from their peers than were their counterparts in the urban triad of the Northeast, the Charleston circle encountered special barriers. Above all, because their city was relatively small and because more than half of its inhabitants were barred by their race from participating in scientific or any other intellectual activity, the Charleston group had a small pool from which to attract a large number to their ranks. While they possessed both the essential understanding of the importance of a scientific association and the will to keep it going, they lacked a sufficient reserve of dedicated naturalists to develop its full potential. Moreover, although they received allocations from the South Carolina legislature during the late 1850s to support three publications in natural history, naturalists in Charleston, indeed, in the entire South, had fewer resources for publishing works in science than did their contemporaries in Boston, New York, and Philadelphia. To the great credit of Charleston, however, the city strongly supported a museum of natural history and fostered the work of its naturalists.

Aside from John Bachman's experimental studies of how vultures locate food and Lewis Gibbes's limited work in experimental physics, the members of the Charleston circle devoted their efforts mainly to collecting specimens and to describing, naming, and classifying species. It was a skill that they honed to perfection and one that made them as good as any of their peers in America and in Europe. Legatees of a long-standing interest in such activities, residents of a region rich in natural history, and followers of the traditional notion that species represent the hand of God in nature, they threw themselves into their tasks with remarkable ardor. Regional pride also served as an incentive, but the Charleston naturalists could not have succeeded had they been devoid of the curiosity and talent essential to the pursuit of science. They were aided by living in a city that possessed a capacity for cultivating the pursuit of natural history, despite its comparatively small size, its relative isolation from the intellectual centers of the country, and its lack of ample resources.

In his prize-winning book *The Launching of Modern American Science, 1846-1876*, however, Robert V. Bruce essentially dismisses the contributions of antebellum southerners to the advancement of science in the United States. While he acknowledges that the "lack of sizable cities surely hampered Southern science," Bruce minimizes even the scientific work done in Charleston, saying the city "could boast of Dr. John E. Holbrook . . . the Reverend John Bachman . . . two or three planters, several physicians, and a couple of professors at the College of Charleston." "Such," he concludes, "was the leading center of Southern science." Asserting that "Southern science depended more heavily on individual amateurs—planters, politicians, lawyers, and especially physicians," Bruce fails not only to recognize the contributions of the Charleston naturalists but also to acknowledge that they were no more amateurs than were their northeastern counterparts. Surely, the favorable views of Agassiz, a host of other American scientists, and a considerable number of European naturalists toward the Charleston circle should have suggested to Bruce that the Charleston naturalists were not mere dabblers in science, but he appears to be content with the conclusion that "Southern scientists . . . faced greater physical hardships than did their Northern counterparts."[1]

Following a common view, Bruce adds that a low level of interest and limited activity in science in the Old South were consequences of the institution of slavery in the region. "The South's peculiar institution," he says, "stunted the life of the mind among masters as well as slaves." Since scientific work is certainly an intellectual activity, however, the case of the Charleston naturalists clearly shows otherwise. Certainly, the faith of the Charleston naturalists in the institution of slavery was terribly misguided and cannot be justifiably defended. However, factors other than slavery account for the comparatively lower output of science in the Old South, not least among them the paucity of large cities in the region, which Bruce admits. As Bruce sees it, the culture of the antebellum South "fostered a pseudo medieval romanticism at odds with the scientific temperament."[2] The success of the Charleston naturalists vitiates that view, however. Without question, all of the Charleston naturalists, two of whom were reared in the Northeast, embraced the cultural system of the South and donned blinders that seriously impaired their moral vision and ultimately brought great suffering to themselves and their region; however, in terms of genuine scientific research, the system hindered them very little, if at all. Indeed, it must be remembered that, despite his acceptance of slavery, John Bachman stood nearly alone among contemporary scientists, both southern and northern, in using science properly to show that all human races belong to one and the same species. To note that the institution of slavery did not unduly handicap the work of the Charleston naturalists is not to condone human bondage, of course. It is only to remind us that good science, at least in many areas, can be done irrespective of cultural values. Science can, of course, be used for evil purposes, but one can hardly accuse the Charleston naturalists of doing that. Slavery was a lamentable legacy passed on to them, and one could only wish they might have abandoned that part of their heritage. But history must deal with what did occur, not with what might have been.

Unquestionably, the impact of the legacy was enduring, and deeply entrenched notions followed the Charleston naturalists to the end of their years, as they did a host of other southerners. But it was not only the legacy—and acceptance of that legacy—that set back science in the South after 1861. A wrecked economy, along with firmly fixed political and social attitudes, generally suppressed the scientific interests of southerners for years to come. While "modern science" took off in the rest of the nation, as Bruce demonstrates, the South lagged behind. The infusion of federal funds boosted the economy elsewhere in the country, but, although the same funds helped many southern universities to promote science, changes of attitude followed more slowly in the South. The postbellum remnant of scientists in Charleston continued to view natural history as the queen of sciences, and their views mirrored those of far too many of their fellow southerners. Half a century elapsed before the South as a whole began to enter the mainstream of scientific research, and another half passed before it began to become an equal in such research. To overlook what went before, however, is to omit a part of the history of the region and a part of the story of the development of science in America.

Abbreviations

A-B/HU	Audubon-Bachman Correspondence, Houghton Library, Harvard University
Acad./ANSP	Academy Correspondence and Records, Academy of Natural Sciences of Philadelphia Library
AJS	*American Journal of Science*
ANSP	Academy of Natural Sciences of Philadelphia Library
APSL	American Philosophical Society Library
ASecr./SIA	Assistant Secretary's Correspondence, Smithsonian Institution Archives
BL/CML	Bachman Letters, Charleston Museum Library
BRB/YU	James J. Audubon Papers, Beinecke Rare Books and Manuscript Library, Yale University
CCL	Special Collections, College of Charleston Library
CLS	Charleston Library Society
CMJR	*Charleston Medical Journal and Review*
CML	Charleston Museum Library
CUL	Clemson University Library
DF/YU	Dana Family Papers, Sterling Memorial Library, Yale University
HU	Harvard University
JHU	Johns Hopkins University Library
Jour. ANSP	*Journal of the Academy of Natural Sciences of Philadelphia*
LRG/LC	Lewis R. Gibbes Papers, Manuscript Division, Library of Congress
McC/McCF	McCrady Family Papers, private collections of the McCrady Family
McC/SCHS	McCrady Family Papers, South Carolina Historical Society
MCZ	Museum of Comparative Zoology, Harvard University
MoHS	Audubon Papers, Missouri Historical Society
Ms/HL	Manuscript Collections, Houghton Library, Harvard University
Proc. AAAS	*Proceedings of the American Association for the Advancement of Science*
Proc. ANSP	*Proceedings of the Academy of Natural Sciences of Philadelphia*
Proc. BSNH	*Proceedings of the Boston Society of Natural History*
Proc. ESNH	*Proceedings of the Elliott Society of Natural History*
RF	Ravenel Family Papers
RL/CML	Edmund Ravenel Letters, Charleston Museum Library
SCHM	*South Carolina Historical and Genealogical Magazine*
SCHS	South Carolina Historical Society, Charleston, S.C.

SCL   Manuscript Collections, South Caroliniana Library
SGM/HSP Samuel G. Morton Papers, Historical Society of Pennsylvania
SHC   Southern Historical Collection
SIA   Smithsonian Institution Archives
USo   University of the South Library
WHL   Waring Historical Library

Chapter One

1. J. Bachman, *Discourse*, 4; Fraser, *Charleston! Charleston!*, 178, 186.

2. J. Bachman, *Discourse*, 4–6; J. Bachman, "Discourse on the 53rd Anniversary," [January 5, 1868], microfiche, CLS; C. L. Bachman, *John Bachman*, 26.

3. J. Bachman, "The Morals of Entomology"; J. Bachman, "An Examination of the Characteristics of Genera and Species," 204; Moore, "Geologists and Interpreters of Genesis."

4. J. Bachman, "Observations on the Changes of Colour"; Audubon and Bachman, *Quadrupeds*, 1:8, 71–72, 110–19, 176–78, 207–8, 258, 310; 2:58–59, 216.

5. Wade, *Slavery in the Cities*, 82, 96–97; Mills, *Statistics of South Carolina*, 421.

6. Wade, *Slavery in the Cities*, 80–100; Fraser, *Charleston! Charleston!*, 199–200; Powers, *Black Charlestonians*, 61–67, 267; Evarts, *Through the South and the West*, 76–77; Clark, *South Carolina*, 104–9, 128.

7. J. Bachman, *Discourse*, 4–6; Fraser, *Charleston! Charleston!*, 190–91; J. Bachman, *An Address Delivered Before the Washington Total Abstinence Society of Charleston*; Wyatt-Brown, *Southern Honor*, 350–61; Buckingham, *Slave States of America*, 1:48–83, 550–76; C. L. Bachman, *John Bachman*, 436 (mention of "Sermon Against Dueling, about 1842").

8. Stearns, *Science in the British Colonies*, 315–18, 599–616; Sanders, "Alexander Garden," 409–37; Sanders and Anderson, *Deep Runs the Heritage*; Beale, "Bosc and the Exequatur"; Johnson, *Scientific Interests*, 126–51; G. A. Rogers, "Elliott, Stephen," 247–49.

9. Fraser, *Charleston! Charleston!*, 180–81, 195–98; Martin, "Ebenezer Kellogg's Visit"; Buckingham, *Slave States*, 47–66; Clark, *South Carolina*, 122–23, 175–77; Mills, *Statistics*, 402–4; Bellows, *Benevolence*, 27–36, 75, 105.

10. Buckingham, *Slave States*, 49–50; *South Carolina: A Guide*, 192–211; G. C. Rogers Jr., *Charleston in the Age of the Pinckneys*, 56–58, 92–94; J. Bachman, *Discourse*, 5–6; Mills, *Statistics*, 404–18. For examples of Bachman's criticisms of Catholic leaders, see J. Bachman, *A Defence of Luther*, and Messmer, *Works of the Right Reverend John England*, 2:57–288.

11. Exceptions among the antebellum Charleston naturalists were Edmund Ravenel, who taught medical chemistry from 1824 to 1835, and Lewis R. Gibbes, who was an able student of chemistry and natural philosophy, or physics. Charles Upham Shepard, an accomplished professor of chemistry and mineralogy at Amherst College, spent one-half of each year as a professor at the medical college in Charleston prior to the Civil War, but his primary association was with New England. The most

accomplished of the Charleston students of chemistry during the antebellum era was J. Lawrence Smith, but he left his native city for good in 1852.

12. Fraser, *Charleston! Charleston!*, 178, 186; Powers, *Black Charlestonians*, 267; Mills, *Statistics*, 422–23.

13. Martin, "Ebenezer Kellogg's Visit," 12–13; Martineau, *Retrospect of Western Travel*, 1:228–29; Mills, *Statistics*, 437, 468; Rhees, *Manual of Public Libraries*, 447–57; Faust, *Sacred Circle*, 7–14; Wallace, *South Carolina: A Short History*, 350.

14. G. C. Rogers Jr., *Charleston in the Age of the Pinckneys*, 89–115, 141–66; Fraser, *Charleston! Charleston!*, 187–246; MacDonald, *Diaries*, 370; C. H. Rogers, *Incidents of Travel in the Southern States*, 49–53.

15. Numbers and Numbers, "Science in the Old South," 9–35; Bachman to John Horlbeck, August 31, 1815, in C. L. Bachman, *John Bachman*, 34–35.

Chapter Two

1. C. L. Bachman, *John Bachman*, 9–19; John Bachman to General John A. Quitman, December 23, 1847, letter tipped in copy of *The Viviparous Quadrupeds of North America*, APSL. The maiden name of Bachman's mother is unknown, but, based upon two references to an uncle in Rhinebeck, it is surmised to be Shop or Shops, likely a corruption of Schopf (see "Journal of Rev. John Bachman, Aug. 9, 1827–Sept. 25, 1827," typescript copy in the CML). In *Heads of Families of New York*, p. 92, there is a Henry Shoop. Eva named one of her sons Henry. A man named Henry Shop, of Dutchess County is listed in *Index to the 1800 Census of New York*, 331.

2. C. L. Bachman, *John Bachman*, 15–17; J. Bachman, "Observations on the Changes of Colour," 231–33; Audubon and Bachman, *Quadrupeds*, 1:8, 71–72, 110–19, 176–78, 207–8, 258, 310, 2:58–59.

3. J. Bachman, "On the Migration of the Birds"; J. Bachman, "Observations on the Changes of Colour," 201.

4. C. L. Bachman, *John Bachman*, 18–20. The statement by Bachman's daughter Catherine that her father attended Williams College is not supported by any firsthand evidence, and records show that he was in Philadelphia in 1804.

5. J. Bachman, "Humboldt Festival"; J. Bachman, *Discourse*, 4.

6. C. L. Bachman, *John Bachman*, 20–25, 412–14; Bost, "The Reverend John Bachman," 5–86.

7. C. L. Bachman, *John Bachman*, 23–25; Bost, "The Reverend John Bachman," 86–91.

8. Clarke, *Wrestlin' Jacob*, passim; Bost, "The Reverend John Bachman," 93–110, 123–32, 197–205; J. Bachman, *Discourse*, 8; C. L. Bachman, *John Bachman*, 29; Neuffer, *Christopher Happoldt Journal*, 32–38; Wilson and Grimes, *Marriage and Death Notices*, 1:7; [J. Bachman], *Funeral Discourse*, 1–23; *History of the Lutheran Church in South Carolina*, 167–203.

9. C. L. Bachman, *John Bachman*, 150–51; Jennie Rose, "John Bachman at Home," typescript, [ca. 1924], CML; Strobel, *Exposition*, 1–2; U.S. Census, *Slave Schedule*, 1850.

10. C. L. Bachman, *John Bachman*, 38–44; John Bachman to Mrs. Jacob F. Roh, August 30, 1823, Ms/HL.

11. C. L. Bachman, *John Bachman*, 44–74; John Bachman to Harriet Bachman, July 24, 26, 28, 1827, Harriet Bachman to John Bachman, August 8, 1827, Maria Martin to Maria Bachman and Harriet Bachman, July 24, 1827, John Bachman to [the Vestry], August 1827, "Journal of the Rev. John Bachman," CML; Arnold, *Four Lives in Science*, 13–35.

12. C. L. Bachman, *John Bachman*, 74–80; "Schirmer Diary," *SCHM* 67 (October 1966): 229.

13. J. Bachman, "Successful Method of Raising Ducks"; J. Bachman, "Reply to the Letter of Samuel George Morton, M.D.," 496–504.

14. Shuler, *Had I the Wings*, 1–17; A. Ford, *John James Audubon*, 283–86; John J. Audubon to Lucy Audubon, October 23, 1831, in Herrick, *Audubon the Naturalist*, 1:5–7.

15. John James Audubon to Lucy Audubon, October 30, November 13, 23, 1831, in Corning, *Letters of Audubon*, 1:145–55; John Bachman to John James Audubon, December 2, 1831, in C. L. Bachman, *John Bachman*, 97–98; Harriet Bachman to John Bachman, December 14, 1831, BL/CML. Audubon refers to "Dr. Henry Ravenel," but it was Dr. Edmund Ravenel.

16. John Bachman to Lewis Reeve Gibbes, May 27, 1834, Botany Department Papers, SHC; Bachman to Gibbes, June 30, 1834, BL/CML; J[ohn] Bachman, *Catalogue, Phaenogamous Plants*, reprinted in *Southern Agriculturalist* 8 (April and June 1835): 189–96, 286–91; "Receipt of Elliot Herbarium" [by Bachman] from "Miss H. H. Elliott," March 8, 1837, Nathaniel Wright Papers, Manuscript Division, Library of Congress; H. H. Elliott to [John Bachman], October 30, 1837, BL/CML.

17. Audubon noted that Bachman was "*well* acquainted" with "the manners of the Southern Birds," in John James Audubon to Victor Audubon, November 4, 1833, in Corning, *Letters of Audubon*, 1:263–67. Bachman could easily have described the two rare species he collected, *Limnothlypis swainsonii* and *Vermivora bachmanii*, but he readily allowed his friend to do so. References to Bachman's assistance appear in all seven volumes of the 1840 edition of Audubon's *Birds of America*, but predominantly in the second and sixth volumes. See also A. Ford, *John James Audubon*, 326.

18. John Bachman to John James Audubon, December 27, 1832, and February 18, March 4, 1833, BRB/YU; Bachman to Audubon, July 21, 1832, and March 3, 1833, MoHS; Audubon to Bachman, January 16, 1832, in Corning, *Letters of Audubon*, 1:172–76; Bachman to Audubon, January 23, February 28, and March 3, 1833, in C. L. Bachman, *John Bachman*, 127–35; Bachman to Audubon, December 23, 1831, in Herrick, *Audubon the Naturalist*, 1:27; Shuler, *Had I the Wings*, 63, 71–72, 94, 122; A. Ford, *John James Audubon*, 294–96, 326.

19. John James Audubon to John Bachman, July 1, 1832, in Corning, *Letters of Audubon*, 1:195–96, 2:1–7, 10–13; J. J. Audubon to Victor Audubon, December 24, 1833, in Herrick, *Audubon the Naturalist*, 2:55; [John Bachman], "Notes on some experiments," December 16–26, 1833, manuscript, BL/CML; John Bachman, "The Vulture," 165–76; Palmer, *Handbook*, 4:13, 20, 23, 38; Terres, *Audubon Society Encyclopedia*, 957.

20. John Bachman, "Remarks in Defence of the Author of 'The Birds of America'"; Herrick, *Audubon the Naturalist*, 2:71–80; [John Bachman], "Review of J. J. Audubon's *Ornithological Biography*, vol. 2, and *Birds of America*, vol. 2," 1835[?], manuscript, MoHS; Bachman to "Mr. Kelly," [editor, *Bucks County Intelligencer*], June 19, 1835, Ms/HL; Shuler, *Had I the Wings*, 135–36.

21. John Bachman to John James Audubon, March 27, 1833, MoHS; Bachman to Audubon, April 1, September 22, 1833, BRB/YU; Bachman to Audubon, December 16, 1834, and March 25, 1835, BL/CML; Audubon to Bachman, May 8, November 5, 19, and December 3, 10, 1834, John E. Thayer Collection, MCZ; Bachman to Audubon, September 14, 1833, in C. L. Bachman, *John Bachman*, 135; J. J. Audubon to Victor Audubon, November 4, 1833, in Corning, *Letters of Audubon*, 1:263–67; Sprunt and Chamberlain, *South Carolina Bird Life*, 70, 95–96, 145–46, 152–53, 222–24, 226, 247, 252–53, 348–49, 363–64, 435–36, 439–42, 533–35, 537–38.

22. Sprunt and Chamberlain, *South Carolina Bird Life*, 6; John Bachman to John James Audubon, January 15 and August 24, 1835, January 22, 1836, and April 24, 1837, BRB/YU; Bachman to Audubon, September 17, 1836, and Bachman to Edward Harris, December 12, 1837, in C. L. Bachman, *John Bachman*, 139–41.

23. John Bachman, "On the Migration of the Birds."

24. John Bachman to John James Audubon, February 18, 1833, BRB/YU; Bachman, "Observations on the Changes of Colour," 197–239; Bachman to Samuel G. Morton, March 30, 1837, Samuel G. Morton Papers, APSL; Bachman to Audubon, March 15, 1837, and March 10, 1838, and George Ord to Bachman, May 26, 1837, BRB/YU.

25. Richard Harlan to John Bachman, February 6 and March 28, 1832, BL/CML.

26. John Bachman to John James Audubon, January 20, 1833, in C. L. Bachman, *John Bachman*, 125–27; Bachman to Audubon, October 4, 1834, and October 17, 1837, BRB/YU.

27. Charles Pickering to John Bachman, January 19, 1836, BL/CML; Bachman, "Description of a New Species of Hare."

28. Richard Harlan to John Bachman, [n.d.], 1834, in C. L. Bachman, *John Bachman*, 136; Minutes, ANSP, March 28, 1837, Acad./ANSP; Charles Pickering to Bachman, June 8, 1836, BL/CML; Bachman to S. G. Morton, February 17 (quotation) and March 15, 17, 1836, Samuel G. Morton Papers, APSL; R[ichard] Harlan, "Description of a new species of Quadrupeds"; Audubon and Bachman, *Quadrupeds*, 2:216; Hamilton and Whitaker, *Mammals*, 179–80.

29. John Bachman, "Observations on the different species of Hares"; Bachman, "Some Remarks on the Genus Sorex"; Thomas Nuttall to Bachman, January 17, 1837, Manuscript Collections, MCZ; Charles Pickering to Bachman, March 25 and December 23, 1837, BL/CML; Bachman to S. G. Morton, April 24, 1837, Pickering to Bachman, April 27, 1837, and Bachman to John Vaughan, May 8, 1837, APS/Arch.; Bachman to Edward Harris, December 12, 1837, in C. L. Bachman, *John Bachman*, 158–59.

30. Bachman, "Some Remarks on the Genus Sorex"; Bachman to S. G. Morton, March 17, 1837, Manuscript Collections, APSL; Audubon and Bachman, *Quadrupeds*, 2:176–77, 3:249–51; Handley and Varn, "Identification of the Carolina Shrews,"

393–94. Handley and Varn venture that a third species may be valid: *Sorex fimbripes* (Bachman, 1837).

31. Bachman, "Observations on the Changes of Colour in Birds," 237; Bachman to Audubon, April 8, 21, 1837, BRB/YU; William Cooper to Bachman, March 28, 1837, BL/CML.

32. John Bachman to Harriet Bachman, May 25 and December 12, 1837, in C. L. Bachman, *John Bachman*, 153–54; Bachman to Edward Harris, January 10, 1838, Ms/HL.

33. *Charleston Mercury*, May 16, 1836; *Charleston Courier*, June 3, 10, 15, 1837; John Bachman, "Notes &c. for Introductory Lecture to the Phil. Society, May 1837," Ms/HL.

34. *Charleston Courier*, June 15, 16, 20, 1837; John Bachman, Review of *Types of Mankind*, *CMJR* 9 (1854): 629; Hovenkamp, *Science and Religion in America*, 119–41.

35. John Bachman to John James Audubon, September 23, 1836, and Bachman to Audubon, October 2, 1837, in C. L. Bachman, *John Bachman*, 141–42, 159–60; Jacob Mintzing to Bachman, May 30, 1838, and John K. Townsend to Bachman, June 6, 1838, BL/CML; Bachman to Audubon, March 10, 1838, BRB/YU; Bachman to Vestry, St. John's, May 18, 1838, in Neuffer, *Christopher Happoldt Journal*, 65; Bachman, *Discourse*, 9; Bachman, "Description of Several New Species of American Quadrupeds";("Read August 7, 1838").

36. Neuffer, *Christopher Happoldt Journal*, entries for June 6–August 15, 1838 (pp. 119–51); John Bachman to John J. Audubon, August 5, 1841, BL/CML.

37. Neuffer, *Christopher Happoldt Journal*, entries for August 16–December 27, 1838 (pp. 151–214); C. L. Bachman, *John Bachman*, 168–74; J. Bachman, "Humboldt Festival"; Bachman to Karl Martin H. Lichtenstein, November 5, 1838, Museum für Naturkunde; Bachman to "The Society of Naturalists & Physicians of Germany, Zoology Department," September 21, 1838, BL/CML.

38. John Bachman to Thomas M. Brewer, March 5, 1839, in Misc. Letters, CML; *Proceedings of the Zoological Society of London* 6 (1839): 85–105; John Bachman, "Monograph of the Genus *Sciurus*."

39. John Bachman to John James Audubon, October 2 and December 4, 1839, January 4, 18, and May 8, 1841, March 14, 1844, March 24 and April 1, 1845, and May 2, 1846, John Bachman to Victor Audubon, May 8, 10, and October 27, 1840, and May 16, 1846, John Bachman to Jane Bachman, December 26, 1840, May 11, 25, 1841, and July 18, 22, 1846, John Bachman to Harriet and Julia Bachman, July 16, 1846, Harriet Bachman to Harriet Eva Bachman Haskell, May 1, 1846, Maria Martin to Eliza Bachman Audubon, April 24, 1840, Maria Martin to John Bachman, March 11 and April 3, 1841, and October 5, 1842, BL/CML; John Bachman to John J. Audubon, April 6, 1840, MoHS; Bachman to Audubon, April 6, 1840, BRB/YU; Audubon to Bachman, April 17, 1840, John E. Thayer Collection, MCZ; C. L. Bachman, *John Bachman*, 212–14; Holcomb, *Marriage and Death Notices from the Charleston Observer*, 154, 168–69.

40. John Bachman to Victor Audubon, July 18, 25, 1846, and April 26, July 10, August 2, 10, 1847, Bachman to John Woodhouse Audubon, July 28, 1847, Bachman

to "My Dear Children," September 3, 6, 8, 1847, Bachman to Jane Bachman, August 15, 1846, and September 10, 1847, Bachman to Dr. William G. Ramsay, September 11, 1847, BL/CML.

41. John Bachman, "Additional Remarks on the Genus Lepus."

42. Thomas M. Brewer to John Bachman, January 9, February 3, April 30, and October 25, 1837, March 5, May 2, 7, 21, July 1, October 10, and December 7, 17, 1838, and February 9, 1841, Ms/HL; John Bachman, "Observations on the Genus Scalops"; John Bachman to Karl Martin H. Lichtenstein, June 18, 1839, and January 6, 1841, Museum für Naturkunde.

Chapter Three

1. John Bachman to John J. Audubon, July 5, 1839, BL/CML.

2. John Bachman to John J. Audubon, September 13 and December 24, 1839, BL/CML; Audubon to Thomas M. Brewer, September 15, 1839, in Brewer, "Reminiscences of Audubon," 674-75.

3. John J. Audubon to John Bachman, January 2, 1840, John E. Thayer Collection, MCZ; Bachman to Audubon, January 13, 1840, and August 5, 1841, BL/CML (the first letter, often quoted, appears in altered form in Herrick, *Audubon the Naturalist*, 2:210-11); Audubon and Bachman, "Descriptions of New Species of Quadrupeds," *Proc. ANSP* 1 (October 1841); Minutes of the Academy of Natural Sciences of Philadelphia, March 21, 1837, Acad./ANSP; Bachman to John Perkins Barratt, March 4, 1842, SCL.

4. Audubon and Bachman, "Descriptions of New Species of Quadrupeds," *Jour. ANSP* 8 (1842).

5. John Bachman to J. J. Audubon, January 26, 1840, and Bachman to Victor Audubon, March 9, 1840, BL/CML; John J. Audubon to Spencer F. Baird, June 13, 1840, in Herrick, *Audubon the Naturalist*, 2:219; Audubon to Baird, June 22, 1840, in Dall, *Spencer Fullerton Baird*, 48; Maria Audubon, *Audubon and His Journals*, 1:467-510, 2:passim.

6. John Bachman to John J. Audubon, June 16 and October 21, 1842, and January 15, 1843, BRB/YU; Bachman to Victor Audubon, June 14 and August 5, 1843, BL/CML.

7. John Bachman to John J. Audubon, November 1[?], 29, 1843, and Bachman to Victor Audubon, February 4, 8, 1844, and October 31, November 24, 27, 29, 1845, BL/CML; Audubon to Bachman, November 12 and December 10, 1843, and March 8, 30, 1845, A-B/HU; Audubon to Bachman, January 8, 1845, in Herrick, *Audubon the Naturalist*, 2:264-67.

8. John Bachman to Victor Audubon, December 5, 1845, and Bachman to John J. Audubon, December 16, 1845, BL/CML; Bachman to Victor Audubon, December 10, 1845, BRB/YU; Bachman to Edward Harris, December 24, 1845, in Herrick, *Audubon the Naturalist*, 2:269-70.

9. Victor Audubon to John Bachman, December 27, 1845, A-B/HU.

10. John Bachman, "An Inquiry Into the Nature and Benefits," *Southern Agricul-*

turalist 3 (February 1843): 49–65 and 3 (March 1843): 81–96; "Agriculture in South-Carolina," *Southern Ladies Book*, n.s., 2 (1843): 200–203; Stephens, "The Mermaid Hoax," *Proceedings of the South Carolina Historical Association, 1983*, 45–55; Bachman to unknown (draft copy), May 8, 1843, BL/CML; Bachman to Lewis R. Gibbes, n.d., 1843, August 8 and September 11, 1843, and John P. Barratt to L. R. Gibbes, March 27, 1843, Lewis R. Gibbes Papers, CML.

11. John Bachman to unknown, March 24, 1846, SCL; Lyell, *Second Visit*, 1:227–28.

12. John Bachman to Victor Audubon, January 3, 10, 1846, BRB/YU.

13. John Bachman to Victor Audubon, January 17, 1846, BRB/YU; Victor Audubon to Bachman, February 2, [10?], 1846, in A-B/HU; Bachman to Victor Audubon, January 23, 25, and February 18, 1846, and Bachman to John J. Audubon, February 5 and March 6, 1846, BL/CML.

14. John Bachman to John J. Audubon, February 28, 1846, and Bachman to Victor Audubon, March 29, May 7, 14, 22, June 16, July 27, August 12, and November 7, 1846, BRB/YU; Victor Audubon to Bachman, March 7, 13, May 13, 19, and August 18, 1846, and John J. Audubon to Bachman, May 31 and July 25, 1846, A-B/HU; Bachman to Victor Audubon, March 22, June 14, July 22, and August 2, 1846, and Bachman to John J. Audubon, June 6 and November 5, 1846, BL/CML; Herrick, *Audubon the Naturalist*, 2:286.

15. John Bachman to Victor Audubon, December 8, 14, 27, and n.d., 1846, BL/CML; Bachman to John J. Audubon, December 23, 1846, BRB/YU.

16. Reviews of *The Quadrupeds of North America* in *American Review* 4 (December 1846): 625–38, *Southern Quarterly Review* 11 (April 1847): 499–503, and *Literary World* 1 (1847): 128–29; Victor Audubon to Bachman, May 31, 1847, A-B/HU; Bachman to John J. Audubon, March 13, 1847, in C. L. Bachman, *John Bachman*, 226–27.

17. John Bachman to Victor Audubon, January 21 and February 10, 1847, BRB/YU; Victor Audubon to Bachman, March 11, 24, July 14, September 1, October 14, November 17, and December 16, 1847, and J. W. Audubon to Bachman, July 20, 1848, A-B/HU; Bachman to Victor Audubon, June 19, 1847, BL/CML.

18. John Woodhouse Audubon to John Bachman, July 20, 1848, and Victor Audubon to Bachman, November 29, 1847, A-B/HU; Bachman to Victor Audubon, March 19, 1847, BRB/YU.

19. John Bachman to Victor Audubon and John Woodhouse Audubon, October 27, 28, 1847, and Bachman to Victor Audubon, November 18, 1847, BL/CML; Adam Gopnik, "Audubon's Passion," *New Yorker*, February 25, 1991, 103.

20. *Charleston Mercury*, November 17, 1847, 2, December 1, 1847, 2, December 2, 1847, December 10, 1847, and December 14, 20, 1847, 2; Victor Audubon to John Bachman, April 4, 1848, A-B/HU.

21. Robert W. Gibbes to Samuel G. Morton, November 30, 1847, SGM/HSP; *Charleston Mercury*, December 21, 1847.

22. John Bachman to Victor Audubon, December 13, 1847, and January 6, 1848, BL/CML; C. L. Bachman, *John Bachman*, 254.

23. Victor Audubon to John Bachman, January 10, February 5, and April 1, 1848; A-B/HU; Bachman to Victor Audubon, February 2, 1848, Ms/HL; Bachman to

Victor Audubon, February 15, 1848, BRB/YU; Bachman to Victor Audubon, February 3, 25, 1848, BL/CML; Bachman to Henry William Ravenel, December 17, 1846, Henry William Ravenel Papers, CUL; White, *Statistics*, Appendix, 1–5; John Bachman, "Notes on the Generation of the Virginia Opossum," *Proc. ANSP* 4 (April 1848): 40–47 (manuscript of same in Acad./ANSP); Myddleton Michel, "Researches," *Proc. AAAS* 3 (1850): 60–65; Hartman, *Possums*, 88, 94.

24. Jacob Schirmer to John Bachman, April 14, 1848, and Mitchell King to Bachman, April 14, 1848, and May 7, 1853, BL/CML; "Reminiscences of the College of Charleston," typescript, n.d., and Faculty Journal, College of Charleston, typescript, April 17, 1848, and May 10, 27, 1850, CCL; [Frederick A. Porcher], manuscript autobiography, ca. 1849, Porcher Family Papers, SCHS.

25. John Bachman to Victor Audubon, December 13, 1847, September 16, October 20, and December 18, 1848, Bachman to Maria Martin, May 11, 1848, Bachman to Jane Bachman, August 27, 1848, and Victor Audubon to "Aunt Maria" Bachman, January 28, 1849, BL/CML; Victor Audubon to Bachman, June 17, August 1, and October 7, 1848, A-B/HU.

26. John Bachman to Victor Audubon, June 30 and September 1, 1849, BL/CML; Bachman to Victor Audubon, August 24, 1849, in C. L. Bachman, *John Bachman*, 274; Bachman to Victor Audubon, June 4, 1849, and Maria Martin Bachman to Victor Audubon, March 30, 1849, BRB/YU.

27. Victor Audubon to John Bachman, August 25, 1851, A-B/HU; *Proc. AAAS* 3 (March 1850): 97; Audubon and Bachman, *Quadrupeds*, 1854, I: 317–28; [John J.] Audubon and [John] Bachman, "Description of a new North American Fox," 114–15; Baird, letter, in *Proc. ANSP* 6 (August 1852): 124.

28. Maria Martin Bachman to "My dear girls," June 18, 1851, BL/CML; Bachman to Victor Audubon, May 31, 1851, A-B/HU.

29. John Bachman to Victor Audubon, July 25, 1851, and July 9, 1852, BRB/YU; Victor Audubon to Bachman, October 25, 1851, A-B/HU; Bachman to Victor Audubon, April 9, June 18, and July 27, 1852, Victor Audubon to "Aunt" Maria Martin Bachman, October 13, 1852, John Bachman to Edward Harris, March 13, 1852, and Bachman to "My Dear Sir," March 25, 1852, BL/CML; Bachman to Victor Audubon, April 3, 1852, in C. L. Bachman, *John Bachman*, 279–81. Dates of the issue of the original and subsequent editions of the volumes are given in Herrick, *Audubon the Naturalist*, 2:405–6; Herrick, "Audubon's Bibliography"; A. Ford, *John James Audubon*, 448–49; and A. Ford, *Audubon's Animals*, 217.

30. John Bachman to [John Chisholm], September 9, 1871, BL/CML. Typical of the references to Bachman in biographies of Audubon published during the 1860s is the brief one that Audubon wrote *The Quadrupeds* "in connection with the Rev. John Backman [*sic*] of South Carolina," Buchanan, *Life and Adventures*, 312; Harlan, *Fauna Americana*; Godman, *American Natural History*; Greene, *American Science*, 314–18.

31. S. S. Smucker to John Bachman, September 18, 1835, and Bachman to John J. Audubon, December 7, 1841, BL/CML; Bachman to "My Dear Sir" [President, South Carolina College], December 7, 1850, SCL. I am indebted to Dr. Oswald Schuette for making a concerted effort to find evidence in Berlin archives of the

supposed Ph.D. degree. Bachman never listed such a degree after his name, whereas he did cite the other honorary degrees conferred upon him.

Chapter Four

1. "The Ravenel Family in France and in America."

2. Ibid., 17–35; Mrs. St. Julien Ravenel [Harriott Horry Rutledge Ravenel], *Charleston*, 311; Holcomb, *Marriage and Death Notices from the (Charleston) Times*, 174; *South Carolina Genealogies*, 2:40, 56, 72–73, 86–87; "Will of Daniel Ravenel," March 11, 1807, in Charleston County Will Transactions, 30:1142–51; *Charleston, South Carolina, Directory and Almanac, 1806*, 67; "Copy Resolution," May 29, 1833, RL/CML; [copy of tombstone inscription] "Sarah Amelia Ford . . . died May 22, 1799 . . . ," and Edmund Ravenel to Benjamin J. Johnson, December 6, 1859, Ford-Ravenel Papers, SCHS; Waring, *History of Medicine*, 287–88; A[nn] K[ing] G[regorie], "Edmund Ravenel."

3. "Diary of Timothy Ford,"132–33; Wilson and Grimes, *Marriage and Death Notices*, 1:101; T. Ford, *An Address to the Literary and Philosophical Society*, CLS; manuscript list of members of Literary and Philosophical Society, SCHS.

4. "Edmund Ravenel 1821," manuscript journal, CML; R[ichardson], "Some Old and New Records of the Sargassum Fish"; Minute Book, Academy of Natural Sciences of Philadelphia, December 3, 31, 1822, Acad./ANSP.

5. "Edmund Ravenel 1821," manuscript journal, CML; Cuvier and Valenciennes, *Histoire naturelle des poissons*, 7:343–51, 9:401, 430, 475, 481.

6. *Charleston Directory and Strangers' Guide for the Year 1822*, 71; Edmund Ravenel to J. Manning, March 16, 1824, Edmund Ravenel Papers, WHL; "Domestic Intelligence"; Waring, *History of Medicine*, 71–78.

7. Wilson and Grimes, *Marriage and Death Notices*, 101, 151; Edmund Ravenel to Louisa C. Ford, May 1, 1827, RL/CML.

8. *South Carolina Genealogies*, 2:73; *Charleston Mercury*, July 15, 1848.

9. Waring, *History of Medicine*, 77–80; Minute Book, 1825–33, September 5, October 1, and December 5, 1831, and July 20, 1832, and "Petition of the Medical College in Reference to Faculty," December 5, 1831 (broadside), WHL; *An Exposition of the Affairs of the Medical Society*; Thomas S. Grimke to Dr. Edmund Ravenel & other curators, May 29, 1833, RL/CML.

10. Edmund Ravenel to Edward W. North, [ca. September or October 1833], Edmund Ravenel Papers, WHL; Waring, *History of Medicine*, 94.

11. Jacob De La Motta to Edmund Ravenel, October 30, 1834, Edward Frost to Edmund Ravenel, March 7, 1835, Edmund Ravenel Papers, WHL; Dickson, *Statements in Reply to Certain Publications*; Ravenel to "Mssrs. King & Durant," [November 20, 1835], Ford-Ravenel Papers, SCHS; Pease and Pease, "Intellectual Life in the 1830s," 246–47.

12. Edmund Ravenel to John Gordon, February 17, 27, 1835, "Register of Mesne Conveyance Office," February 23, 1835, and "Bond," May 25, 1835, St. Julien Ravenel Family Papers, SCHS; "List of the Gordon Gang," n.d., mortgage of "Negroes,"

May 25, 1835, "Bill of Sale," February 10, 1836, and Sale of Patent right to "brick making machine," August 11 and October 30, 1838, Ford-Ravenel Papers, SCHS; Irving, *A Day on Cooper River*, 23.

13. "Edmund Ravenel 1821," manuscript journal containing lists of donations of shells and seventy-six entries between January 18, 1829, and June [n.d.], 1837, Amos Binney to Edmund Ravenel, April 14, 1831, March 24, 1832, and May 17, 1834, T. A. Conrad to Ravenel, February [n.d.], 1833, Isaac Lea to Ravenel, May 5, and September 28, 1834, W. H. Lindsay to Ravenel, March 5, July 11, and October 17, 1831, and April 20, 1832, C. A. Poulson to Ravenel, November 17, 1831, and February 17, 1832, Thomas Say to Ravenel, July 13, 1832, D. Humphreys Storer to Ravenel, March 7 and October 5, 1831, June 17, 1832, and October 2, 1834, [J. G.] F. Wurdemann to Ravenel, [from] "Paris, 1833," Baron de Ferussac to Ravenel, January 25, 1833, and Malthus A. Ward to Ravenel, December 2, 1830, RL/CML; E. Ravenel Jr., *Catalogue of Recent and Fossil Shells in the Cabinet of the Late Edmund Ravenel*.

14. Edmund Ravenel to Amos Binney, November 21, 1830, Manuscript Collections, MCZ; Isaac Lea to Ravenel, April 14 and November 29, 1831, and January 17, 1832, T. A. Conrad to Ravenel, October 18, 1834, Thomas Say to Ravenel, October 16, 1832, RL/CML; Gillespie, "Preparing for Darwin," 96–110.

15. Gillespie, "Preparing for Darwin," 104–10; John McCrady Diary, McC/McCF.

16. Ravenel to S. G. Morton, February 13, 1833, Samuel G. Morton Papers, APSL.

17. Edmund Ravenel to S. G. Morton, June 29, 1832, Acad./ANSP; Morton to Ravenel, September 21, 1832, and Ravenel to Morton, August 7, 1837, RL/CML; Conrad, *New Fresh Water Shells*, 39–40.

18. Edmund Ravenel, "Description of Two Fossil Scutella Sent to S. G. Morton of Phila. For publication—July 1841," manuscript, Edmund Ravenel to S. G. Morton, January 2, 1843, and June 13, 1845, and Morton to Ravenel, November 23, 1842, and May 1, 1844, RL/CML; Ravenel to Henry W. Ravenel, April 8, 1844, Henry William Ravenel Papers, SCL; Edmund Ravenel, "Description of some new species of Fossil Organic remains, from South Carolina," manuscript, Acad./ANSP; Edmund Ravenel, "Description of two new species of Fossil Scutella from South Carolina," *Proc. ANSP* 1 (September 1841): 81–82, and reprinted with additional paragraph and two figures in *Jour. ANSP* 8 (1842): 333–36; Edmund Ravenel, "Description of some new species of Fossil Organic remains, from the Eocene"; Edmund Ravenel, "Description of a new recent species of Scutella." Reference to various notes, comments, and specimens sent by Ravenel to the ANSP are mentioned in *Proc. ANSP* 1 (September 1841): 89; 1 (December 1841): 131; 1 (September 1842): 210; 1 (November 1842): 221; 1 (January 1843): 234; 2 (May 1844): 57–58; 2 (June 1845): 253.

19. Edmund Ravenel to S. G. Morton, March 26, 1844, and April 28, 1845, SGM/HSP; Charles Lyell to Ravenel, December 30, [1841], Lewis R. Gibbes to Ravenel, January 21, 1847, Ravenel to Gibbes, January 25, 1847, and January 8, 1848, Henry W. Ravenel to Edmund Ravenel, May 23, [1848?], Michael Tuomey to Edmund Ravenel, January 17, 1846, and Isaac Lea to Ravenel, November 19, [1843?], RL/CML; Tuomey to L. R. Gibbes, January [n.d.], 1846, LRG/LC; Lyell, "On the Newer Deposits of the Southern States of North America"; Lyell, *Travels*, 1:138–41; Wilson, *Lyell in America*, 76–85; E. Ravenel, *Echinidae, Recent and Fossil*; Tuomey

and Holmes, *Pleiocene Fossils* (reprint ed.), vii, 2–11, 21, 27–28; Jellison and Swartz, "Scientific Interests of Robert W. Gibbes." Acknowledgments of Ravenel's donations and contributions to the ANSP during the period 1844–1847 are mentioned in *Proc. ANSP* 2 (May–June 1844): 57–58; 2 (May 1845): 232; 2 (August 1845): 260; and 3 (December 1847): 326.

20. L. R. Gibbes to Edmund Ravenel, February 18, 1845, and Ravenel to Gibbes, February 5 and April 7, 12, 1847, RL/CML; Edmund Ravenel to Henry W. Ravenel, December 14, 1846, CUL; H. A. De Saussure to Edmund Ravenel, February 10, 1845, Ford-Ravenel Papers, SCHS; L. R. Gibbes to S. G. Morton, March 12, 1845, SGM/HSP.

21. Edmund Ravenel, "On the Medical Topography of St. John's"; E. Ravenel, "On the Advantages of a Sea-Shore Residence";*Charleston Mercury*, July 15, 1848.

22. Edmund Ravenel, "On the Recent Squalidae of the Coast of South-Carolina"; Edmund Ravenel, "Illustration of several New Animals," cited in the *Charleston Courier*, March 15, 1850; Ravenel, manuscript draft of AAAS paper on Squalidae, 1850, and L. R. Gibbes, P. C. Gaillard, and F. S. Holmes to Edmund Ravenel, March 20, 1850, John McCrady to Ravenel, July 18, August 30, and October 8, 1853, RL/CML.

23. John Edwards Holbrook, *North American Herpetology*, 3:109, 4:v, 5:105; Edmund Ravenel to L. R. Gibbes, January 8, 15, March 20, April 2, 19, May 11, November 26, and December 11, 1857, Gibbes to Ravenel, August 3, 1857, and February 2, 1858, Spencer F. Baird to Ravenel, June 4, 1855, Oscar M. Lieber to Ravenel, February 17 and September 16, 28, 1859, F. S. Holmes to Ravenel, October 5, 1853, and October 21 [1856?], Isaac Lea to Ravenel, February 13, 1855, Temple Prime to Ravenel, December 1, 1851, Francis A. Sauvelle to Ravenel, December 9, 1857, and E. R. Showalter to Ravenel, November 28, 1859, and February 21, 1860, RL/CML; Ravenel to Baird, April 13 and June 14, 1860, ASecr./SIA; Ravenel to Henry W. Ravenel, December 14, 1846, Henry William Ravenel Papers, CUL; Tuomey and Holmes, *Pleiocene Fossils* (reprint ed.), vii, 10–11, 21. McCrady named the genus for Ravenel in an informal presentation before the ESNH; see *Proc. ESNH* 1 (1858): 282–83.

24. "Tax Return of Edmund Ravenel, 1844, St. Thomas Parish," "Edmund Ravenel, Taxable Property, City of Charleston, 1852," "List of Negroes," Alexander H. Mazÿck to Edmund Ravenel, March 4, 10, 1859, J. P. DeVeaux to Ravenel, January 26, 1859, P. D. Lorre[?] to Ravenel, February 3, 1846, unknown U.S. Treasury Department Official to Ravenel, February 11, 1859, Ford-Ravenel Papers, SCHS; [Edmund Ravenel Plantation Account Book], entries from 1851–61, "Bill of Sale, 1854," "List of 72 Rice Field Negroes . . . to Be Sold," February 22, [1849], and Thomas Farr Capers to Ravenel, February 4, March 13, and April 3, 1851, RF, SCHS; U.S. Census, *Slave Schedule*, 1860, 397–98; *List of the Taxpayers of the City of Charleston for 1859*, 286–87; F. A. Ford, *Census of the City of Charleston, South Carolina, for the Year 1861*.

25. [Edmund Ravenel], "Suggestions in relation to Yellow Fever on the[?] Island SC, 1858," manuscript, and another untitled version of the same, [Edmund Ravenel], "U. States Hospitals, at Ft. Moultrie, To Edmund Ravenel Dr.," September 1858,

William E. Martin to Ravenel, September 20, 1858, and Edward McCrady to Ravenel, September 20, 1858, RF, SCHS; Ravenel to John H. Honour, April 26, 1859, Ford-Ravenel Papers, SCHS; Ravenel to "Mr. Rt. Adger," [1858?], RL/CML.

26. Isaac Lea to Edmund Ravenel, April 28, 1857, April 26 and September 29, 1858, December 26, 1859, and January 11, 1860, Ravenel to Lea, January 6, 1860, Ravenel to unknown [of the ANSP], [1860], S. P. Woodward to Ravenel, March 31, 1857, Robert McAndrew to Ravenel, March 31, 1858, and September 26, 1859, Ravenel to McAndrew, March 16, 1860, Philip P. Carpenter to Ravenel, June 27, 1859, and February 5, 1860, Felipe Poey to Ravenel, February 10, 1858, William Stimpson to Ravenel, December 21, 1857, January 30 and April 29, 1858, June 1, July 15, and December 24, 1860, and April 14, 1861, Ravenel to Stimpson, January [n.d.], 1861, L. R. Gibbes to Ravenel, February 2, 1858, and February 9, 1860, Ravenel to Gibbes, February 3, 4, 15, April 2, and November 19, 27, 1858, October 20 and November 7, 1859, February 6, June 6, July 5, November 27, and December 13, 1860, Stephen Elliott Jr. to Ravenel, August 17, 1857, RL/CML; Edmund Ravenel, "Description of three new species of Univalves"; Edmund Ravenel, "Tellinidae of South Carolina," (read before the Society on December 15, 1860); Isaac Lea, *A Synopsis*, 33, 64; Isaac Lea, "Rectification of Mr. T. A. Conrad's 'Synopsis,' published in the 'Proceedings' of the Academy of Natural Sciences of Philadelphia, February, 1853"; and Lea's marginal notes on page proof of same, Acad./ANSP.

27. William Stimpson to Ravenel, January 30, 1861, Ravenel to L. R. Gibbes, February 13, June 10, and October 24, 1861, RL/CML; Edmund Ravenel, "Descriptions of New Recent Shells from the Coast of South Carolina." A thorough study of Stimpson and of his relationship to the Charleston naturalists appears in Ronald Vasile's biography of Stimpson (manuscript in preparation).

28. W. G. Binney to Edmund Ravenel, September 27, October 3, 18, November 2, and December 24, 1859, and January 14, April 12, and December 24, 1860, Edmund Ravenel to L. R. Gibbes, February 13 and November 9, 1861, and Henry W. Ravenel to Edmund Ravenel, January 15, 1861, RL/CML; [Edmund Ravenel], "Lincoln vs. Cotton," October 24, 1860, and Ravenel to B. Johnson, November 10, 1860, Ford-Ravenel Papers, SCHS; Ravenel to Thomas Bland, January 21, 1861, Acad./ANSP; M. L. Ravenel De Saussure to L. R. Gibbes, September 8, 1865, LRG/LC.

29. Tryon, "A Sketch of the History of Conchology in the United States," 179–80; Edmund Ravenel Jr., *Catalogue of Recent and Fossil Shells in the Cabinet of the Late Edmund Ravenel*; Turgeon et al., *Common and Scientific Names*; Lee, "Edmund Ravenel." I am grateful to Dr. Harry Lee for sharing with me his knowledge of Ravenel taxa and patronyms.

Chapter Five

1. Morse, *Genealogical Register*, 145; *Second Annual Report of the Receipts and Expenditures of the Town of Norfolk*, 15; Richardson, "Dr. Anthony Cordes and Some of His Descendants," 225, 228–29; Harriott H. Ravenel to Marcus Benjamin, July 25, [1890s], Marcus Benjamin Papers, SIA.

2. Clipping from unidentified newspaper, [ca. September 10, 1871], and "Brown University Graduate Records," [ca. 1913], John Hay Library Archives, Brown University; *Columbian Centinel*, March 19, 1800, 2; Adler, "Holbrook, John Edwards"; Worthington and Worthington, "John Edwards Holbrook."

3. Clipping from unidentified newspaper [ca. September 10, 1871] and "Brown University Graduate Records" [ca. 1913], John Hay Library Archives, Brown University.

4. John Edwards Holbrook to Silas Pinckney Holbrook, February 3, 1819, SCL.

5. Ibid.

6. Ibid., December 1819, SCL.

7. Ibid.

8. Ibid.

9. John Edwards Holbrook to Mary Holbrook, February 2, 1820, in possession of Bertram Holbrook Holland, Brookline, Mass. (copy supplied by courtesy of Richard D. Worthington, Department of Biological Sciences, University of Texas at El Paso); Ogier, *Memoir*, 8–9.

10. Minute Book, Medical Society of South Carolina, 1810–25, December 9, 1822, typescript, WHL; Ogier, *Memoir*, 5–6; Cuvier and Valenciennes, *Histoire Naturelle des Poissons*, 7:454–67.

11. Minute Book, Medical Society of South Carolina, 1810–25, February 2, April 12, and June 24, 1824, October 1, 1825, September 1, 1831, and July 5, 24, September 2, and October 2, 1833, WHL; F. Peyre Porcher to [Marcus Benjamin], June 15, 1891, Marcus Benjamin Papers, SIA.

12. "List of Members of the Literary and Philosophical Society, Charleston, February 8th 1832," typescript, SCHS; Frederick A. Porcher, "Memoir," typescript, CCL; *Constitution and By-Laws of the Elliott Society of Natural History*, 15.

13. Weber, "Dr. John Rutledge and His Descendants," 94–95; Smith, "Charleston and Charleston Neck," 33–35; Holman, "Charleston in the Summer of 1841"; Bremer, *Homes of the New World*, 1:270–73, 498, 505–7; E. C. Agassiz, *Louis Agassiz*, 2:495–96, 509.

14. Holbrook, *North American Herpetology* (reprint ed.), 1:vii–xi, 2:v–vi, 3:12, 5:vi; Rutledge, "Artists in the Life of Charleston," 148, 217; Blum, *Picturing Nature*, 147–48, 150, 195, 363–64.

15. Holbrook, *North American Herpetology* (reprint ed.), 1:x–xi, 2:v–vi, 5:48, 66, 110; John M. B. Harden, manuscript journal, June 19, 21, 1841, in possession of Mrs. George Buroughs, Hinesville, Ga.

16. John Edwards Holbrook to Samuel S. Haldeman, June 25, July 16, [1839?], and November 13, 1839, Acad./ANSP; Holbrook, *North American Herpetology* (reprint ed.), 1:xi, 2:v–vi, 4:136, 5:v, 82.

17. Holbrook, *North American Herpetology* (reprint ed.), 1:x–xi, 3:32, 4:v–vi, 5:v–vi; Holbrook to D. Humphreys, August 25 [1839?], October 1, 1839, and November 29, [1840?], Boston Museum of Science.

18. John Edwards Holbrook to Samuel G. Morton, April 19, 1834, and November 28, 1835, APSL; Holbrook to Morton, April 14, 1844, Acad./ANSP; Holbrook, *North American Herpetology*, 4 vols. (Philadelphia: J. Dobson, 1836–40) and ex-

panded versions of vols. 1 and 2 issued with original title page, ca. 1839; Holbrook, *North American Herpetology*, 1842 ed., 1:v ("Publisher's Notice"), vii ("Errata" slip), xiii–xiv ("Preface"); Holbrook, *North American Herpetology*, reprint of 1842 ed. (1976), "Editor's Note," by Kraig Adler, v–vi; Shimek, "Holbrook's *North American Herpetology*, First Edition," *Proceedings of the Iowa Academy of Science* 31 (1924): 427–30.

19. John Edwards Holbrook to D. Humphreys Storer, November 29 and December 25, [1840?], Boston Museum of Science; Holbrook to Storer, n.d., Manuscripts, Massachusetts Historical Society.

20. Louis Agassiz, [Eulogy on John Edwards Holbrook], 348.

21. Holbrook, *North American Herpetology*, 1842 ed., 1:ix; Adler, "Holbrook, John Edwards"; Adler, "New Genera and New Species," xxix–xliii; Girard, "On a New American Saurian Reptile," 200; DeKay, *Zoology of New York*, pt. 3:iv; *Southern Rose* 5 (February 18, 1837): 103, 6 (June 9, 1838): 336, 7 (December 12, 1838): 144.

22. *North American Review* 49 (July 1839): 145–55; *AJS* 35 (1839): 186–87; Holbrook, *North American Herpetology* (reprint ed.), 2:70–71, 3:19.

23. Holbrook, *North American Herpetology*, 1842 ed.

24. Ogier, *Memoir*, 8–9; G. Bibron to Holbrook, May 25, 1845, St. Julien Ravenel Childs Papers, SCHS; Holbrook, *North American Herpetology*, 1842 ed., 1:title page.

25. *CMJR* 15 (September 1860): 705–6; *Charleston Courier*, May 1, 1849; "Memoirs of Frederick Adolphus Porcher," *SCHM* 47 (1946): 226; "Officers of the South-Carolina Historical Society, Elected June, 1855," printed circular, William J. Rivers Papers, SCL; Way, *History of the New England Society*, 137–42; Holbrook, "Education of Farmers."

26. John Edwards Holbrook to [George Ord?], August 15, 1846, and to George Ord, January 8, 1847, Ms./APSL; Holbrook to Spencer F. Baird, June 15 [1850?], ASecr./SIA.

27. Stephens, "John Edwards Holbrook," 451–52; Holbrook, *Southern Ichthyology*, pts. 2 and 3.

28. John Edwards Holbrook, "Reptiles" and "Fish"; Gill, "John Edwards Holbrook," 59–60; Ogier, *Memoir*, 12; Spencer F. Baird to Holbrook, November 15, 1849, and March 10, October 19, and December 21[?], 1850, and Holbrook to Baird, January 13, 1850, ASecr./SIA; Holbrook to D. Humphreys Storer, [1840], Boston Museum of Science.

29. Edward Rüppell to John Edwards Holbrook, March 15, 1846, Ms/HL; L. Agassiz, [Eulogy], 348.

30. Lurie, *Louis Agassiz*, 142–44; Lurie, "Louis Agassiz and the Races of Man," 234–35; *Charleston Courier*, December 6 and 13, 1847.

31. L. Agassiz, [Eulogy], 348; E. C. Agassiz, *Louis Agassiz*, 2:495–97.

32. Lurie, *Louis Agassiz*, 162–65; *Charleston Courier*, February 19, 1849.

33. *Proc. AAAS* 3 (1850): v–viii, xxi–xxiii.

34. John Edwards Holbrook, "An account of several species of Fish"; Stephens, "John Edwards Holbrook," 451–52; Gabriel E. Manigault, manuscript autobiography, Manigault Family Papers, SHC.

35. Stephens, "John Edwards Holbrook," 451–53; Holbrook, *Ichthyology of South Carolina*, 1855 ed.

36. Holbrook, *Ichthyology of South Carolina*, 1855 ed.; Gill, "Review of Holbrook's *Ichthyology of South Carolina*"; Holbrook, *Ichthyology of South Carolina*, 1860 ed.

37. John Edwards Holbrook to Joseph Leidy, February 12, 1860, Francis S. Holmes to Leidy, March 30, 1860, Holbrook to John Lawrence LeConte, March 12, 1860, and Harriott P. Holbrook to John Lawrence LeConte, April 25, 1860, Manuscript Collections, APSL; *CMJR* 15 (July 1860): 568; *CMJR* 15 (September 1860): 705–6.

38. "Will of Harriott Pinckney Holbrook," in Charleston County Will Transactions, 50:151–52, South Carolina Department of Archives and History; U.S. Census, *Population Schedule*, 1860, and *Slave Schedule*, 1860.

39. *List of the Taxpayers of the City of Charleston for 1859*, 158.

Chapter Six

1. *South Carolina Genealogies*, 2:223; *[Tributes to] Lewis R. Gibbes*, copy in CCL; Smallwood and Smallwood, *Natural History and the American Mind*, 111–12; Sarah P. Gibbes, manuscript sketch of life of Lewis Reeve Gibbes, written on blank pages in a copy of *The Life and Exploits of Don Quixote de la Mancha*, Hinson Collection, CLS.

2. Sarah P. Gibbes, manuscript sketch of Lewis Reeve Gibbes; Stevenson, *Diary of Clarissa Adger Brown*, 35–36; Klosky, *Pendleton Legacy*, 57–58; LaBorde, *History of the South Carolina College*, 194–95; Sargent, *Silva of North America*, 12:70.

3. *South Carolina Genealogies*, 2:223; LaBorde, *History*, 195; Faculty Minute Book, 1814–33, December 27, 1827, and May 19 and November 27, 1829, SCL; Lewis R. Gibbes to "Aunt" [Henrietta A. Drayton], January 10, 1825, Gibbes-Gilchrist Papers, SCHS.

4. Holcomb, *Marriage and Death Notices from the Charleston Observer*, 31; Lewis R. Gibbes to John Gibbes, April 12, 1828, LRG/LC; R[obert] Anderson et al. to [South Carolina College faculty?], in Pendleton Academy Letters/SCL; Lewis R. Gibbes to Board of Trustees, South Carolina College, November 30, 1831, SCL.

5. Warner, "The Idea of Southern Medical Distinctiveness," and "A Southern Medical Reform"; Lewis R. Gibbes to Board of Trustees, South Carolina College, November 30, 1831, and Proceedings of the Board of Trustees, November 30 and December 3, 1831, University Archives, University of South Carolina; LaBorde, *History*, 195.

6. LaBorde, *History*, 195; L. R. Gibbes, *Catalogue of the Phoenogamous Plants*; L. R. Gibbes, "Observations on Solar Eclipse of 30 Nov. 1834," *Charleston Mercury*, December 17, 1834, in collection of newspaper articles by Lewis R. Gibbes, Lewis R. Gibbes Papers, College of Charleston Library; L. R. Gibbes, "On the General Principles of the Resistance of Fluids."

7. Malone, *Public Life of Thomas Cooper*, 337–67; A. Blanding to Board of Trustees, June 3, 1835, Proceedings of the Board of Trustees [of South Carolina College],

November 24, 1831, November 26 and December 5, 9, 12, 1834, and November 27, 1837, University Archives, University of South Carolina; Robert W. Gibbes to Lewis R. Gibbes, November 20, 1835, LRG/LC; LaBorde, *History*, 195–96.

8. *Catalogue of the Trustees . . . 1836*, 7, 37, 46; Lewis R. Gibbes to "My dear Aunt" [Henrietta A. Drayton], July 18, 1837, Gibbes-Gilchrist Papers, SCHS; Joel R. Poinsett to David Baillie Warden, March 25, 1836, David B. Warden Papers, Maryland Historical Society; John Edwards Holbrook to "My dear friend" [Achille Valenciennes], April 20, 1836, Manuscript Collections, APSL; Stevenson, *Diary of Clarissa Adger Brown*, 35–36; L. R. Gibbes, [On Eugene Chevruel], 223.

9. Lewis R. Gibbes, folder containing dried plants and notes, in the Charleston Museum; E. Chevruel to Gibbes, October 27, 1836, CML; L. R. Gibbes, [On Eugene Chevruel], 223; L. R. Gibbes "Early Life of Francis Arago," *Charleston College Magazine* 1: 225–31, 2: 41; Gibbes to "My dear Aunt," May 8 and July 18, 1837, Gibbes-Gilchrist Papers, SCHS; Gibbes, manuscript notebook "1836 or 1837," CCL.

10. A. Blanding to Lewis R. Gibbes, December 26, 1837, Blanding to Board of Trustees, College of Charleston, December 26, 1837, David Johnson to Board of Trustees, December 28, 1837, Secretary of Board of Trustees to Gibbes, February 3, 1838, Thomas Cooper to Gibbes, May 11, 1838, W. F. DeSaussure to Gibbes, July 23, 1838, Alister[?] Garden to Gibbes, June 1, 1838, and R. W. Gibbes to Gibbes, May 20, 1838, November 28, 1845, LRG/LC; L. R. Gibbes to Adele Janvier, March 20, 1838, in Gibbes's French letter book, CCL; R. W. Gibbes to L. R. Gibbes, December 3, 1846, Robert W. Gibbes Papers, SCL.

11. Lewis R. Gibbes to Amelie Moret, August 16, 1839, Gibbes to Louisa I. Gibbes, February 6, 1838, and Gibbes to Marie M. Burrill, November 26, 1845, all in Gibbes's French letter book, CCL; Snowden, "The Late Seventies"; "Memoirs of Frederick Adolphus Porcher," *SCHM* 47 (1946): 153.

12. Faculty Reports to the President of the College of Charleston, March 7, 1871, March 12, 1874, and March 11, 1876, CCL; Lewis R. Gibbes to John H. Means, Chairman of Board of Visitors of South Carolina College, [November 22, 1859], SCL; Snowden, "The Late Seventies."

13. James P. Espy to Lewis R. Gibbes, December 4, 1841, July 2 and September 5, 1842, February 26 and March 2, 27, 1843, and February 12, 1844, Thomas Parker to Gibbes, April 30, 1839, and [unknown] to Gibbes, November 4, 1840, LRG/LC; Gibbes, "The Natchez Tornado," *Charleston Courier*, September 3, 1841; Gibbes, "Meteorological Register for the City of Charleston," *Charleston Courier*, March 28, 1843; Gibbes, "Atmospherical Phenomena," *Charleston Mercury*, January 22, 1844.

14. Joseph Henry to Lewis R. Gibbes, December 6, 1843, and A. D. Bache to Gibbes, April 7, 1847, LRG/LC; Gibbes to Henry, March 13, 1844, and March 1 and June 30, 1845, and Henry to Gibbes, May 31, 1845, in Henry, *Papers of Joseph Henry*, 4:53–56, 236–44, 280–84, 303–8. Copies of the newspaper articles written by Gibbes are contained in his scrapbook, CCL.

15. Minute Book of Board of Trustees of the College of Charleston, May 27, 1847, and January 10, 1848, CCL; Mitchell King to Lewis R. Gibbes, July 17, 1847, and January 24, February 10, April 29, June 15, and October 11, 1848, Gibbes to Elias Loomis, October 24, 1854, LRG/LC; Gibbes to Bache, November 18 and Decem-

ber 14, 1853, and April 19, June 12, and [August?] 1854, Manuscripts, Huntington Library; Gibbes to Bache, June 30, 1855, Alexander D. Bache Papers, SIA.

16. Lewis R. Gibbes to J. D. Dana, June 4, 1849, DF/YU; Thomas Parker to Gibbes, March 21 and August 20, 1840, LRG/LC.

17. John Bachman to Lewis R. Gibbes, January 18, 1842, November 30, 1845, and October 27, 1849, Lewis R. Gibbes Papers, CML; Audubon and Bachman, *Quadrupeds*, 1:vii.

18. R. T. Brumby to Lewis R. Gibbes, July 23, 1842, and August 5, 1843, J. Hamilton Couper to Gibbes, April 22 and June 19, 1843, Robert W. Gibbes to Gibbes, November 7, 1844, January 27, 1845, Michael Tuomey to Gibbes, November 22, 1843, and November 29, 1844, and Isaac Lea to Gibbes, July 30, 1850, LRG/LC; Gibbes to A. A. Gould, March 15, 1845, Ms/HL; Robert W. Gibbes, "Memoir on the Fossil genus *Basilosaurus*," 8; Tuomey, "Notice of the Discovery," 152; Lea, "Description of New Species," 254; Ruffin, *Report*, 31–32, 45–48; Tuomey, *Report*, iii, 139, 166, 186; L. R. Gibbes, "Catalogue of the Fauna of South Carolina."

19. B. A. Gould to Lewis R. Gibbes, November 12, 1849, Stephen Alexander to Gibbes, September 2, 1846, O. M. Mitchell to Gibbes, July 15, 1847, Sears Walker to Gibbes, December 9, 1850, J. B. DeBow to Gibbes, March 18 and October 29, 1846, and July 17, 1847, and F. Peyre Porcher to Gibbes, October 27, 1849, LRG/LC; W. G. Simms to Gibbes, May 8, 1849, and April 15, 1850, Simms, *Letters of William Gilmore Simms*, 2:514–15, 3:34.

20. Collection of newspaper articles by Lewis R. Gibbes, CCL; Gibbes to J. D. Dana, June 4, 1849, DF/YU; Dana to Gibbes, January 3, 1851, LRG/LC.

21. Lewis R. Gibbes to J. D. Dana, November 21 and December 16, 1851, DF/YU; copy of newspaper article in the collection of newspaper articles by Lewis R. Gibbes, CCL; L. R. Gibbes, "Aurora Borealis."

22. J. D. Dana to Lewis R. Gibbes, March 31, 1849, and J. G. F. Wurdemann to Gibbes, November 20, 1844, and January 9, 1845, LRG/LC; E. C. Agassiz, *Louis Agassiz*, 493–94; Gibbes to Philippe [Felipe] S. Poey, February 28, 1844, and February 25, 1845, and Gibbes to Don José Yradi, September 29, 1845, in Gibbes's French letter book, CCL.

23. Lewis R. Gibbes to Louisa I. Gibbes, September 4, 1845, August 21, 24, 1846, and August 16 and September 7, 8, 1847, and Gibbes to Gabriel Bibron, August 28, 1846, in Gibbes's French letter book, CCL; Amos Binney to Gibbes, April 29 and July 9, 1846, Robert H. Browne to Gibbes, December 15, 1847, and J. E. Hilgard to Gibbes, May 18 and July 21, 1849, LRG/LC; Gibbes to Robert Bridges, July 21 and December 8, 1847, and April 25, 1849, Acad./ANSP; Gibbes to A. A. Gould, July 20, 1845, and June 2, 1847, Ms/HL; Gibbes to D. Humphreys Storer, September 12 and October 17, 1845, Boston Museum of Science; L. R. Gibbes, "Of the Collection of Crustaceans"; L. R. Gibbes, "Catalogue of the Crustacea"; L. R. Gibbes, "Crustacea."

24. Lewis R. Gibbes to A. A. Gould, April 25, 1849, Ms/HL; Gibbes to Jacob W. Bailey, October 29, 1850, Boston Museum of Science; Emma Gibbes to "The Chief of the Department of Natural History," April 21, 1900, SIA; L. R. Gibbes, "On the Carcinological Collections."

25. Williams, *Shrimps*, 88, 127, 321, 347, 391, 459; Williams et al., *Common and Scientific Names*, passim; Lunz, *Rediscovery of* Squilla Neglecta *Gibbes*; Manning and Heard, "Stomatopod Crustaceans"; Lewis R. Gibbes to J. D. Dana, March 8, 1852, DF/YU.

26. Dana, Comment on "On the Carcinological Collections"; J. D. Dana to Lewis R. Gibbes, December 24, 1850, and May 26 and December 8, 1853, and Joseph Leidy to Gibbes, March 1, 1851, LRG/LC; Gibbes to Leidy, June 7, 1852, Acad./ANSP.

27. L. R. Gibbes, "Description, with figures, of six species of Porcellana"; L. R. Gibbes, "Monograph of the Genus Cryptopodia" (read before the ESNH in June 1856); L. R. Gibbes, "Description of Ranilia Muricata" (read before the ESNH in July 1857); Gibbes to Spencer F. Baird, July 21, 1855, ASecr./SIA; Williams, *Shrimps*, 244–48, 389, 462.

28. *[Tributes to] Lewis R. Gibbes*, 12; L. R. Gibbes, "Description (with figure) of Menobranchus punctatus"; L. R. Gibbes, "Description of a New Species of Salamander"; Stephens, "John Edwards Holbrook," 455–56.

29. Emma Gibbes to "The Chief of Department of Natural History," April 21, 1900, SIA; Lewis R. Gibbes to J. D. Dana, December 10, 16, 1851; L. R. Gibbes, "Xenotime"; G. Robinson, "Charles Upham Shepard."

30. Lewis R. Gibbes to James Hall, August 19, 1856, James Hall Papers, New York State Library; L. R. Gibbes, "Remarks on Niagara Falls"; L. R. Gibbes, "On Some Points" (read before the ESNH in January 1857).

31. Lewis R. Gibbes, "Notice of the Phenomena" (read before the ESNH in September 1858); L. R. Gibbes, "Report of the Society's Commission to collect information"; L. R. Gibbes, "Personal observations," 153–63.

32. Lewis R. Gibbes to John Torrey, May 27, 1839, February 22, 1844, October 13, 1851, May 9, and September 21, 28, 1852, April 22 and May 31, 1853, and October 16, 1854, Asa Gray to Gibbes, April 18, June 6, July 26, 1842, March 7 and August 7, 1844, March 17, 1852, August 6, 1856, January 22, May 3, and September 4, 1858, March 22, 1859, and February 12, 1882, Gibbes to Gray, February 8, 12, and March 23, 1882, J. W. Bailey to Gibbes, September [n.d.], 1847, and July 24, 1854, LRG/LC; Gibbes to Torrey, October 31, 1851, and March 9, 1852, New York Botanical Gardens Library; Gibbes to Dana, November 21, 1851, DF/YU; Gibbes to Torrey, May 20, 1853, John Torrey Correspondence, Gray Herbarium Library, HU; Maria H. Gibbes to [Librarian, Harvard University], December 31, 1897, and March 7, 1914, Gray Herbarium Library, HU; "Catalogue of Herbarium of Professor L. R. Gibbes, Late of College of Charleston, S.C.," Acad./ANSP; Henry W. Ravenel to Gibbes, July 17, 1855, and April 10, 1856, Botany Department Papers, SHC; Gibbes to Edmund Ravenel, February 2, 1858, CML; Gibbes, "Ravenel's Fungi of Carolina," *Charleston Courier*, June 25, 1855, January 9, 1856, and October 12, 1860; L. R. Gibbes, "Botany of Edings' Bay" (read before the ESNH in October 1857); L. R. Gibbes, "On the representatives" (read before the ESNH in January 1858); Bailey, "Microscopical Observations," 4; Sargent, *Silva of North America*, 12:69–70.

33. Joseph Henry to Lewis R. Gibbes, October 13, 1851, June 10, 1852, June 27, 1854, and August 4 and October 8, 1857, LRG/LC.

34. Joseph Henry to Lewis R. Gibbes, September 7 and December 30, 1859, LRG/LC.

35. Slotten, *Patronage*, 112–46; J. D. Runkel to Lewis R. Gibbes, March 2 and November 2, 1859, LRG/LC; L. R. Gibbes, "On a Convenient Form of Aspirator"; L. R. Gibbes, "Portable and easily made Heliotrope"; L. R. Gibbes, "Note on the Cycloid"; L. R. Gibbes, "Note on Maxima and Minima"; L. R. Gibbes, "An Easy Mode of Approximating" (reprints of the three articles containing Gibbes's corrections are in LRG/LC).

36. Stephens, "Mermaid Hoax"; A. A. Gould to Lewis R. Gibbes, November 21, 1845, LRG/LC; Gibbes to Gould, September 12 and December 2, 1845, Augustus A. Gould Papers, Houghton Library, HU.

37. A. A. Gould to Lewis R. Gibbes, July 24, 1846, LRG/LC; Gibbes to Gould, August 3, 1846, Ms/HL.

38. Lewis R. Gibbes to Spencer F. Baird, March 29, 1851, and Baird to Gibbes, April 3, 1851, ASecr./SIA; L. R. Gibbes, *Rules for the Accentuation of Names*.

39. Francis Markoe Jr. to Lewis R. Gibbes, July 26, 1842, February 27, May 13, and August 26, 1843, and February 13 and March 4, 1844, and Benjamin Silliman Jr. and J. Lawrence Smith to Gibbes, February 28, 1845, LRG/LC; Gibbes to Silliman, March 24, 1845, and June 22, 1846, and Gibbes to Jeffries Wyman, July 7, 1847, Acad./ANSP; Gibbes to A. A. Gould, January 6, 1848, Ms/HL; Gibbes to Markoe, *National Intelligencer*, June 2, 1843; Kohlstedt, *Formation of the American Scientific Community*, 54–77.

40. Walter R. Johnson to Lewis R. Gibbes, September 24, 1844, diploma indicating Gibbes's election to membership in the Academy of Natural Sciences of Philadelphia, May 22, 1847, Edward C. Herrick to Gibbes, [n.d. but soon after August 29, 1850], LRG/LC; Gibbes to Samuel G. Morton, July 6, 1848, SGM/HSP; membership diploma for Gibbes, Boston Society of Natural History, LRG/LC; list of committee members for the third Meeting of the American Association for the Advancement of Science, in *Proc. AAAS* 3 (1850): v–viii; "Circular of the Local Committee of the American Association for the Advancement of Science. Thirteenth Annual Meeting, Springfield, Wednesday, August 3, 1859, 10 o'clock A.M.," Ford-Ravenel Papers, SCHS.

41. *South Carolina Genealogies*, 2:229; Anna Barnwell [Mrs. Lewis] Gibbes to "My dear Aunt," October 5, 1848, Gibbes-Gilchrist Papers, SCHS; I. L. Engnis [?] to Lewis R. Gibbes, October 4, 1848, M. L. Ravenel to Gibbes, December 26, 1858, Alexander Glennin to Gibbes, March 9, 1859, Asa Gray to Gibbes, March 9, 1859, Joseph Henry to Gibbes, March 18, 1859, and Gibbes to John LeConte, July 6, 1868, LRG/LC; U.S. Census, *Population Schedule*, 1850 (p. 292) and 1860 (p. 472); U.S. Census, *Slave Schedule*, 1840 and 1850; *List of the Taxpayers of the City of Charleston for 1859*, 126.

Chapter Seven

1. *Charleston Yearbook, 1882*, 335–37; *Cyclopedia of Eminent and Representative Men*, 508–10; *Charleston Courier*, November 23, 1824; [Mrs. George F. von Kolnitz],

"Francis Simmons Holmes, First Curator of the museum of the College of Charleston," 1907, typescript in possession of Dwight G. von Kolnitz, Spartanburg, South Carolina. This chapter draws in part upon Stephens, *Ancient Animals*.

2. Wilson and Grimes, *Marriage and Death Notices*, 1:104; Holcomb, *Marriage and Death Notices from the Charleston Observer*, 124; *Rules of the South Carolina Society*, 11–66, 121; *Charleston Directory and Stranger's Guide for the Year 1822*, 48; [von Kolnitz], "Francis Simmons Holmes"; Francis S. Holmes Scrapbook, CML.

3. Deeds of Conveyance, St. Andrew's Parish, South Carolina, Book M12: 181 and Book N11: 528, 530; U.S. Census, *Population Schedule*, 1840, 145; *Proceedings of the Agricultural Convention*, 52, 78, 263–71, 293–305, 313–19; *Charleston Courier*, June 3, 1840; *Charleston Mercury*, January 10, 1843; *Southern Quarterly Review* 2 (1842): 275; Ruffin, *Essay*; Holmes, *Southern Farmer*.

4. Holmes, *Phosphate Rocks*, 56–57; Michael Tuomey to Lewis R. Gibbes, November 22, 1843, and July 2, 27, 1844, LRG/LC; *Cyclopedia of Eminent and Representative Men*, 1:109.

5. Francis S. Holmes to Samuel G. Morton, July 21 and August 11, 1845, and January 6, 1846, SGM/HSP; Holmes to J. E. Gray, June 27, 1846, J. E. Gray Collection, APSL; Holmes, "Description of a Bezoar Stone"; Holmes, "Notes on the Geology of Charleston."

6. Tuomey, "Notice of the Discovery," 152; Tuomey, *Report*, 166, 186; R. W. Gibbes, "New Species of *Myliobates*"; R. W. G[ibbes], Review of "Report on the Geology of South Carolina"; R. W. Gibbes to Lewis R. Gibbes, August 4, 1846, and July 11, 1847, LRG/LC; R. W. Gibbes to S. G. Morton, March 10, 1847, SGM/HSP. For an account of Leidy's paleontological work, see Warren, *Joseph Leidy*.

7. Minute Book, Board of Trustees, College of Charleston, October 18, 1847, and March 6, 28, and May 4, 1850, and Journal of the Faculty, College of Charleston, January 21, 1848, CCL; F. S. Holmes Scrapbook, CML; Holmes to S. G. Morton, June 15 and August 24, 1848, SGM/HSP.

8. Holmes, "Observations on the Geology of Ashley River"; Robert W. Gibbes, "Remarks on the Fossil Equus"; Holmes, *Remains of Domestic Animals Discovered Among Post-Pleiocene Fossils*; Holmes, "Remarks on Post-Pliocene Fossils"; Robert W. Gibbes to L. R. Gibbes, November 10, 20, 1853, LRG/LC.

9. *Charleston Courier*, March 16, 1850, 2; *Charleston Mercury*, May 25, 1850, 2; Minute Book, Board of Trustees, May 4, June 14, 15, July 30, November 25, 30, and December 2, 1850, CCL; Charleston Museum Accession Book entries for March 28, July 15, July 30, and November 25, 1850 [reconstructed by F. S. Holmes ca. 1866], CML; Mazÿck, *The Charleston Museum*, 25; Cardozo, *Reminiscences*, 53–55.

10. Minute Book, Board of Trustees, July 29 and November 1, 1851, CCL; *Charleston Mercury*, February 1, 1851; Holmes, "Museum of Natural History, South-Carolina, July 25, 1852," circular addressed to "Professor Baird, Smithsonian Institution Department of Natural History," ASecr./SIA.

11. *Charleston Courier*, January 26, 1852; Holmes, "Museum of Natural History" circular; "Natural History in Charleston."

12. *Report of the Curator*.

13. Tuomey and Holmes, *Pleiocene Fossils*, reprint ed., i–xvi, 1–53, 30 plates;

Francis S. Holmes to Gabriel E. Manigault, June 13, 1855, Miscellaneous Letters and Diaries, CML; *Reports and Resolutions of the General Assembly*, 1855, 307–8; Holmes to Joseph Leidy, January 27, 1857, Joseph Leidy Papers, ANSP. Brief notices or reviews of the parts of *Pleiocene Fossils* appear in *AJS*, 2nd ser., 19 (1855): 452; 20 (1855): 128, 301–2; 24 (1857): 159, 447; and 25 (1858): 146.

14. Francis S. Holmes, printed circular, [ca. 1857], Manuscript Collections, CLS.

15. Ibid.

16. Holmes, "Maffitt Channel Borings"; Francis S. Holmes to Henry W. Ravenel, September 26, 1853, Henry William Ravenel Papers, CUL; Holmes, L. A. Frampton, and Francis T. Miles to Lewis R. Gibbes, October 5, 1853, LRG/LC; Minute Book, Board of Trustees, October 17, 1853, CCL; Stephens, "Scientific Societies," 55–66; *Charleston Mercury*, April 20, 1859, 2.

17. Holmes, "Descriptions of New Fossil Balani" (read in March 1855); Holmes, "Description of a New Species of Ostrea" (read in June 1856); Holmes, "Contributions to the Natural History of the American Devil fish" (read in July 1856); *Charleston Mercury*, June 24, 1854; Stephens, *Ancient Animals*, 16.

18. *Catalogue of the Trustees, Faculty and Students of the College of Charleston . . . November 1855*, 26; *Catalogue of the Trustees, Faculty and Students of the College of Charleston . . . January, 1859*, 30–32; Minute Book, Board of Trustees, July 24, 1857, and June 23, 1858, February 26 and July 23, 1859, and January 7, March 22, and April 21, 1860, CCL; manuscript proceedings of the ESNH, December 15, 1856, CML; Mitchell King to Lewis R. Gibbes, March 23, 1858, LRG/LC; Stephens, *Ancient Animals*, 16–17.

19. Mitchell King to Lewis R. Gibbes, August 3, 1859, LRG/LC; Lewis R. Gibbes et al., "Memorial to the Honorable President and members of the Board of Trustees of the College of Charleston," July 1857, with marginalia by Gibbes, Minute Book, Board of Trustees, June 23, 1858, CCL.

20. Francis S. Holmes to Joseph Leidy, January 11, 1855, January 24, [1856?], January 14, March 29, July 1, and September 1, 1857, and April 28, September 8, and n.d., 1858, and n.d., Joseph Leidy Papers, ANSP; John McCrady to Lewis R. Gibbes, March 20, 1855, LRG/LC; Warren, *Joseph Leidy*.

21. Holmes, *Post-Pleiocene Fossils*; *Reports and Resolutions of the General Assembly*, 1857, 444; [Holmes], form for subscribers, January 1, 1858, Langdon Cheves Papers, SCHS; *Annual Report of the Board of Regents of the Smithsonian Institution*, 1858 ed., 20–22, and 1859 ed., 29; Holmes to Joseph Leidy, December 24 [1857], Joseph Leidy Papers, ANSP.

22. Tuomey, *Report*, 53–59; R. W. Gibbes, *The Present Earth*, 6; Holmes, *Post-Pleiocene Fossils*, preface; Hovenkamp, *Science and Religion in America*, 132–45.

23. Holmes, *Remains of Domestic Animals Discovered Among Post-Pleiocene Fossils*, 1–16; Leidy, "Descriptions of Vertebrate Fossils," in Holmes, *Post-Pleiocene Fossils*, 100–105; Ray and Sanders, "Pleistocene Tapirs," 288–98; Holmes to Leidy, three letters, all with n.d., Joseph Leidy Papers, ANSP.

24. S[timpson], Review of *Post-Pliocene Fossils*; Stimpson to Edmund Ravenel, January 30, 1861, RL/CML; Review of *Post-Pleiocene Fossils of South-Carolina*.

25. Francis S. Holmes to Edmund Ravenel, September 12, 1859, RL/CML. I am indebted to Albert E. Sanders for this assessment of *Post-Pleiocene Fossils.*

26. Campbell and Campbell, "Revision"; Turgeon et al., *Common and Scientific Names . . . Mollusks,* 35, 40, 44, 93, 104; *Charleston Courier,* March 20, 24, 25, 27, and December 4, 1858, and January 29, May 5, and October 1, 1859.

27. Francis S. Holmes to Joseph Leidy, March 12, 1859, and January 7, 1861, in Acad./ANSP; U.S. Census, *Population Schedule,* 1860, 442.

Chapter Eight

1. McCrady, "Description of *Oceania* (*Turritopsis*) *nutricula*"; Stephens and Calder, "John McCrady of South Carolina." McCrady's paper, read before the ESNH in December 1856, was published in mid-1857 in the third number of the *Proc. ESNH*; the whole volume, however, the first of the *Proc. ESNH,* was not published until May 1859. See Calder, Stephens, and Sanders, "Comments on the Date of Publication."

2. Edward McCrady, "Biography Notes on Prof. John McCrady, 1831–1881," typescript, and manuscript poems, McC/McCF; "S. H. P." to "My dear Friend" [Mrs. John McCrady], October 21, 1881, McCrady Family Papers, USo; Louis de Bernière McCrady and Mary de Bernière Barnwell, "Mrs. Edward McCrady, 2nd and her De Bernière Papers," typescript, in possession of John McCrady, Charleston, S.C.; Easterby, *History of the College of Charleston,* 125–27, 141.

3. Agassiz, *Introduction*; alumni card for John McCrady, University Archives, HU; *List of Students of the Lawrence Scientific School,* 16, 20; Edward McCrady, "Biography Notes," McC/McCF; [John McCrady], handwritten sketch of work with Agassiz and studies of medusae, McC/SCHS.

4. John McCrady to Edmund Ravenel, July 18, August 30, October 8, 1853, RL/CML; McCrady to "my dear Father, Wed. [October] 27th [1852?]," and to "My dear Mother," February 25, 1853, McC/McCF; Tuomey and Holmes, *Pleiocene Fossils,* vii, 4, 6–11.

5. John McCrady to "Dear Sister," May 8, 1853, and January 23, 1855, McCrady to "My dear Father," September 29, 1853, and McCrady to "My dear Mother," n.d. [1853?] and January 23 [1855], McC/McCF; McCrady to Lewis R. Gibbes, March 20, 1855, LRG/LC.

6. John McCrady to "Dear Sister," January 23, 1855, McCrady to "My dear Mother," January 23 [1855], and McCrady, "Scene—A Court in Heaven; Recording Angel (writes)," manuscript, McC/McCF; Lewis R. Gibbes to A. D. Bache, November 18 and December 14, 1853, LRG/LC. On the relationship of Agassiz to his students, see Winsor, *Reading the Shape of Nature,* 20–61.

7. McCrady, "Gymnopthalmata of Charleston Harbor."

8. Ibid.

9. Ibid.; Stephens and Calder, "John McCrady of South Carolina."

10. Stephens and Calder, "John McCrady of South Carolina"; Stimpson, Review

of "Description of *Oceania* (*Turritopsis*) *nutricula* . . ." and "Gymnopthalmata of Charleston Harbor."

11. McCrady, "Gymnopthalmata," 135, 160–65.

12. McCrady, "On the development of two species of Ctenophora"; McCrady, "Remarks on the embryology of a species of Bolina"; McCrady, "Remarks on the Zoological Affinities of Graptolites"; McCrady, "Remarks on the Eocene Formation"; reference to McCrady's paper on "Larva of Brachiopods," in *Proc. ESNH* 2 (1891): 25, and to his comment on *Oculina*, in *Proc. ESNH* 2 (1891): 8; manuscript titled "Remains of old paper on Annelids of Charleston Harbor," 6 pp. and figures, dated "August 22, 1860," and manuscript note, "Lent to Prof. Edward S. Morse, Salem, Mass., Sept. 24, 1873, Notice and Drawing from *Mercury* of a Brachiopod Larva discovered 1859," McC/McCF; McCrady, "Letter from John McCrady on the *Lingula pyramidata*"; Mayer, "Descriptions," 1–9; Fortey, *Fossils*, 116. *Oculina arbuscula* was described by A. E. Verrill in 1864.

13. Manuscript proceedings of the ESNH, May 15, 1857, 222–23, CML.

14. Stephens, "Scientific Societies in the Old South," 55–66; Calder, Stephens, and Sanders, "Comments on the Date of Publication"; John McCrady to Edmund Ravenel, January 7, 1858, RL/CML.

15. McCrady, *A System of Independent Research*, 7–8.

16. Ibid., 15–16.

17. Ibid., 17–18.

18. Ibid., 19–23.

19. McCrady, "A Few Thoughts on Southern Civilization."

20. Ibid., 1: 227, 339–49, 546, 554.

21. Ibid., 2: 214–15, 223–24.

22. McCrady, *Home Education*, 11–13.

23. Ibid., 23–27.

24. On American racial views of the time, see, for example, Jordan, *White Over Black*, pts. 4 and 5, and Fredrickson, *The Black Image in the White Mind*, 1–42.

25. McCrady, "The Study of Nature," 588–90, 595–98.

26. Ibid., 602–6.

27. Various comments on evolution made by McCrady before the ESNH in 1860 appear in *Proc. ESNH* 2 (1891): 20, 23, 26; [Lewis R. Gibbes], manuscript sheet, "February 15th 1860," and [John McCrady], manuscript sheet beginning "One of the most important works in Natural Science is that of Mr. Darwin," McC/McCF.

28. McCrady, "Law of Development by Specialization," read before the ESNH on October 1, 1860.

29. Ibid., 102.

30. Ibid., 114. For a useful overview of the context in which McCrady formulated his notions on embryonic development, see Coleman, *Biology in the Nineteenth Century*, 35–56.

31. Burton, *Siege of Charleston*, 1–4, 14–15; Fraser, *Charleston! Charleston!*, 250–51; John McCrady to "Lassie [Sarah Dismukes]," n.d., and McCrady to Paul Dismukes, July 2, 1859, McC/McCF.

Chapter Nine

1. Stanton, *The Leopard's Spots*, 122–44.

2. John Bachman to "Rev. J. D.," n.d., 1837, in C. L. Bachman, *John Bachman*, 358; Payne, *Recollections*, 23–25, 35; Bost, "The Reverend John Bachman," 390–425. The U.S. Census shows that the Bachmans owned twelve slaves in 1840, but, for reasons unknown, by 1850 the number had dropped to six (see U.S. Census, *Population Schedule*, 1840 and 1850).

3. J. Bachman, "Agricultural Labor"; Morton, *Crania Americana*; Morton, *Crania Aegyptiaca*; Meigs, *Catalogue of Human Crania*.

4. Morton, "Hybridity in Animals," 39–50.

5. Ibid., 203–12.

6. Louis Agassiz to Samuel G. Morton, September 20, 1845, cited in *Early American Literature*, 56–57; Lurie, *Louis Agassiz*, 256–65; Lurie, "Louis Agassiz and the Races of Man." Bachman refers to the discussions of the Literary Club on page 3 of his *Doctrine of the Unity*, as does Thomas Smyth in the preface of his *The Unity of the Human Races*, but Smyth erroneously cites Agassiz's presentation as occurring in 1846 instead of 1847. Comments on many of the members of the club are contained in "Memoirs of Frederick Adolphus Porcher," *SCHM* 47 (1946): 218–21. James Moultrie invited Lewis R. Gibbes to meet with the club on December 15, 1847, when the topic of discussion would be "the Unity of the Races," but it is not known whether Gibbes attended (see James Moultrie to Gibbes, LRG/LC). See also Smyth, *Autobiographical Notes*, 211–71.

7. Nott, "The Mulatto a Hybrid"; Horsman, *Josiah Nott*, 6–103.

8. John Bachman to Samuel G. Morton, October 15, 1849, SGM/HSP.

9. Ibid.

10. Robert W. Gibbes to Samuel G. Morton, January 21, 1850, SGM/HSP; Luker, *Southern Tradition*, 76–80.

11. Nott, "An examination," *Proc. AAAS* 3 (1850): 98–106, with comments by Louis Agassiz, ibid., 106–7; Smyth, *Unity of the Human Races*, 353–54; Josiah Nott to S. G. Morton, March 1, 1850, and Robert W. Gibbes to Morton, March 31 and April 10, 1850, SGM/HSP; Agassiz, "Geographical Distribution"; Agassiz, "The Diversity of Origin." The daguerreotypes are located in the Peabody Museum of Archaeology and Ethnology, Harvard University.

12. J. Bachman, *Doctrine of the Unity*, preface.

13. Ibid., 7–8.

14. Ibid., 19–21; Mayr, *Populations*, 55–57; Mayr, *Growth*, 251–97.

15. J. Bachman, *Doctrine of the Unity*, 21–35.

16. Ibid., 41–81.

17. Ibid., 84–114.

18. Ibid., 115–25, 288.

19. Ibid., 127–82.

20. Ibid., 67, 184–209.

21. Ibid., 209–46.

22. Ibid., 295–300.

23. J. Bachman, "An Investigation," 168–69.

24. Ibid., 172–86.

25. Ibid., 186.

26. Ibid., 170–76; Edward Griffith et al., *Animal Kingdom*.

27. D. J. Cain and F. P. Porcher to Samuel G. Morton, March 25 and April 6, 1850, SGM/HSP; Morton, "Letter to the Rev. John Bachman."

28. Morton, "Letter to the Rev. John Bachman," 329–40.

29. Ibid., 340–44.

30. Josiah C. Nott to Samuel G. Morton, May 4, 26, 1850, SGM/HSP.

31. J. Bachman, "A Reply"; F. P. Porcher to Samuel G. Morton, June 24, 1850, D. J. Cain and F. P. Porcher to Morton, July 10, 1850, R. W. Gibbes to Morton, June 17, 1850, SGM/HSP.

32. J. Bachman, "A Reply," 466–74.

33. Ibid., 474–508.

34. Josiah C. Nott to Samuel G. Morton, July 25 and August 26, 1850, SGM/HSP.

35. Morton, "Additional Observations," 755–77.

36. Ibid., 760–61, 763, 787.

37. D. J. Cain and F. P. Porcher to Samuel G. Morton, April 22 and December 28, 1850, SGM/HSP; Morton, "Notes on Hybridity, designed as a Supplement," *CMJR* 6 (1851): 145–52.

38. Morton, "Notes on Hybridity, designed as a further Supplement," *CMJR* 6 (1851): 301–8; Cull, "On the Recent Progress of Ethnology," 304; Browne, *Trichologia Mammalium*.

39. Morton, "Notes on Hybridity, and some collateral subjects," *CMJR* 6 (1851): 373–83.

40. Josiah C. Nott to Samuel G. Morton, April 6, 1851, and D. J. Cain and F. P. Porcher to Morton, April 10, 1851, SGM/HSP; McCord, "Diversity of the Races." A generally favorable, but much shorter, review of *Doctrine of the Unity* had appeared in *Southern Quarterly Review* 17 (April 1850): 250. Sympathetic reviews also appeared in *Southern Medical and Surgical Journal*, n.s., 6 (August 1850): 504, and in *AJS*, 2nd ser., 11 (March 1851): 302. For an important sketch of Louisa McCord, see the entry for her in Duyckinck and Duyckinck, *Cyclopedia of American Literature*, 2: 251–53.

41. J. Bachman, "Additional Observations," 385–86.

42. J. Bachman, "Letter from Rev. John Bachman," *CMJR* 6 (1851): 598; R. W. Gibbes, "Death of Samuel George Morton, M.D."

Chapter Ten

1. Nott and Gliddon, *Types of Mankind*.

2. Ibid., 733–38; Nott, "Diversity of the Human Race," 113.

3. J. Bachman, Review of *Types of Mankind*, *CMJR* 9 (1854): 627–59.

4. Ibid., 631–46, 649.

5. Ibid., 657–58.

6. Nott contended that Agassiz simply chose to ignore Bachman's criticisms. See Nott, "Reply,", 18–54.

7. Winsor, "Louis Agassiz and the Species Questions," 89–117; Agassiz, "Essay on Classification," 3–232, and Edward Lurie, "Editor's Introduction," xx–xxxiii, both in Agassiz, *Contributions to the Natural History of the United States of America*, reprint ed.

8. Agassiz, "Natural Provinces," lvii–lxxviii; J. Bachman, "An examination of a few of the statements of Prof. Agassiz," 790–96; Nott, "Letter of Dr. Nott."

9. J. Bachman, "An examination of a few of the statements of Prof. Agassiz," 797–806; A. L., "On the Unity of the Human Race"; W. C. Robinson, *Columbia Theological Seminary*, 18–54.

10. J. Bachman, "An Examination of the Characteristics of Genera and Species." 204–7.

11. Ibid., 206–12. See also Goodman, "Bred in the Bone?"

12. J. Bachman, "An examination of Prof. Agassiz's Sketch of the Natural Provinces," 482–83, 486.

13. Ibid., 492–93, 497, 499.

14. Ibid., 501–2, 509.

15. Ibid., 519–20, 530, 533.

16. Nott, "Communication from Dr. Nott." 862.

17. Ibid., 862–64; Asa Gray to John Bachman, January 15 and October 16, 1855, Ms/HL; Dana, "Thoughts on Species"; Rossiter, "Portrait of James Dwight Dana," 121–22.

18. Nott, "Reply to Dr. Bachman's Review," 753, 755–56.

19. Ibid., 758–59, 765, 767.

20. Nott, ["Thoughts on Race"]; Burke, "Strictures on Dr. Bachman's paper," 433–34, 441–42, 448.

21. Cull, "On the Recent Progress of Ethnology."

22. Messmer, *Works of the Right Reverend John England*, 1: 57–288.

23. Advertisement in *Charleston Courier*, March 13, 1852, reprinted in J. Bachman, *A Defence of Luther*, 1.

24. J. Bachman, *A Defence of Luther*, 1–21.

25. Ibid., 22–55, 197 (quotation); *United States Catholic Miscellany* 21 (March 20–May 29, 1852): 286–87, 294–95, 300, 311, 314, 326–27, 335, 345–46, 351, 358, 362.

26. John Bachman to [Henry Summer?], September 6, 1851, SCL.

27. John Bachman to Victor Audubon, [September 11, 1851], BL/CML; Victor Audubon to Bachman, September 18, 1851, A-B/HU.

28. John Bachman to Victor Audubon, [1857], BL/CML; Bachman to Edmund Ruffin, January 18 and May 23, 1860, Edmund Ruffin Papers, Virginia Historical Society.

29. John Bachman to Lucy Audubon, March 30, 1856, and Bachman to Victor Audubon, April 14, 1856, and March 2, 9, August 15, and December 17, 1857, BL/CML; C. L. Bachman, *John Bachman*, 343; Holcomb, *Marriage and Death Notices from the Lutheran Observer*, 180–81.

30. Edward Harris to John Bachman, July 16, 1860, and [Maria Bachman] to [Catherine L.] Kate [Bachman], November 12, 1860, BL/CML.

Chapter Eleven

1. Fraser, *Charleston! Charleston!*, 247–53; Burton, *Siege of Charleston*, 44–51; John McCrady to Sarah McCrady, January 18, 20, March 29, and April 7, 13, 1861, McCrady to Edward McCrady, March 11, 1861, McC/McCF; McCrady to "Dear Sister," McCrady Family Papers, SCL; Survivors Association Records, manuscript note, SCHS.

2. Trustees Minute Book, March 12, July 20, September 29, and October 31, 1861, CCL; John McCrady to Edward McCrady, March 18, 1861, McC/McCF; *Charleston Mercury*, November 19, 1861, 2.

3. Nepveux, *George Alfred Trenholm*, 73; Childs, *Journal of Henry William Ravenel*, 103, 105, 110, 114; *Charleston Mercury*, December 12, 14, 1861; Rosen, *Confederate Charleston*, 86.

4. Maria Bachman to [Catherine L.] Kate [Bachman], November 16, 1861, BL/CML; John Bachman to Henry Summer, November 12, 1861, and Bachman to John C. Faber, May 30, 1862, SCL; *Charleston Daily Courier*, May 10, 1862.

5. John Bachman to Edmund Ruffin, November 22, 1862, BL/CML; C. L. Bachman, *John Bachman*, 369, 375; Bachman to "Mrs. J. W. E.," September 8, 1863, in C. L. Bachman, *John Bachman*, 371–72; *History of the Lutheran Church in South Carolina*, 284–85.

6. Childs, *Journal of Henry William Ravenel*, 121, 141, 148, 196; Cauthen, *Journals*, 169, 196, 233, 256–57, 285; *War of the Rebellion*, Series 4, 3:698, 702; Easterby, *History of the College of Charleston*, 148–49; N. R. Middleton to Lewis R. Gibbes, October 10, 1863, and Gibbes to John LeConte, December 1, 1863, LRG/LC.

7. John McCrady to Sarah McCrady, February 19, 28, March 9, April 25, June 7, and September 20, 1863, J. P. Benjamin to John McCrady, March 13, 1862, Edward McCrady Jr. to John McCrady, September 23 and December 26, 1862, John McCrady to Edward McCrady, July 10, October 14, and November 18, 1862, and April 18, 21, 1863, John McCrady to "Your Excellency," n.d., and "Confederate States Engineer's Office," June 24, 1862 [printed circular], McC/McCF.

8. S. Dismukes to "Brother," August 24, [1864], E. A. Dismukes to "Mrs. McCrady," October 8, 1864, John McCrady to Edward McCrady, November 20, 29, 1864, John McCrady to Sarah McCrady, November 28 and December 13, 1864, and James I. Waring, M.D., to [McCrady's commanding officer], September 20, 1864, McC/McCF.

9. Edward McCrady Jr. to his wife, December 10, 1864, John McCrady to [Sarah McCrady], December 14, 30, 1864, and February 27 and March 14, 1865, McC/McCF; Francis S. Holmes to Lewis R. Gibbes, September 14, 1864, and Daniel Ravenel to Gibbes, October 19, 1865, LRG/LC; J. S. Coles to "Office of Superintendent of Field Transportation," October 26, 1864, in SCL; Jones, "Siege and Evacuation of Savannah," 60–85.

10. John Bachman to Edmund Ruffin, November 15, 1864, BL/CML; John Le-Conte to Lewis R. Gibbes, April 10, 1865, LRG/LC; "Evacuation of Charleston: Flight of a Family; Notes from Cousin Rose Ravenel's Account," typescript, Mrs. St. Julien Ravenel Family Papers, SCHS.

11. John Bachman to Robert G. Chisolm, February 7, 1865, BL/CML; Catherine L. Bachman, [Address to Daughters of the Confederacy, 1898], manuscript, SCHS.

12. Catherine L. Bachman, [Address to the Daughters of the Confederacy], SCHS; C. L. Bachman, *John Bachman*, 384–85.

13. *Edgefield Advertiser*, September 6 and October 13, 1865; Francis S. Holmes to Joseph Leidy, December 9, 1865, Joseph Leidy Papers, ANSP; Lewis R. Gibbes to John McCrady, November 13, 1865, McC/McCF; McCrady to Gibbes, September 25, 1865, St. Julien Ravenel to Gibbes, October 15, 1865, Gibbes to A. T. Porter, May 28, 1865, Daniel Ravenel to Gibbes, November 21, 29, and December 5, 1865, and J. W. Miles to Gibbes, Nov[ember 18]65, LRG/LC; "Report of the Committee on the organization of the College. Filed 16 Novr 1865," and Trustees Minute Book, November 17, 1865, CCL.

14. Pocket notebook diary of John McCrady, entries from January 1 through October 15, 1866, McC/McCF; John McCrady to Lewis R. Gibbes, September [n.d.] 1865, and May 23 and October 18, 19, 1866, and Gibbes to McCrady, May 24, 1866, LRG/LC.

15. Trustees Minute Book, August 27, 1866, Francis S. Holmes to Daniel Ravenel, April 20, 1866, CCL; Holmes to Joseph Henry, October 18, 1868, Secretary's Correspondence, SIA.

16. Francis S. Holmes to N. R. Middleton, "Report of the Professor of Geology & Zoology, Etc., and Curator of the Museum of Natural History . . . for the year ending March 1868," Trustees Minute Book, April 22, 1868, Faculty Journal, College of Charleston, 1860–70, May 1, 1868, CCL; John McCrady Diary, May 8, 13, 16, 1868, McC/McCF; Notice of Holmes's resignation from ESNH, *Proc. ESNH* 2 (1891): 48.

17. McCrady Diary, May 16, August 28, and September 15, 17, 28, 1868, McC/McCF; Trustees Minute Book, August 22, 31, and September 21, 28, 1868, CCL.

18. Trustees Minute Book, letter from Holmes, [n.d., but entered following September 28, 1868], and report of the Standing Committee, September 28, 1868, CCL.

19. Stephens, *Ancient Animals*, 41–45; Trustees Minute Book, September 28, October 8, and November 2, 1868, Holmes to Trustees, October 2, 1868, and report of the Committee on the Library and Museum, October 8, 1868, CCL.

20. Trustees Minute Book, December 28, 1868, and January 25, 1869, CCL; Stephens, *Ancient Animals*, 47–48.

21. Stephens, *Ancient Animals*, 37–41.

22. John McCrady Diary, January 22, 23, 25, and March 4, 6, 1869, and John McCrady, "Journal of Researches in Natural History and on Kindred Subjects," manuscript, 1868–75, McC/McCF; Faculty Journal, March 26, 1869, and Trustees Minute Book, March 6, 1869, CCL. In his diaries of 1869–72, McCrady commented frequently on his experience in trying to raise oysters. In 1873, he published "Observations on the Food and the Reproductive Organs of Ostrea Virginia, with some Account of Bucephalus Cuculus Nov. Spec." This was an excellent, detailed, scien-

tific description, though the oyster described was not, in fact, a new species, and near the end of the article McCrady deviated from his report to criticize the Darwinian theory and to tout his law of development, none of which was directly related to his research on the oyster.

23. Trustees Minute Book, March 19, 1869, CCL; McCrady Diary, March 20, 30, and May 14, 1869, March 15, April 8, 18, July 26, and October 12, 1870, May 21, 1871, and June 13, 1872, in McC/McCF; McCrady to Lewis R. Gibbes, August 20, 1870, LRG/LC.

24. McCrady Diary, August 19, 20, 21, 26, September 14, and October 13, 14, 16, 1869, April 23, 30, August 27, and October 4–6, 1870, and March 9, 12, April 30, August 3, and November 2, 1872, McC/McCF; *Charleston Daily Courier*, August 20, 21, and September 15, 1869.

25. McCrady Diary, April 1, 3, August 13, and September 10, 1868, and March 1–5, April 8, 10, 12, 1872, McCrady, manuscript draft of report on yellow fever, 1871, McCrady, "Journal of Researches," McCrady to Spencer F. Baird, August 12 and September 14, 1870, McCrady to "Editor" of *Nature*, September 13, 1870, and McCrady to James H. Coffin, September 14, 1870, McC/McCF; McCrady to Joseph Henry, May 27, 1870, McCrady to James H. Coffin, June 1, 1870, and McCrady to Spencer F. Baird, ASecr./SIA; McCrady, Ravenel, and Pankin, "Report," 17–41; Stephens, "Scientific Societies in the Old South," 64–65. McCrady also presented a long paper on pisciculture, which reflected considerable knowledge of fishes, a reasonable plan of economic development, and a thoughtful commentary on scientific experimentation. Originally titled "Address Delivered Before the Agricultural Society of South Carolina, at Their Seventy-Sixth Annual Meeting, January 12, 1871," it was reprinted as *Essay by John McCrady*.

26. McCrady Diary, May 7 and June 29, 1868, May 29, 1869, May 3, 5, 6, 13, June 19, and December 26, 1870, and February 25 and December 8, 1871, and McCrady to Louis Agassiz, May 12, 1869 (letterpress book), McCrady, "Journal of Researches," and McCrady, "Geographical Distribution of the Races of Men," n.d. (44-page manuscript), McC/McCF; *Charleston Daily Courier*, May 4, 6, 11, 1870.

27. McCrady Diary, April 3 and October 14, 30, 31, 1868, January 8, 11, April 22, 28, 30, May 1, 4, 5, August 3, September 28, and October 28, 29, 1869, April 3, May 20, July 8, September 17, 20, 21, November 12, 27, and December 31, 1870, February 6, March 8, 10, April 28, May 10, 11, June 17, July 16, September 11, 15, and October 9, 1871, January 6, October 5, November 25, 27, 29, and December 3, 29, 1872, and January 1, 3, 17, 1873 [final entry for 1873, as McCrady was too poor to buy another blank book], and McCrady to Max Müller, [n.d.], 1869 [draft copy], McCrady to "Mssrs. Editors" [of *American Journal of Science*], September [1870], McCrady to "Mssrs. Silliman and Dana," September 23, 1870, McCrady to "Prof. Venable," September 28, 1870, McCrady to Max Müller, July 28, 1870, and February 8, 1871, McCrady to Charles Venable, February 20, 1871, [all in McCrady's letterpress book], and McCrady to "Editor of Appleton's Popular Science Monthly," April 14, 1873 [draft copy], McCrady, "Journal of Researches," and McCrady, "Geographical Distribution of the Races of Men"—all in McC/McCF; McCrady to Lewis R. Gibbes, August 20, 1870, LRG/LC.

28. Louis Agassiz to John McCrady, March 29, April 13, 18, May 30, and June 18, 1873, and McCrady to Agassiz, April 5, 7, 17, 24, and June 13, 1873, McC/McCF; Harvard University Corporation Minutes, Scientific School, 142:221, 223, University Archives, HU.

Chapter Twelve

1. John McCrady to Gabriel E. Manigault, October 25, 1873, McC/McCF; McCrady to St. Julien Ravenel, [1873], Harriott Horry Rutledge Ravenel Papers, SCHS.

2. John McCrady, "Memorial Address, Delivered in Agassiz Lecture Room the day after his Burial," [December 15, 1873], manuscript, and McCrady Diary, January 1, February 6, March 30, and April 13, 1874, John McCrady to Edward McCrady, December 7, 21, 1873, and E[lizabeth] C. Agassiz to John McCrady, n.d., McC/McCF.

3. John McCrady to Edward McCrady, December 15, 25, 1873, Edward McCrady to John McCrady, January 25, 1874, and McCrady Diary, January 13, 16–20, 28, 31, February 2, March 13, April 27, May 2, 13, 24, and October 3, 1874, McC/McCF; James, *Correspondence*, 4:580.

4. McCrady Diary, January 8, 13, 29, February 6, March 1, April 15, May 7, 13, 14, June 2, 3, 8, and July 16, 1874, McC/McCF; McCrady to Edward McCrady, April 26, 1874, McC/SCHS.

5. John McCrady to Charles W. Eliot, July, 24, 1874 [draft copy], and McCrady Diary, August 8–11, 17, 25, and September 1, 1874, McC/McCF; Eliot to McCrady, July 11 and August 8, 1874, and McCrady to Eliot, July 29, 1874, University Archives, HU.

6. McCrady Diary, October 3, 9, and December 3, 18, 21, 1874, and Edward McCrady to John McCrady, January 25, 1874, McC/McCF.

7. Morrill Wyman to Edward McCrady, January 7, 1876, McC/SCHS; McCrady Diary, January 14, 15, 21, 31, February 12, 19, 23, 24, March 1, 3, 4, 6, 7, 14, 26, 28, 31, April 2, 8, 9, 11, 24, May 10, 17, 21, and June 8, 1875, McC/McCF; John McCrady to Charles W. Eliot, June 3, 1875, University Archives, HU.

8. McCrady Diary, February 28, March 7, April 1, May 22, 27–30, and June 3–5, 1875, McC/McCF.

9. Ibid., August 20, September 2, 22 23, October 20, 21, and November 11, 15–29, 1875; John McCrady to Charles W. Eliot, September 29, 1875, University Archives, HU. An extant copy of one of his examinations in Natural History 4, or the "Elementary" course, is lengthy and exacting. The document is in McC/McCF.

10. McCrady Diary, December 5, 20–31, 1875, McC/McCF.

11. Ibid., January 1–13, April 13, and June 27, 1876, and John McCrady, "Notice of Cheiropsalmus Carolinensis" (manuscript), McC/McCF; Alexander Agassiz to Daniel C. Gilman, January 12, 1876, Daniel C. Gilman Papers, JHU.

12. McCrady Diary, September 28, October 3, 4, 8–18, November 10, 29, and December 3, 4, 8, 11, 1876, and John McCrady to Sarah McCrady, October 1, 2, 1876, and John McCrady to Edward McCrady, Jr., October 12, November 1, and December 3, 1876, McC/McCF.

13. McCrady Diary, January 12, 16, 27, 31, and February 6, 8, 12, 17, 19–21, 24, 26–28, 1877, in McC/McCF; John McCrady to Charles W. Eliot, February 28, 1877, University Archives, HU. Meanwhile, the father and the eldest brother of John McCrady were leading the opposition to representation by blacks in the Episcopal Diocese of South Carolina, as noted in Tyler, "Drawing the Color Line."

14. John McCrady to Edward McCrady, March 26 and April [n.d.], 1877, McC/SCHS.

15. McCrady Diary, March 1, 16, 20, 23–26, 1877, McC/McCF; John McCrady to Edward McCrady, March 26, April 2, April [n.d.], and April [n.d.], 1877, and Charles W. Eliot to John McCrady, March 27, 1877, in McC/SCHS; McCrady to Eliot, March 26, 1877, University Archives, HU.

16. McCrady Diary, March 27–30, April 1–2, 1877, McC/McCF.

17. Ibid., April 4–9, 12, 15 20–30, 1877, and John McCrady to Edward McCrady, Jr., April 8, 11, 1877, McC/McCF; William Henry Trescott to Edward McCrady, Jr., April 19, 1877, and John McCrady to Edward McCrady, May 4, 31, 1877, McC/SCHS; *Boston Daily Advertiser*, April 6, 1877. Before he left Cambridge, McCrady also presented a paper at a meeting of the Boston Society of Natural History, which was published in its proceedings as "A Provisional Theory of Generation." Although the paper reflected McCrady's admirable knowledge of embryology, it indicated that he was wedded to the flawed notion of embryogenesis as tantamount to a type of evolution, hence to his law of development.

18. McCrady Diary, June 1, 9, 11, 20–21, 26, and July 2, 4, 12, 17, 24, 31, 1877, and John McCrady to Sarah McCrady, July 11, 17, 25, 29, 1877, McC/McCF; George Z. Gray to "Rt. Rev. & Dear Sir," June 25, 1877, and N. S. Shaler to "the Rt. Reverend the Bishop of Tennessee, etc.," July 21, 1877, University Archives, USo.

19. John McCrady to Sarah McCrady, August [3], 1877, and McCrady Diary, June 4, 11, and August 3–7, 1877, in McC/McCF; John McCrady to Edward McCrady, August 5, 1877, McC/SCHS; *Proceedings of the Board of Trustees of the University of the South, 1877*, 7, 13, 30–31; Allman, Comments on Resignation of John McCrady.

20. McCrady Diary, August 30, September 3–6, 1877, McC/McCF; John McCrady to Edward McCrady, September 10, 1877, McC/SCHS.

21. McCrady Diary, September 25, October 3, 4, 9, 16, 27, November 18, December 10, 1877, McC/McCF.

22. Ibid., January 23, 25, 28, 30, February 6, 8, March 2, 31, and April 2, 4–6, 1878, and McCrady to John Kershaw, February 11, 1878, and McCrady, "The Forge of Thought" (manuscript), McC/McCF; John McCrady to "Messr. Macmillan & Co.," April 5, 1878, McC/SCHS.

23. McCrady, "Cook and His Biology"; McCrady, Review of *Boston Monday Lecture: Biology*, by Joseph Cook.

24. McCrady Diary, April 13–14, May 8, 17, July 4, August 6, 16, September 4, 10, 17, 22, December 31, 1878, and May 31, July 1, October 9, 13, 15, December 31, 1879, and John McCrady to D. F. Boyd, August 26, 1878, McCrady to William S. Atkinson, September 2, 1878, McCrady to W. M. Green, May 31, 1879, and McCrady to J. L. Tucker, December 17, 1879, McC/McCF; McCrady to Alexander Gregg, March 24,

1879, and McCrady to W. M. Green, June 5, 1879, and August [n.d.], 1879, McCrady Family Papers, USo; List of the Officers and Faculty of the University of the South; Fairbanks, *History of the University of the South*, 174–201.

25. John McCrady to "His Honor Judge Kershaw," February 11, 1878, John Mc-Crady to Edward McCrady, May 5, June 19, September 1, 15, October 6, and December 29, 1878, and June 12, August 17, and December 22, 1879, McC/SCHS; McCrady Diary, June 4, 6, 8, 14, 23, 26, 28, 1880; John McCrady to Telfair Hodgson, June 23, 29, 1880, and Hodgson to McCrady, June 23, 1880, University Archives, USo; Mc-Crady to Lewis R. Gibbes, December 21, 1880, LRG/LC.

26. McCrady Diary, January 12, March 3, 30, 1881, John McCrady to Sarah Mc-Crady, March 28, April 1, 2, 5, 11, and n.d., 1881, McC/McCF; John McCrady to Daniel C. Gilman, March 8, 20, 1881, Special Collections Department, JHU.

27. John McCrady to Albert T. McNeal, June [n.d.], 1880, McCrady to Telfair Hodgson, June 4, 24, 28, and July 3, 1880, McCrady to Edward McCrady, July 11, 21, 1880, March 6, 28, April 5, 11, 15, May 1, August 8, 11, 19, and October 9, 1881, McC/SCHS; John McCrady to St. Julien Ravenel, February 25, 1881, Harriott Horry Rutledge Ravenel Papers, SCHS; John McCrady to J. L. Tucker, December 17, 1879, McCrady to Sarah McCrady, March 28, April 1, 2, 5, 11, and n.d., 1881, Sarah McCrady to "My dearest Friend," [n.d., 1881], McCrady Diary, January 12, March 3, 31, and August 12–26, 1881, and two unidentified newspaper clippings, John Mc-Crady to Telfair Hodgson, September 23, 1881, McC/McCF; Fanny G. Hodgson to Mrs. John McCrady, October 28, 1881, "Burning of Otey Hall," Wilmer, "In Memoriam," and *Proceedings of the Board of Trustees of the University of the South, 1881*, 7, 8, 22, all in University Archives, USo; John McCrady to Daniel C. Gilman, March 8, 20, 1881, and undated clipping from the *Baltimore American*, Special Collections Department, JHU.

28. McCrady Diary, July 6–8, 28, and September 15, 1871, McC/McCF; and *Charleston Daily Courier*, July 31, 1871 (for Ravenel); clipping from unidentified newspaper, in John Hay Library Archives, Brown University (for Holbrook); *Charleston Mercury*, February 25, 1874, J. T. Wightman to J. B. Haskell, July 1, 1874, BL/CML; E. T. Horn, "St. John's Evangelical Lutheran Church," *Charleston Yearbook, 1884*, 267–79; C. L. Bachman, *John Bachman*, 424–34 (for Bachman); Stephens, *Ancient Animals*, 49–50.

29. Faculty Reports to the President, College of Charleston, 1875, and "Proceedings of the Elliott Society of Science and Arts" (manuscript) May 13–December 9, 1869, May 10–June 9, 1870, November 5, 1875, CCL; James D. Dana to Lewis R. Gibbes, November 7 and December 4, 1875, Arthur Mazÿck to Gibbes, October 27, 1882, C. B. Colson to Gibbes, January 30, 1885, William G. Mazÿck to Gibbes, June 8, 1885, Gibbes to Colson, June 27, 1885, and Gibbes to Edward Burgess, May 29, 1886, LRG/LC; *Charleston News and Courier*, October 21 and 27, 1882; Taylor, "Lewis Reeve Gibbes." The second part of the ESNH proceedings, the part published in 1885, is the section on pages 75 to 125 of the bound volume, that is, *Proc. ESNH* 2 (1891): 75–125.

30. Faculty Reports, 1876, CCL; ms. "Proceedings of the Elliott Society," January 22, April 9, June 11, July 9, October 5, and November 12, 1885, April 22 and Novem-

ber 25, 1886, November 17, 1887, and July 26 and October 25, 1888, CML; Lewis R. Gibbes to Richard Rathbun, January 15, 1881, SIA; Asa Gray to Gibbes, February 12, 1882, Gibbes to Gray, February 8 and March 23, 1882, Gibbes to C. S. Sargent, July 4, 1890, LRG/LC.

31. President, University of South Carolina, to Lewis R. Gibbes, July 1, 1890, LRG/LC; Gibbes to Board of Trustees, College of Charleston, June 22, 1888, June 16, 1890, June 23 and October 20, 1891, and July 13, 1892, CCL.

32. Fraser, *Charleston! Charleston!*, 301–14.

33. Ibid., 314–41; "Proceedings of the Elliott Society," December 7, 1900, and January 17, 1901, CML; Fraser, *Charleston! Charleston!*, 355–56.

Epilogue

1. Bruce, *Launching*, 57–63; Stephens, Review of *Launching*. Errors mentioned in the review were corrected in the paperback edition of Bruce's book, but others not cited in the review remain uncorrected.

2. Bruce, *Launching*, 61–62; Numbers and Numbers, "Science in the Old South."

BIBLIOGRAPHY $\cdot\;\cdot$

Primary Sources

MANUSCRIPT COLLECTIONS

Academy of Natural Sciences of Philadelphia Library, Philadelphia, Pa.
  Academy Correspondence and Records
  Isaac Lea Papers
  Joseph Leidy Papers
American Museum of Natural History Archives, New York, N.Y.
American Philosophical Society Library, Philadelphia, Pa.
  American Philosophical Society Archives
  Asa Gray Papers
  John Edward Gray Correspondence
  Miscellaneous Manuscript Collections
  Samuel G. Morton Papers
Boston Museum of Science, Boston, Mass.
  Boston Society of Natural History Manuscripts
Charleston Library Society, Charleston, S.C.
  Hinson Collection
  Manuscript Collections
Charleston Museum Library, Charleston, S.C.
  John Bachman Letters
  Lewis R. Gibbes Letters
  Edmund Ravenel Letters
  Miscellaneous Letters and Diaries, South Carolina Collection
Clemson University Library, Special Collections, Clemson, S.C.
  Henry William Ravenel Papers
College of Charleston Library, Special Collections, Charleston, S.C.
  Board of Trustees Minutes
  Chrestomathic Literary Society Papers
  Faculty Journal
  Faculty Reports
  Lewis R. Gibbes Papers
  Museum Reports
  Frederick A. Porcher Papers
Harvard University, Cambridge, Mass.
  Gray Herbarium Library
    Asa Gray Correspondence

John Torrey Correspondence
Harvard University Archives
Houghton Library
    Audubon-Bachman Correspondence
    Augustus A. Gould Papers
    Miscellaneous Manuscript Collections
Museum of Comparative Zoology
    John E. Thayer Collection
    Miscellaneous Collections
Peabody Museum of Archaeology and Anthropology
    Collection of J. T. Zealy Daguerreotypes
Historical Society of Pennsylvania, Philadelphia, Pa.
Samuel G. Morton Papers
Huntington Library, San Marino, Calif.
John Bachman Manuscripts
Johns Hopkins University Library, Baltimore, Md.
Daniel C. Gilman Papers
Miscellaneous Manuscripts
Library of Congress, Manuscript Division, Washington, D.C.
Lewis R. Gibbes Papers
Nathaniel Wright Papers
Maryland Historical Society, Baltimore, Md.
David B. Warden Papers
Massachusetts Historical Society, Boston, Mass.
Missouri Historical Society, St. Louis, Mo.
John James Audubon Papers
Museum für Naturkunde, Zentralinstitut der Humboldt-Universität zu Berlin,
    Berlin, Germany
Lichtenstein Correspondence
New York Botanical Gardens Library, New York, N.Y.
John Torrey Correspondence
New York State Library, Albany, N.Y.
James Hall Papers
Smithsonian Institution Archives, Washington, D.C.
Alexander D. Bache Papers
Marcus Benjamin Papers
Isaac Lea Papers
Assistant Secretary, Incoming and Outgoing Correspondence
Secretary, Incoming and Outgoing Correspondence
South Carolina Historical Society Library, Charleston, S.C.
Bacot-Huger Papers
Langdon Cheves Papers
St. Julien Ravenel Childs Family Papers
Ford-Ravenel Papers
Gibbes-Gilchrist Papers

McCrady Family Papers
Manigault Family Papers
Porcher Family Papers
Harriott Horry Rutledge Ravenel Papers
Mrs. St. Julien Ravenel Family Papers
South Caroliniana Library, University of South Carolina, Columbia, S.C.
John P. Barratt Papers
Robert W. Gibbes Papers
McCrady Family Papers
Manigault Family Papers
Gabriel Manigault Papers
Pendleton Academy Letters
Porcher Family Papers
Francis Peyre Porcher Papers
Ravenel Family Papers
Henry W. Ravenel Papers
William J. Rivers Papers
Southern Historical Collection, University of North Carolina, Chapel Hill, N.C.
Botany Department Papers
Manigault Family Papers
University of South Carolina, Columbia, S.C.
University Archives
University of the South Library, Sewanee, Tenn.
University Archives
John McCrady Papers
Virginia Historical Society Library, Richmond, Va.
Edmund Ruffin Papers
Waring Historical Library, Medical University of the State of South Carolina,
Charleston, S.C.
John Edwards Holbrook Papers
Medical Society of South Carolina Records
Francis Peyre Porcher Papers
Edmund Ravenel Papers
Yale University, New Haven, Conn.
Beinecke Rare Books and Manuscript Library
John James Audubon Papers
Sterling Memorial Library, Manuscript Collections
Dana Family Papers
Private Collections
Mrs. George Buroughs, Hinesville, Ga.
Bertram Holbrook, Brookline, Mass.
John McCrady, Charleston, S.C.
McCrady Family Papers, Sewanee, Tenn.
John McCrady Correspondence
John McCrady Diaries and Journals

John McCrady Manuscripts
Dwight G. von Kolnitz, Spartanburg, S.C.

NEWSPAPERS AND MAGAZINES

Boston Daily Advertiser, April 6, 1877
Charleston Courier
Charleston Daily Courier
Charleston Mercury
Columbian Centinel, March 19, 1800
Edgefield (S.C.) Advertiser
National Intelligencer, June 2, 1843
Southern Rose
United States Catholic Miscellany

CENSUS REPORTS

United States Census. *Population Schedule*, 1790–1860.
United States Census. *Slave Schedule*, 1820–1860.

PUBLISHED WORKS

A[gassiz] L[ouis]. "The Diversity of Origin of the Human Races." *Christian Examiner*, 4th ser., 14 (July 1850): 110–45.
———. "Essay on Classification." In *Contributions to the Natural History of the United States of America*, edited and with introduction by Edward Lurie, i–xxxiii. Reprint ed. Cambridge: Harvard University Press, 1969.
———. [Eulogy on John Edwards Holbrook]. *Proceedings of the Boston Society of Natural History* 14 (1871): 348–50.
———. "Geographical Distribution of Animals." *Christian Examiner*, 4th ser., 13 (March 1850): 181–204.
———. *Introduction to the Study of Natural History*. New York: Greeley & McElrath, 1847.
———. "Sketch of the Natural Provinces of the Animal World and Their Relation to the Different Types of Man." In *Types of Mankind; Or Ethnological Researches, Based Upon the Ancient Monuments, Paintings, Sculptures, and Crania of Races and Upon Their Natural, Geographical, Philological and Biblical History: Illustrated by Selections from Inedited Papers of Samuel George Morton, M.D., And by Additional Contributions from Prof. L. Agassiz, LL. D.; U. Usher, M.D.; and Prof. H. S. Patterson, M.D.* Philadelphia: Lippincott, Grambo & Co., 1854, lviii–lxviii.
"Agriculture in South-Carolina." *Southern Ladies Book*, n.s., 2 (1843): 200–203.
A. L. "On the Unity of the Human Race." *Southern Quarterly Review*, n.s., 10 (October 1854): 273–304.
[Allman, George J.] Comments on the resignation of John McCrady. *Nature* 16 (May 10, 1877): 31.
Annual Report of the Board of Regents of the Smithsonian Institution. Washington, D.C.: Smithsonian Institution, 1858 and 1859.

Audubon, John James. *The Birds of America*. 7 vols. New York: J. J. Audubon, 1840–44.

Audubon, [John J.], and [John] Bachman. "Description of a new North American Fox. Genus Vulpes. Cuv." *Proceedings of the Academy of Natural Sciences of Philadelphia* 6 (June 1852): 114–15.

———. "Descriptions of New Species of Quadrupeds inhabiting North America." *Proceedings of the Academy of Natural Sciences of Philadelphia* 1 (October 1841): 92–103, and *Journal of the Academy of Natural Sciences of Philadelphia* 8 (1842): 280–323.

———. *The Quadrupeds of North America*. 3 vols. New York: V. G. Audubon, 1851–54.

Audubon, Maria R., ed. *Audubon and His Journals*. 2 vols. New York: Dover, 1960.

[Bachman, Catherine L.]. *John Bachman*. Charleston: Walker, Evans & Cogswell, 1888.

Bachman, John. "Additional Observations on Hybridity in Animals." *Charleston Medical Journal and Review* 6 (1851): 383–96.

———. "Additional Remarks on the Genus Lepus, with corrections of a former paper, and descriptions of other species of Quadrupeds found in North America." *Journal of the Academy of Natural Sciences of Philadelphia* 8 (1839): 75–103.

———. *An Address Delivered Before the Total Abstinence Society of Charleston, S.C. on Wednesday Evening, July 27th, 1842*. Charleston: Burgess & James, 1842.

———. "Agricultural Labor, As One of the First Conditions of a National Existence (From an Agricultural Oration, delivered in South-Carolina in 1840)." *Magnolia*, n.s., 2 (1843): 16–20.

———. *Catalogue, Phaenogamous Plants and Ferns, Native and Naturalized, Found Growing in the Vicinity of Charleston, South Carolina*. Charleston: A. E. Miller, 1834.

———. *A Defence of Luther Against John Bellinger, M.D., and others*. Charleston: William Y. Paxton, 1853.

———. "Description of a New Species of Hare found in South Carolina." *Journal of the Academy of Natural Sciences of Philadelphia* 7, pt. 2 (1837): 194–99.

———. "Description of several New Species of American Quadrupeds." *Journal of the Academy of Natural Sciences of Philadelphia* 8, pt. 1 (1839): 57–73.

———. *A Discourse Delivered on the Forty-third Anniversary of His Ministry in Charleston*. Charleston: A. J. Burke, 1858.

———. *The Doctrine of the Unity of the Human Race Examined on the Principles of Science*. Charleston: C. Canning, 1850.

———. "An examination of a few of the statements of Prof. Agassiz, in his 'Sketch of the natural provinces of the animal world, and their relation to the different types of men.'" *Charleston Medical Journal and Review* 9 (1854): 790–806.

———. "An examination of Prof. Agassiz's Sketch of the Natural Provinces of the Animal World, and their relation to the different Types of Man, with a Tableau accompanying the Sketch." *Charleston Medical Journal and Review* 10 (1855): 482–534.

———. "An Examination of the Characteristics of Genera and Species as applicable to the doctrine of the Unity of the Human Race." *Charleston Medical Journal and Review* 10 (1855): 201–22.

[——]. *The Funeral Discourse of the Rev. John G. Schwartz, Delivered on September 11, 1831.* Charleston: James S. Burgess, 1831.

——. "The Humboldt Festival." *Charleston Daily Courier*, September 15, 1869, 4.

——. "An Inquiry Into the Nature and Benefits of an Agricultural Survey of the State of South-Carolina." *Southern Agriculturist* 3 (February 1843): 49–65 and (March 1843): 81–96.

——. "An Investigation of the Cases of Hybridity in Animals on Record, considered in reference to the Unity of the Human Species." *Charleston Medical Journal and Review* 5 (January 1850): 168–97.

——. "A Letter from Rev. John Bachman." *Charleston Medical Journal and Review* 6 (1851): 598.

——. "Monograph of the Genus *Sciurus*, with Descriptions of new Species and their Varieties, as existing in North America." *Magazine of Natural History* 3 (1839): 113–23, 154–62, 220–27, 330–37, 378–90.

——. "The Morals of Entomology." In *The Charleston Book: A Miscellany in Prose and Verse*, edited by William Gilmore Simms, 30–40. Charleston: Samuel Hart, 1845.

——. "Notes on the Generation and Development of the Opossum (Didelphis virginiana)." *Proceedings of the Academy of Natural Sciences of Philadelphia* 4 (April 1848): 40–47.

——. "Observations on the Changes of Colour in Birds and Quadrupeds." *Transactions of the American Philosophical Society*, n.s., 6 (1838): 197–239.

——. "Observations on the different species of Hares (genus Lepus) inhabiting the United States and Canada." *Journal of the Academy of Natural Sciences of Philadelphia* 7 (1837): 282–361.

——. "Observations on the Genus Scalops, (Shrew Moles,) with Descriptions of the Species Found in North America." *Journal of the Boston Society of Natural History* 4 (1842–43): 26–35.

——. "On the Migration of the Birds of North America." *American Journal of Science* 30 (July 1836): 81–100.

——. "Remarks in Defense of the Author of 'The Birds of America.'" *Boston Journal of Natural History* 1 (1834): 15–31, and *Loudon's Magazine of Natural History* 7 (1834): 164–75.

——. [Remarks on species of *Sciurus* in the collections of the Zoological Society of London.] *Proceedings of the Zoological Society of London* 6 (1839): 85–105.

——. "A Reply to the Letter of Samuel George Morton, M.D., on the question of Hybridity in Animals, considered in reference to the Unity of the Human Species." *Charleston Medical Journal and Review* 5 (1850): 466–508.

——. Review of *Types of Mankind*, by Josiah C. Nott and George R. Gliddon. *Charleston Medical Journal and Review* 9 (1854): 627–59, 790–806; ibid. 10 (1855): 201–22, 482–534.

——. "Some Remarks on the Genus Sorex, with a monograph of the North American Species." *Journal of the Academy of Natural Sciences of Philadelphia* 7 (1837): 362–403.

——. "A Successful Method of Raising Ducks." *Farmer's Register* 1 (1833): 356–59.

——. "The Vulture." *Scientific Tracts and Family Lyceum*, March 15, 1834.

Bailey, J. W. "Microscopical Observations Made in South Carolina, Georgia, and Florida." *Smithsonian Contributions to Knowledge* 2, pt. 8 (1851): 1–48 and 3 plates.

Bremer, Fredericka. *The Homes of the New World: Impressions of America.* 2 vols. New York: Harper and Brothers, 1853.

Brewer, Thomas M. "Reminiscences of John James Audubon." *Harper's New Monthly Magazine*, October 1880, 665–75.

Browne, Peter A. *Trichologia Mammalium, or a Treatise on the Organization, Properties and Uses of Hair and Wool.* Philadelphia: Jones, 1853.

Buchanan, Robert. *The Life and Adventures of Audubon the Naturalist.* New York: E. P. Dutton, 1869.

Buckingham, J. S. *Slave States of America.* 2 vols. London: Fisher, Son & Co., 1842.

Burke, Luke. "Strictures on Dr. Bachman's paper entitled 'An Examination of Prof. Agassiz's Sketch of the Natural Provinces of the Animal World.'" *Charleston Medical Journal and Review* 11 (1856): 433–58.

"The Burning of Otey Hall." *Cap and Gown* [journal of the University of the South] 1 (October 1881): 4–5.

Cardozo, J. N. *Reminiscences of Charleston.* Charleston: Joseph Walker, 1866.

Catalogue of Herbarium of Professor L. R. Gibbes. N.p., n. pub., n.d. [ca. 1898]. Copy in Academy of Natural Sciences of Philadelphia Library.

Catalogue of the Trustees, Faculty and Students of the College of Charleston, South Carolina, with the Regulations for the Government of the College Library and Museum of Natural History, November 1855. Charleston: A. E. Miller, 1856.

Catalogue of the Trustees, Faculty and Students of the College of Charleston, South Carolina, with the Regulations for the Government of the College Library and Museum of Natural History, January, 1859. Charleston: Walker, Evans & Co., 1859.

Catalogue of the Trustees, Faculty and Students of the Medical College of the State of South-Carolina, 1836. Charleston: Medical College of the State of South Carolina, 1836.

Cauthen, Charles E., ed. *Journals of the South Carolina Executive Councils of 1861 and 1862.* Columbia: South Carolina Archives Department, 1956.

Charleston Directory and Stranger's Guide for the Year 1822.

Charleston, South Carolina, Directory and Almanac, 1806.

Charleston Yearbook, 1882, 1883.

Childs, Arney Robinson, ed. *The Journal of Henry William Ravenel.* Columbia: University of South Carolina Press, 1947.

Clark, Thomas D., ed. *South Carolina: The Grand Tour, 1780–1865.* Columbia: University of South Carolina Press, 1973.

Conrad, Timothy A. *New Fresh Water Shells of the United States.* Philadelphia: Judah Dobson, 1834.

Constitution and By-Laws of the Elliott Society of Natural History. Charleston: Walker, Evans and Co., 1857.

"Cook and His Biology." *Popular Science Monthly* 12 (1878): 495–500.

Corning, Howard, ed. *Letters of John James Audubon, 1826–1840*. 2 vols. Boston: The Club of Odd Volumes, 1930.

[Cull, Richard]. "On the Recent Progress of Ethnology." *Journal of the Ethnological Society of London* 4 (1856): 298–304.

Cuvier, Georges, and Achille Valenciennes. *Histoire naturelle des poissons*. 22 vols. Paris: Levrault, 1828–49.

[Dana, James Dwight]. Comment on "On the Carcinological Collections," by Lewis R. Gibbes. *American Journal of Science*, 2nd ser., 11 (January 1851): 128.

——. "Thoughts on Species." *American Journal of Science*, 2nd ser., 24 (1857): 305–16.

DeKay, James E. *Zoology of New York, or the New-York Fauna*. 6 pts. in 5 vols. Albany: W. & A. White & J. Visscher, 1842.

"Diary of Timothy Ford, 1785–1786." *South Carolina Historical and Genealogical Magazine* 13 (1912): 132–47, 181–204.

Dickson, Samuel Henry. *Statements in Reply to Certain Publications from the Medical Society of South-Carolina*. Charleston: n.p., 1834.

"Domestic Intelligence." *Carolina Journal of Medicine, Science, and Agriculture* 1 (1825): 392–94.

Evarts, Jeremiah. *Through the South and the West, 1826*. Edited by J. Orin Oliphant. Lewisburg, Pa.: Bucknell University Press, 1956.

An Exposition of the Affairs of the Medical Society of South-Carolina, So Far as They Appertain to the Establishment of a Medical College in Charleston, and the Subsequent Division of the Latter Into Two Schools of Medicine. Charleston: The Medical Society, 1831.

Ford, Frederick A. *Census of the City of Charleston, South Carolina, for the Year 1861*. Charleston: Evans & Cogswell, 1861.

Ford, Timothy. *An Address to the Literary and Philosophical Society on Physical Science*, 1818.

Gibbes, Lewis R. "The Annular Phase of Venus." *Proceedings of the Elliott Society of Natural History* 2 (1891): 246–48.

——. "Aurora Borealis of September 29th, 1851." *American Journal of Science*, 2nd ser., 13 (May 1852): 128.

——. "Botany of Edings' Bay." *Proceedings of the Elliott Society of Natural History* 1 (1859): 241–48.

——. "Catalogue of the Crustacea in the Cabinet of the Academy of Natural Sciences of Philadelphia, August 20th, 1847, with notes on the most remarkable." *Proceedings of the Academy of Natural Sciences of Philadelphia* 5 (1850): 22–30.

——. "Catalogue of the Fauna of South Carolina." In *Report on the Geology of South Carolina*, by M. Tuomey, appendix, i–xxiv. Columbia: A. S. Johnston, for the State, 1848.

——. *A Catalogue of the Phoenogamous Plants of Columbia, S.C., and Its Vicinity*. Columbia: "Printed at the Telescope Office," 1835.

———. "Crustacea." In *Statistics of the State of Georgia*, by George White, appendix, 21–24. Savannah: W. Thorne Williams, 1849.

———. "Description of a New Species of Salamander." *Boston Journal of Natural History* 5 (1845–47): 89–90.

———. "Description of Ranilia Muricata Milne Edwards." *Proceedings of the Elliott Society of Natural History* 1 (1859): 225–29 and 1 plate.

———. "Description (with figure) of Menobranchus punctatus." *Boston Journal of Natural History* 6 (1853): 369–73.

———. "Description, with figures, of six species of Porcellana, inhabiting Eastern Coast of North America." *Proceedings of the Elliott Society of Natural History* 1 (1859): 6–13.

———. "Early Life of Francis Arago." *Charleston College Magazine* 1 (January 1, 1855): 225–31, 272–80, 289–99; 2 (May 1, 1855): 33–41.

———. "An Easy Mode of Approximating to the Time of Vibration in a Circular Arc." *Mathematical Monthly* 3 (November 1860): 40–46.

———. "The Identity of the Comets 1886b, 1844b, and 1678." *Proceedings of the Elliott Society of Natural History* 2 (1891): 126–28.

———. "Monograph of the Genus Cryptopodia." *Proceedings of the Elliott Society of Natural History* 1 (1859): 32–38.

———. "Note on Maxima and Minima." *Mathematical Monthly* 1 (July 1859): 335–42.

———. "Note on the Cycloid." *Mathematical Monthly* 1 (June 1859): 297–300.

———. "Notice of Stalactites formed in Artificial Structures." *Proceedings of the Elliott Society of Natural History* 2 (1891): 224.

———. "Notice of the Phenomena attending the Shock of the Earthquake of Dec. 19, 1857." *Proceedings of the Elliott Society of Natural History* 1 (1859): 288–89.

———. "Of the Collection of Crustaceans in the Cabinet of the Boston Society of Natural History." *Proceedings of the Boston Society of Natural History* 2 (1845): 69–70.

———. "On a Convenient Form of Aspirator." *Proceedings of the Elliott Society of Natural History* 1 (1859): 291–94.

———. [On Eugene Chevruel]. *Proceedings of the Elliott Society of Natural History* 2 (1891): 223–24.

———. "On Some Points Which Have Been Overlooked in the Past and Present Condition of Niagara Falls." *Proceedings of the Elliott Society of Natural History* 1 (1859): 91–100.

———. "On the Carcinological Collections of the Cabinets of Natural History in the United States." *Proceedings of the American Association for the Advancement of Science* 3 (1850): 167–201.

———. "On the General Principles of the Resistance of Fluids, in a notice of the Fifth Article of no. XV of the Southern Review." *American Journal of Science* 27 (1835): 135–39.

———. "On the representatives of the genus Cactus in this State." *Proceedings of the Elliott Society of Natural History* 1 (1859): 272–73.

———. "Personal observations and notes on the displacement of monuments." *Proceedings of the Elliott Society of Natural History* 2 (1891): 153–71.

——. "A portable and easily made Heliotrope." *Proceedings of the Elliott Society of Natural History* 2 (1891): 173–76.

——. "Remarks on Niagara Falls." *Proceedings of the American Association for the Advancement of Science* 10 (1856): 69–76.

——. "Report of the Society's Commission to collect information concerning the Charleston Earthquake of August 31, 1886." *Proceedings of the Elliott Society of Natural History* 2 (1891): 139–52.

——. *Rules for the Accentuation of Names in Natural History, with Examples Zoological and Botanical.* Charleston: Elliott Society of Natural History, 1860.

——. "Synoptical Table of the Chemical Elements." *Proceedings of the Elliott Society of Natural History* 2 (1891): 77–90.

——. "Xenotime from the Gold Region of Georgia." *American Journal of Science*, 2nd ser., 13 (May 1852): 142–43.

G[ibbes], R[obert] W. "Death of Samuel George Morton, M.D." *Charleston Medical Journal and Review* 6 (1851): 594–98.

——. "Memoir on the Fossil genus *Basilosaurus*, Harlan, (*Zeuglodon*, Owen), with a notice of Specimens from the Eocene Green Sand of South Carolina." *Journal of the Academy of Natural Sciences of Philadelphia*, 2nd ser., 1 (1847–50): 1–13 and 5 plates.

——. "New Species of *Myliobates* from the Eocene of South Carolina, with other genera not heretofore observed in the United States." *Journal of the Academy of Natural Sciences of Philadelphia*, 2nd ser., 1 (1847–50): 299–300.

——. *The Present Earth, The Remains of a Former World: A Lecture Delivered Before the South Carolina Institute, September 6, 1849.* Columbia: A. S. Johnston, 1849.

——. "Remarks on the Fossil Equus." *Proceedings of the American Association for the Advancement of Science* 3 (1850): 66–69.

——. Review of *Report on the Geology of South Carolina*, by Michael Tuomey. *Southern Quarterly Review* 31 (October 1849): 161–78.

Girard, Charles. "On a New American Saurian Reptile." *Proceedings of the American Association for the Advancement of Science* 4 (1851): 200–202.

Godman, John. *American Natural History.* 3 vols. Philadelphia: H. C. Carey and I. Lea, 1826–28.

Gongaware, George J., comp. *The History of the German Friendly Society of Charleston, South Carolina, 1766–1916.* Richmond, Va.: Garrett & Massie, Publishers, 1935.

Griffith, Edward, et al. *The Animal Kingdom Arranged in Conformity with Its Organization by the Baron Cuvier, with Supplementary Additions to Each Order.* 15 vols. London: G. B. Whittaker, 1827–32.

Harlan, R[ichard]. "Description of a new species of Quadrupeds of the order Rodentia, inhabiting the United States." *American Journal of Science* 31 (January 1837): 385–86.

——. *Fauna Americana: Being a Description of the Mammiferous Animals Inhabiting North America.* Philadelphia: A. Finly, 1825.

Heads of Families at the First Census of the United States Taken in the Year 1790, New York. Spartanburg, S.C.: Reprint Company, 1981.

Henry, Joseph. *The Papers of Joseph Henry*, Vol. 4. Edited by Nathan Reingold et al. Washington, D.C.: Smithsonian Institution Press, 1981.

Holbrook, John Edwards. "An account of several species of Fish observed in Florida, Georgia, &c." *Journal of the Academy of Natural Sciences of Philadelphia*, 2nd ser., 3 (1855): 47–58.

——. "Education of Farmers." *Southern Agriculturist* 12 (July 1839): 394–97.

——. *Ichthyology of South Carolina*. Charleston: John Russell, 1855. [Rev. ed.], Charleston: Russell and Jones, 1860.

——. *North American Herpetology*. 4 vols. Philadelphia: J. Dobson, 1836–1840.

——. *North American Herpetology*. 5 vols. [2nd ed.], Philadelphia: J. Dobson, 1842. Reprint, Athens, Ohio: Society for the Study of Amphibians and Reptiles, 1976.

——. "Reptiles" and "Fish." In *Statistics of the State of Georgia*, by George White, appendix, 13–15 and 16–19, respectively. Savannah: W. J. Williams, 1849.

——. *Southern Ichthyology; Or a Description of the Fishes Inhabiting the Waters of South Carolina, Georgia, and Florida*. New York: Wiley and Putnam, pt. 2, 1847, and pt. 3, 1848.

Holcomb, Brent H., comp. *Marriage and Death Notices from the Charleston Observer, 1827–45*. Columbia: The Author, 1980.

——. *Marriage and Death Notices from the (Charleston) Times*. Baltimore: Genealogical Publishing Co., 1979.

——. *Marriage and Death Notices from the Lutheran Observer, 1831–1861 and the Southern Lutheran, 1861–1865*. Easley, S.C.: Southern Historical Press, 1979.

Holman, Harriett R. "Charleston in the Summer of 1841: The Letters of Harriott Horry Rutledge." *South Carolina Historical and Genealogical Magazine* 46 (1945): 1–14.

Holmes, Francis S. "Contributions to the Natural History of the American Devil fish, with Descriptions of a New Genus from the Harbour of Charleston, South Carolina." *Proceedings of the Elliott Society of Natural History* 1 (1859): 39–46.

——. "Description of a Bezoar Stone, found in the stomach of a Buck, 'cervus virginianus.'" *Southern Journal of Medicine and Pharmacy*, 2 (1847): 527–30. Reprinted in *American Journal of Science*, 2nd ser., 7 (1849): 187–201.

——. "Description of a New Species of Ostrea, Found Living in the Waters of the Coast of South Carolina." *Proceedings of the Elliott Society of Natural History* 1 (1859): 29.

——. "Descriptions of New Fossil Balani, from the Eocene Marl of Ashley River, S.C." *Proceedings of the Elliott Society of Natural History* 1 (1859): 21.

——. "Maffitt Channel Borings." *Southern Quarterly Review*, n.s., 8 (April 1853): 509–13.

——. "Notes on the Geology of Charleston." *Charleston Medical Journal and Review* 3 (1848): 655–71.

——. "Observations on the Geology of Ashley River, South-Carolina." *Proceedings of the American Association for the Advancement of Science* 3 (1850): 201–4.

——. *Phosphate Rocks of South Carolina and the "Great Carolina Marl Bed".* Charleston: Holmes' Book House, 1870.

——. *Post-Pleiocene Fossils of South Carolina.* Charleston: Russell & Jones, 1860. [Includes "Description of Vertebrate Fossils," by Joseph Leidy, 99–122.]

——. *Remains of Domestic Animals Discovered Among Post-Pleiocene Fossils in South Carolina.* Charleston: James and Williams, 1858.

——. [Remarks on Post-Pliocene Fossils of South Carolina.] *Proceedings of the Academy of Natural Sciences of Philadelphia* 11 (1859): 177–86.

——. *The Southern Farmer and Market Gardener.* Charleston: Burgess and James, 1842.

Index to the 1800 Census of New York. Baltimore: Genealogical Publishing Co., 1984.

James, William. *The Correspondence of William James.* Edited by Ignas K. Skrupskelis and Elizabeth M. Berkeley. 6 vols. Charlottesville: University Press of Virginia, 1992–98.

Klosky, Beth Ann. *The Pendleton Legacy.* Columbia, S.C.: Sandlapper Press, 1971.

Lea, Isaac. "Description of New Species of the Family Unionidae." *Transactions of the American Philosophical Society,* n.s., 10 (1847): 253–94.

——. "Rectification of Mr. T. A. Conrad's 'Synopsis of the Family Naiades of North America,' published in the 'Proceedings' of the Academy of Natural Sciences of Philadelphia, February, 1853." *Proceedings of the Academy of Natural Sciences of Philadelphia* 7 (December 1854): 236–50.

——. *A Synopsis of the Family of Naiades.* Philadelphia: Blanchard and Lea, 1852.

Leidy, Joseph. "Description of Vertebrate Remains, chiefly from the Phosphate Beds of South Carolina." *Journal of the Academy of Natural Sciences of Philadelphia,* 2nd ser., 8 (1874–81): 209–61.

List of Students of the Lawrence Scientific School, 1847–1900. Cambridge: Harvard University, 1901.

[List of the Officers and Faculty of the University of the South]. *The University Record,* n.s., 1 (March 1879): 1–2.

List of the Taxpayers of the City of Charleston for 1859. Charleston: Walker, Evans & Company, 1860.

Lyell, Charles. "On the Newer Deposits of the Southern States of America." *Quarterly Journal of the Geological Society of London* 2 (1846): 405–7.

——. *A Second Visit to the United States of America.* 2 vols. New York: Harper & Brothers, 1849.

——. *Travels in North America in the Years 1841–1842.* 2 vols. New York: Wiley and Putman, 1845.

M[cCord], L[ouisa] S. "Diversity of the Races: Its Bearing Upon Negro Slavery." *Southern Quarterly Review* 19 (April 1851): 392–419.

McCrady, John. [Address on Humboldt]. *Charleston Daily Courier,* September 15, 1869, 2.

——. "Chinese Immigration: Should We Favor It at the Present Time?" *Charleston Daily Courier,* May 11, 1870, 2.

——. [Comments on Chinese Immigration]. *Charleston Daily Courier,* May 6, 1870, 2.

——. "Description of *Oceania* (*Turritopsis*) *nutricula* nov. spec. and the embryo-logical history of a singular Medusan Larva, found in the Cavity of the Bell." *Proceedings of the Elliott Society of Natural History* 1 (1859): 55–90.

——. *Essay [on Pisiculture] by John McCrady.* Charleston: Walker, Evans & Cogswell, 1872.

——. "A Few Thoughts on Southern Civilization." *Russell's Magazine* 1 (June 1857): 224–28, (July 1857): 338–49, (September 1857): 546–56, and 2 (December 1857): 212–26.

——. "Gymnopthalmata of Charleston Harbor." *Proceedings of the Elliott Society of Natural History* 1 (1859): 103–221.

——. *Home Education: A Necessity of the South. An Oration Delivered Before the Chrestomathic Society of the College of Charleston on Friday, March 2d, 1860.* Charleston: Walker, Evans & Co., 1860.

——. "Joseph Cook on Biology." *Literary World* 8 (November 1877): 87–89.

——. "The Law of Development by Specialization. A Sketch of Its Probable Universality." *Journal of the Elliott Society* 1 (1860): 101–14.

——. "Letter from John McCrady, Esq., Charleston, on the *Lingula pyramidata* described by Mr. W. Stimpson." *American Journal of Science*, 2nd ser., 30 (1860): 157–58.

——. "Observations on the Food and the Reproductive Organs of Ostrea Virginia, with some Account of Bucephalus Cuculus Nov. Spec." *Proceedings of the Boston Society of Natural History* 16 (December 3, 1873): 170–92.

——. "On the development of two species of Ctenophora found in Charleston Harbor." *Proceedings of the Elliott Society of Natural History* 1 (1859): 254–71.

——. "A Provisional Theory of Generation." *Proceedings of the Boston Society of Natural History* 19 (April 8, 1877): 171–85.

——. "Remarks on the embryology of a species of Bolina found in Charleston Harbor." *Proceedings of the Elliott Society of Natural History* 1 (1859): 223–24.

——. "Remarks on the Eocene formation in the neighborhood of Alligator, Florida, with descriptions of three new species (Ravenelia gen. nov.) of fossil echinoderms." *Proceedings of the Elliott Society of Natural History* 1 (1859): 282–83.

——. "Remarks on the Zoological Affinities of Graptolites." *Proceedings of the Elliott Society of Natural History* 1 (1859): 229–36.

——. "The Study of Nature and the Arts of Civilized Life." *DeBow's Review*, n.s., 5 (May–June 1861): 579–606.

——. *A System of Independent Research, the Chief Educational Want of the South. Address Delivered Before the Society of the Alumni of the College of Charleston, at the Inauguration of the Charleston College Library.* Charleston: A. J. Burke, 1856.

McCrady, John, St. Julien Ravenel, and C. F. Pankin. "Report." *Transactions of the Howard Association of Charleston, S.C. during the Epidemic of 1871.* Charleston: Walker, Evans & Cogswell, [1871].

MacDonald, Donald. *The Diaries of Donald MacDonald, 1824–1826.* Introduction by Caroline Dale Snedeker. Clifton, N.J.: A. M. Kelley, 1973.

Martineau, Harriet. *Retrospect of Western Travel.* 2 vols. London: Saunders and Otley, 1838.

Meigs, James Aitken. *Catalogue of Human Crania, in the Collections of the Academy of Natural Sciences of Philadelphia*. Philadelphia: J. B. Lippincott & Co., 1857.

"Memoirs of Professor Frederick Adolphus Porcher." *South Carolina Historical and Genealogical Magazine* 45 (1944): 30–40, 80–98, 146–56, 200–216; ibid. 46 (1945): 25–39, 78–92, 140–58, 198–208; ibid. 47 (1946): 32–52, 83–108, 150–62, 214–27; ibid. 48 (1947): 20–25.

Messmer, Sebastian G., ed. *The Works of the Right Reverend John England, First Bishop of Charleston*. 7 vols. Cleveland: The Arthur Clark Company, 1908.

Michel, Myddleton. "Researches on the Generation of the Virginia Opossum—Didelphys Virginiana." *Proceedings of the American Association for the Advancement of Science* 3 (1850): 60–65.

Mills, Robert. *Statistics of South Carolina, Including a View of Its Natural, Civil and Military History*. Charleston: Hurlburt and Lloyd, 1826.

Morse, Abner. *Genealogical Register of the Descendants of the Early Planters of Sherborn, Holliston, and Midway, Massachusetts*. Boston: Damrell and Moore, 1855.

Morton, Samuel George. "Additional Observations on Hybridity in Animals." *Charleston Medical Journal and Review* 5 (1850): 755–805.

——. *Crania Aegyptiaca; or, Observations on Egyptian Ethnography, Derived from Anatomy, History and the Monuments*. Philadelphia: J. Penington, 1844.

——. *Crania Americana; or, a Comparative View of the Skulls of Various Aboriginal Nations of North and South America To Which is Prefixed an Essay on the Varieties of the Human Species*. Philadelphia: J. Dobson, 1839.

——. "Hybridity in Animals, considered in reference to the question of the Unity of the Human Species." *American Journal of Science*, 2nd ser., 3 (1847): 39–50, 203–12.

——. "Letter to the Rev. John Bachman, D.D., on the question of Hybridity in Animals, considered in reference to the Unity of the Human Species." *Charleston Medical Journal and Review* 5 (1850): 328–44.

——. "Notes on Hybridity, and some collateral subjects." *Charleston Medical Journal and Review* 6 (1851): 373–83.

——. "Notes on Hybridity, designed as a further Supplement to a Memoir on that subject in the last number of this Journal." *Charleston Medical Journal and Review* 6 (1851): 301–8.

——. "Notes on Hybridity, designed as a supplement to the memoir on that subject in a former number of this Journal." *Charleston Medical Journal and Review* 6 (1851): 145–52.

"Natural History in Charleston." *Charleston Medical Journal and Review* 7 (1852): 140.

Neuffer, Claude Henry, ed. *The Christopher Happoldt Journal*. Contributions from the Charleston Museum, No. 13, 1960.

[Nott, Josiah]. "Communication from Dr. Nott." *Charleston Medical Journal and Review* 9 (1854): 862–64.

[——]. "Diversity of the Human Race." *DeBow's Review*, 3rd ser., 2 (February 1851): 113–32.

——. "An examination of the Physical History of the Jews, in its bearing on the

Question of the Unity of the Races." *Proceedings of the American Association for the Advancement of Science* 3 (1850): 98–106.

[——]. "Letter of Dr. Nott." *Charleston Medical Journal and Review* 10 (1855): 883–84.

——. "The Mulatto a Hybrid—Possible Extermination of the Two Races if the Whites and Blacks Are Allowed to Intermarry." *American Journal of Science*, n.s., 6 (1843): 252–53.

——. "Reply to Dr. Bachman's Review of Agassiz's Natural Provinces." *Charleston Medical Journal and Review* 10 (1855): 753–67.

——. ["Thoughts on Race"]. In *The Moral and Intellectual Diversity of Races, with Particular Reference to Their Respective Influence in the Civil and Political History of Mankind, from the French of Count A. de Gobineau: with an Analytical Introduction and Copious Notes*, edited by H. Hotz, appendix B, 473–512. Philadelphia: J. B. Lippincott, 1856.

Nott, Josiah C., and George R. Gliddon. *Types of Mankind; or Ethnological Researches, Based Upon the Ancient Monuments, Paintings, Sculptures, and Crania of Races and Upon Their Natural, Geographical, Philological and Biblical History: Illustrated by Selections from Inedited Papers of Samuel George Morton, M.D., and by Additional Contributions from Prof. L. Agassiz, LL.D.; U. Usher, M.D.; and Prof. H. S. Patterson, M.D.* Philadelphia: Lippincott, Grambo & Co., 1854.

Ogier, Thomas L. *A Memoir of Dr. John Edwards Holbrook.* Charleston: Walker, Evans, & Cogswell, 1871.

Payne, Daniel Alexander. *Recollections of Seventy Years.* Nashville: A.M.E. Sunday School Union, 1888.

Proceedings of the Agricultural Convention and of the State Agricultural Society of South Carolina, from 1839–1845—Inclusive. Columbia: Summer & Carroll, 1846.

Proceedings of the Board of Trustees of the University of the South, 1877–81.

Proceedings of the Elliott Society of Natural History 1 (1859) and 2 (1891).

Ravenel, Edmund. *Catalogue of Recent Shells in the Cabinet of Edmund Ravenel, M.D.* Charleston: Privately printed, 1834.

——. "Description of a new recent species of Scutella." *Proceedings of the Academy of Natural Sciences of Philadelphia* 2 (June 1845): 253–54.

——. "Description of some new species of Fossil Organic remains, from the Eocene of South Carolina." *Proceedings of the Academy of Natural Sciences of Philadelphia* 2 (October 1844): 96–98.

——. "Description of three new species of Univalves, recent and fossil." *Proceedings of the Elliott Society of Natural History* 1 (1859): 280–82.

——. "Description of two new species of Fossil organic Remains, from South Carolina." *Proceedings of the Academy of Natural Sciences of Philadelphia* 1 (September 1841): 81–82.

——. "Description of two new species of Fossil Scutella, from South Carolina." *Journal of the Academy of Natural Sciences of Philadelphia* 8 (1842): 333–36.

——. "Descriptions of New Recent Shells from the Coast of South Carolina." *Proceedings of the Academy of Natural Sciences of Philadelphia* 13 (1861): 41–44.

——. *Echinidae, Recent and Fossil of South Carolina*. Charleston: Burges & James, 1848.

——. "Hirudo?" *Proceedings of Elliott Society of Natural History* 1 (1859): 24.

——. "On the Advantages of a Sea-Shore Residence in the Treatment of Certain Diseases, and the Therapeutic Employment of Sea-Water." *Charleston Medical Journal and Review* 3 (1848): 619–31.

——. "On the Medical Topography of St. John's, Berkeley, S.C., and its relation to Geology." *Charleston Medical Journal and Review* 4 (1849): 697–704.

——. "On the Recent Squalidae of the Coast of South-Carolina, and Catalogue of the Recent and Fossil Echinoderms of South-Carolina." *Proceedings of the American Association for the Advancement of Science* 3 (1850): 159–61.

——. "Tellinidae of South Carolina." *Proceedings of the Elliott Society of Natural History* 2 (1891): 33–40.

Ravenel, Edmund, Jr. *Catalogue of Recent and Fossil Shells in the Cabinet of the Late Edmund Ravenel, M.D., Charleston, S.C., December 1874*. Charleston: Walker, Evans & Cogswell, 1874.

"The Ravenel Family in France and in America." *Transactions of the Huguenot Society of South Carolina* 6 (1889): 38–54.

Report of the Curator of the Museum of Natural History of the College of Charleston, S.C., March, 1854. Charleston: A. J. Burke, 1854.

Reports and Resolutions of the General Assembly of the State of South Carolina Passed at the Annual Session of 1855. Columbia: E. H. Britton & Co., State Printers, 1855.

Reports and Resolutions of the General Assembly of the State of South Carolina Passed at the Annual Session of 1857. R. W. Gibbes, State Printers, 1857.

Review of *Post-Pleiocene Fossils of South-Carolina*, by Francis S. Holmes. *Charleston Medical Journal and Review* 15 (January 1860): 106–7.

Review of *The Doctrine of the Unity*, by John Bachman. *American Journal of Science*, 2nd ser., 11 (March 1851): 302.

Review of *The Doctrine of the Unity*, by John Bachman. *Southern Medical and Surgical Journal*, n.s., 6 (August 1850): 504.

Review of *The Doctrine of the Unity*, by John Bachman. *Southern Quarterly Review* 17 (April 1850): 250.

Review of *The Quadrupeds of North America*, by Audubon and Bachman. *American Review* 4 (December 1846): 625–38.

Review of *The Quadrupeds of North America*, by Audubon and Bachman. *Literary World* 1 (1847): 128–29.

Review of *The Quadrupeds of North America*, by Audubon and Bachman. *Southern Quarterly Review* 11 (April 1847): 499–503.

Rhees, William J. *Manual of Public Libraries, Institutions, and Societies in the United States, and British Provinces of North America*. Philadelphia: J. B. Lippincott & Co., 1859.

Roos, Rosalie. *Travels in America, 1851–1855*. Edited and translated by Carl L. Anderson. Carbondale: Southern Illinois University Press, 1982.

Ruffin, Edmund. *An Essay on Calcareous Manures*. Petersburg, Va.: J. W. Campbell, 1832.

——. *Report of the Commencement and Progress of the Agricultural Survey of South Carolina, for 1843*. Columbia: A. H. Pemberton, State Printer, 1843.

The Rules of the South Carolina Society Established at Charleston in the Said Province, Sept. 11, 1837, originally incorporated, May 1, 1751. 17th ed. Charleston: The Society, 1937.

"The Schirmer Diary." *South Carolina Historical Magazine* 44 (1943): 54, 113, 182, 237; 67 (1966): 167, 229; 68 (1967): 37, 97; 69 (1968): 59, 139, 204, 262; 70 (1969): 59, 122, 196.

Second Annual Report of the Receipts and Expenditures of the Town of Norfolk, Including the List of Officers and Other Statistics for the Year Ending 31 Jan. 1872. Norfolk, Mass.: W. H. Rockwood, 1872.

Simms, William Gilmore. *The Letters of William Gilmore Simms*. 6 vols. Edited by Mary Simms Oliphant, Alfred Taylor Odell, and T. C. Duncan Eaves. Columbia: University of South Carolina Press, 1954.

Smith, Henry A. M. "Charleston and Charleston Neck: The Original Grantees and the Settlements Along the Ashley and Cooper Rivers." *South Carolina Historical and Genealogical Magazine* 19 (January 1918): 3–76.

Smyth, Thomas. *Autobiographical Notes, Letters and Reflections*. Edited by Louisa Cheves Stoney. Charleston: Walker, Evans & Cogswell Company, 1914.

——. *The Unity of the Human Races Proved to be the Doctrine of Scripture, Reason, and Science*. New York: George P. Putnam, 1850.

Snowden, Yates. "The Late Seventies." *Charleston College Magazine* 27 (June 1924): 18–26.

South Carolina: A Guide to the Palmetto State. New York: Oxford University Press, 1941.

South Carolina Genealogies. 2 vols. Spartanburg, S.C.: The Reprint Company, 1983.

Stevenson, Mary, comp. *The Diary of Clarissa Adger Brown, Ashtabula Plantation, 1865*. Pendleton, S.C.: Foundation for Historic Preservation in Pendleton Area, 1973.

S[timpson], W[illiam]. Review of "Description of *Oceania* (*Turritopsis*) *nutricula* . . ." and "Gymnopthalmata of Charleston Harbor," by John McCrady. *American Journal of Science*, 2nd ser., 29 (May 1860): 130–32.

——. Review of *Post-Pliocene Fossils*, by Francis S. Holmes. *American Journal of Science*, 2nd ser., 33 (1862): 298–300.

Strobel, Martin. *Exposition of the Relationship Existing between Jacob Martin and Elizabeth Pennington, Residing Together in Philadelphia*. Charleston: Privately printed, 1825.

[Tributes to] Lewis R. Gibbes. N.p., n.d. 36 pp. Copy in College of Charleston Library.

Tuomey, Michael. "Notice of the discovery of a Cranium of the Zeuglodon." *Proceedings of the Academy of Natural Sciences of Philadelphia* 3 (February 1847): 151–53. Reprinted in *American Journal of Science*, 2nd ser., 4 (September 1847): 283–85.

——. *Report on the Geology of South Carolina*. Columbia: A. S. Johnston, for the State, 1848.

Tuomey, M[ichael], and F[rancis] S. Holmes. *Pleiocene Fossils of South Carolina.* Charleston: James & Williams, 1857. Reprint ed., Ithaca, N.Y.: Paleontological Research Foundation, 1974, with addendum by Druid Wilson.

Way, William. *History of the New England Society of Charleston, South Carolina.* Charleston: The Society, 1920.

The War of the Rebellion: A Compilation of the Official Records of the Union and Confederate Armies. Ser. 4, Vol. 3. Washington, D.C.: Government Printing Office, 1880–1901.

Weber, Mabel L. "Dr. John Rutledge and His Descendants." *South Carolina Historical and Genealogical Magazine* 31 (1930): 93–106.

White, George. *Statistics of the State of Georgia: Including an Account of Its Natural, Civil and Ecclesiastical History.* Savannah: W. Thorne Williams, 1849.

Wilmer, G. T. "In Memoriam." *Cap and Gown* [journal of the University of the South] 1 (November 1881): 1.

Wilson, Teresa, and Janice L. Grimes, comp. *Marriage and Death Notices from the Southern Patriot, 1815–1830.* 2 vols. Easley, S.C.: Southern Historical Press, 1982.

Secondary Sources

Adler, Kraig. *A Brief History of Herpetology in North America Before 1900.* N.p.: Society for the Study of Amphibians and Reptiles, 1979.

——. "Holbrook, John Edwards (1794–1871)." In *Contributions to the History of Herpetology,* Society for the Study of Amphibians and Reptiles Contributions to Herpetology No. 5, edited by Kraig Adler, 33–34. N.p.: n.p., 1989.

——. "New Genera and New Species Described in Holbrook's *North American Herpetology.*" In John Edwards Holbrook, *North American Herpetology,* reprint ed., introductory essay, xxix–xliii. Athens, Ohio: Society for the Study of Amphibians and Reptiles, 1976.

Agassiz, Elizabeth C., ed. *Louis Agassiz, His Life and Correspondence.* 2 vols. Boston: Houghton-Mifflin Company, 1886.

Agassiz, Louis. [Eulogy on John Edwards Holbrook]. *Proceedings of the Boston Society of Natural History* 4 (1871): 348–51.

Arnold, Lois Barber. *Four Lives in Science: Women's Education in the Nineteenth Century.* New York: Schocken Books, 1984.

Beale, George Robinson. "Bosc and the Exequatur." *Prologue* 10 (Fall 1978): 133–51.

Bellows, Barbara L. *Benevolence Among Slaveholders: Assisting the Poor in Charleston, 1670–1860.* Baton Rouge: Louisiana State University Press, 1993.

Blum, Ann Shelby. *Picturing Nature: American Nineteenth-Century Zoological Illustration.* Princeton: Princeton University Press, 1993.

Bost, Raymond Morris. "The Reverend John Bachman and the Development of Southern Lutheranism." Ph.D. diss., Yale University, 1963.

Bozeman, Theodore Dwight. *Protestants in an Age of Science: The Baconian Ideal and Antebellum Religious Thought.* Chapel Hill: University of North Carolina Press, 1977.

Bruce, Robert V. *The Launching of Modern American Science, 1846–1876*. New York: Alfred A. Knopf, 1987.

Burton, E. Milby. *The Siege of Charleston, 1861–1865*. Columbia: University of South Carolina Press, 1970.

Calder, Dale R., Lester D. Stephens, and Albert E. Sanders. "Comments on the Date of Publication of John McCrady's Hydrozoan Paper "Gymnopthalmata of Charleston Harbor.'" *Bulletin of Zoological Nomenclature* 49 (December 1992): 287–89.

Campbell, Lyle D., and Sarah C. Campbell. "Revision of Tuomey and Holmes' Pleiocene Fossils of South Carolina." [*South Carolina*] *Geologic Notes* 20 (Fall 1976): 101–14.

Clarke, Erskine. *Wrestlin' Jacob: A Portrait of Religion in the Old South*. Atlanta: John Knox Press, 1979.

Coleman, William. *Biology in the Nineteenth Century: Problems of Form, Function, and Transformation*. Baltimore: John Wiley and Sons, 1971.

Craven, Avery O. *Edmund Ruffin, Southerner: A Study in Secession*. Baton Rouge: Louisiana State University Press, 1932.

Cyclopedia of Eminent and Representative Men of the Carolinas: South Carolina. Madison, Wisc.: Brant and Fuller, 1892.

Dall, William H. *Spencer Fullerton Baird: A Biography*. Philadelphia: J. B. Lippincott, 1915.

Daniels, George H. *American Science in the Age of Jackson*. New York: Columbia University Press, 1968.

Duyckinck, Evert, and George L. Duyckinck. *Cyclopedia of American Literature*. 2 vols. New York: C. Scribner, 1856.

Early American Literature, Medicine and Science and Thought Reform, and History. Providence, R.I.: M & S Rare Books, Inc. Catalogue Sixty-Three, 1997.

Easterby, J. H. *A History of the College of Charleston*. Charleston: Trustees of the College of Charleston, 1935.

Fairbanks, George R. *History of the University of the South, at Sewanee, Tennessee*. Jacksonville, Fla.: The H. & W. B. Drew Company, 1905.

Faust, Drew Gilpin. *A Sacred Circle: The Dilemma of the Intellectual in the Old South, 1840–1860*. Baltimore: Johns Hopkins University Press, 1977.

Ford, Alice, ed. *Audubon's Animals*. New York: Studio Publications, 1951.

———. *John James Audubon*. Norman: University of Oklahoma Press, 1964.

Fortey, Richard. *Fossils: The Key to the Past*. Cambridge: Harvard University Press, 1991.

Fraser, Walter J., Jr. *Charleston! Charleston! The History of a Southern City*. Columbia: University of South Carolina Press, 1989.

Fredrickson, George M. *The Black Image in the White Mind: The Debate on Afro-American Character and Destiny, 1817–1914*. Middletown, Conn.: Wesleyan University Press, 1987.

Gill, Theodore. "John Edwards Holbrook." *National Academy of Sciences Biographical Memoirs* 5 (1905): 47–77.

———. "Review of Holbrook's *Ichthyology of South Carolina*." *American Journal of Science*, 2nd ser., 37 (1864): 89–94.

Gillespie, Neal C. "Preparing for Darwin: Conchology and Natural Theology in Anglo-American Natural History." In *Studies in History of Biology*, edited by William Coleman and Camille Limoges, 93–145. Baltimore: Johns Hopkins University Press, 1984.

Goodman, Alan H. "Bred in the Bone?" *The Sciences* 37 (March–April 1997): 20–25.

Gopnik, Adam. "Audubon's Passion." *New Yorker*, February 25, 1991, 103.

Greene, John C. *American Science in the Age of Jefferson*. Ames: Iowa State University Press, 1984.

G[regorie], A[nn] K[ing]. "Edmund Ravenel." *Dictionary of American Biography*, 15:394–95.

Haller, John S., Jr. *Outcasts from Evolution: Scientific Attitudes of Racial Inferiority, 1859–1900*. Urbana: University of Illinois Press, 1971.

Hamilton, William J., and John O. Whitaker, Jr. *Mammals of the Eastern United States*. 2nd ed. Ithaca: Cornell University Press, 1979.

Handley, Charles O., and Merrill Varn. "Identification of the Carolina Shrews of Bachman." In *Advances in the Biology of Shrews*, edited by J. F. Merritt, G. L. Kirkland, and R. K. Rose, 393–400. Pittsburgh: Carnegie Museum of Natural History, 1994.

Hartman, Carl G. *Possums*. Austin: University of Texas Press, 1952.

Haygood, Tamara Miner. *Henry William Ravenel, 1814–1887: South Carolina Scientist in the Civil War Era*. Tuscaloosa: University of Alabama Press, 1987.

Herrick, Francis Hobart. *Audubon the Naturalist*. 2 vols. New York: D. Appleton and Company, 1917.

———. "Audubon's Bibliography." *Auk* 44 (1919): 372–80.

A History of the Lutheran Church in South Carolina. Columbia: South Carolina Synod of the Lutheran Church, 1971.

Holifield, E. Brooks. *The Gentlemen Theologians: American Theology in Southern Culture, 1795–1860*. Durham: Duke University Press, 1978.

Horn, E. T. "St. John's Evangelical Lutheran Church." In *The Charleston Yearbook, 1884*. [Charleston]: n.p., [1885].

Horsman, Reginald. *Josiah Nott of Mobile: Southerner, Physician, and Racial Theorist*. Baton Rouge: Louisiana State University Press, 1987.

Hovenkamp, Herbert. *Science and Religion in America, 1800–1860*. Philadelphia: University of Pennsylvania Press, 1978.

Irving, John B. *A Day on Cooper River*. Edited by Louisa Cheves Stoney. 3rd ed. Columbia, S.C.: R. L. Bryan Company, 1969.

Jellison, Richard M., and Phillip S. Swartz. "The Scientific Interests of Robert W. Gibbes." *South Carolina Historical Magazine* 66 (1965): 77–97.

Johnson, Thomas Cary, Jr. *Scientific Interests in the Old South*. New York: D. Appleton-Century Company, 1936.

Jones, Charles C., Jr. "The Siege and Evacuation of Savannah, Georgia, in December, 1864." *Southern Historical Society Papers*. Richmond, Va.: The Society, 1889.

Jordan, Winthrop D. *White Over Black: American Attitudes Toward the Negro, 1550–1812*. Chapel Hill: University of North Carolina Press, 1968.

Klosky, Beth Ann. *The Pendleton Legacy*. Columbia, S.C.: Sandlapper Press, 1971.

Kohlstedt, Sally Gregory. *The Formation of the American Scientific Community: The American Association for the Advancement of Science, 1848–1860*. Urbana: University of Illinois Press, 1976.

LaBorde, M[aximilian]. *History of the South Carolina College*. Charleston: Walker, Evans & Cogswell, 1874.

Lee, Harry G. "Edmund Ravenel, Eminent Conchologist of Antebellum Charleston." Presented before the annual meeting of the American Malacological Union, Charleston, S.C., June 21, 1988.

Luker, Ralph. *A Southern Tradition in Theology and Social Criticism, 1830–1930*. New York: Edwin Mellen Press, 1984.

Lunz, G. Robert, Jr. *The Rediscovery of* Squilla Neglecta *Gibbes*. Leaflet No. 5. Charleston: The Charleston Museum, 1933.

Lurie, Edward. *Louis Agassiz: A Life in Science*. Chicago: University of Chicago Press, 1960.

——. "Louis Agassiz and the Races of Man." *Isis* 45 (1954): 227–42.

Malone, Dumas. *The Public Life of Thomas Cooper, 1783–1839*. Columbia: University of South Carolina Press, 1966.

Manning, Raymond B., and Richard W. Heard. "Stomatopod Crustaceans from the Carolinas and Georgia, Southeastern United States." *Gulf Research Reports* 9 (January 1997): 313–14.

Martin, Sidney Walter. "Ebenezer Kellogg's Visit to Charleston, 1817." *South Carolina Historical and Genealogical Magazine* 49 (January 1948): 1–14.

Mayer, Alfred G. "Descriptions of New and Little-known Medusae from the Western Atlantic." *Bulletin of the Museum of Comparative Zoology at Harvard College* 37 (1900): 1–9.

Mayr, Ernst. *The Growth of Biological Thought: Diversity, Evolution, and Inheritance*. Cambridge: Harvard University Press, 1982.

——. *Populations, Species, and Evolution*. Cambridge: Harvard University Press, 1970.

Mazÿck, William G. *The Charleston Museum: Its Genesis and Development*. Charleston: Walker, Evans & Cogswell Co., 1908.

Mitchell, Betty L. *Edmund Ruffin, A Biography*. Bloomington: Indiana University Press, 1981.

Moore, James R. "Geologists and Interpreters of Genesis in the Nineteenth Century." In *God and Nature: Historical Essays on the Encounter Between Christianity and Science*, edited by David C. Lindberg and Ronald L. Numbers, 322–50. Berkeley: University of California Press, 1986.

Nepveux, Ethel S. *George Alfred Trenholm and the Company that Went to War*. Charleston: Privately published, 1973.

Numbers, Ronald L., and Janet S. Numbers. "Science in the Old South: A Reappraisal." In *Science and Medicine in the Old South*, edited by Ronald L. Numbers and Todd L. Savitt, 9–35. Baton Rouge: Louisiana State University Press, 1989.

Palmer, Ralph S., ed. *Handbook of North American Birds*. 5 vols. New Haven: Yale University Press, 1988.

Pease, Jane H., and William H. Pease. "Intellectual Life in the 1830's: The Institutional Framework and the Charleston Style." In *Intellectual Life in Antebellum Charleston*, edited by Michael O'Brien and David Moltke-Hansen, 233–54. Knoxville: University of Tennessee Press, 1986.

Powers, Bernard E., Jr. *Black Charlestonians, A Social History, 1822–1885*. Fayetteville: University of Arkansas Press, 1994.

Ravenel, Mrs. St. Julien [Harriott Horry Rutledge Ravenel]. *Charleston: The Place and the People*. New York: Macmillan Company, 1925.

Ray, Clayton E., and Albert E. Sanders. "Pleistocene Tapirs in the Eastern United States." In *Contributions in Quaternary Vertebrate Paleontology*, edited by H. H. Genoways and M. R. Dawson, 283–313. Pittsburgh: Carnegie Museum of Natural History, 1984.

R[ichardson] E. M. "Some Old and New Records of the Sargassum Fish (*Antennarius ocellatus*)." *Charleston Museum Quarterly* n.v. (1932): 13.

Richardson, H. M., comp., "Dr. Anthony Cordes and Some of His Descendants," *South Carolina Historical and Genealogical Magazine* 43 (1942): 133–55, 219–42; 44 (1943): 17–42, 115–23, 184–95.

Robinson, Gloria. "Charles Upham Shepard." In *Benjamin Silliman and His Circle*, edited by Leonard G. Wilson, 85–103. New York: Science History Publications, 1979.

Robinson, William Childs. *Columbia Theological Seminary and the Southern Presbyterian Church*. Decatur, Ga.: [Dennis Lindsey Printing Co.], 1931.

Rogers, C. H. *Incidents of Travel in the Southern States and Cuba*. New York: R. Craighead, Printer, 1862.

Rogers, George A. "Elliott, Stephen." In *Biographical Dictionary of American and Canadian Naturalists and Environmentalists*, edited by Keir B. Sterling et al., 247–49. Westport, Conn.: Greenwood Press, 1997.

Rogers, George C., Jr. *Charleston in the Age of the Pinckneys*. Rev. ed. Columbia: University of South Carolina Press, 1980.

Rosen, Robert N. *Confederate Charleston: An Illustrated History of the City and the People During the Civil War*. Columbia: University of South Carolina Press, 1994.

Rossiter, Margaret W. "A Portrait of James Dwight Dana." In *Benjamin Silliman and His Circle*, edited by Leonard G. Wilson, 105–27. New York: Science History Publications, 1979.

Rutledge, Anna Wells. "Artists in the Life of Charleston." *Transactions of the American Philosophical Society* 39, pt. 2 (1949): 101–260.

Sanders, Albert E. "Alexander Garden (1730–1791), Pioneer Naturalist in Colonial America." In *Collection Building in Ichthyology and Herpetology*, American Society of Ichthyologists and Herpetologists (ASIH) Special Publication No. 3, edited by Theodore W. Pietsch and William D. Anderson Jr., 409–37. [Lawrence, Kans.]: ASIH, 1997.

Sanders, Albert E., and William D. Anderson Jr. *Deep Runs the Heritage: The Story of Natural History Investigations in South Carolina*. Columbia: University of South Carolina Press, 1999.

Sanders, Albert E., and Warren Ripley. *Audubon: The Charleston Connection*. Contributions from the Charleston Museum No. 16, 1986.

Sargent, Charles Sprague. *The Silva of North America*. 14 vols. Reprint ed. New York: Peter Smith, 1947. Originally published 1890–1902.

Shimek, B. "Holbrook's *North American Herpetology*, First Edition." *Proceedings of the Iowa Academy of Science* 31 (1924): 427–30.

Shuler, Jay. *Had I the Wings: The Friendship of Bachman and Audubon*. Athens: University of Georgia Press, 1995.

Slotten, Hugh Richard. *Patronage, Practice, and the Culture of American Science: Alexander Dallas Bache and the U.S. Coast Survey*. Cambridge: Cambridge University Press, 1994.

Smallwood, William Martin, and Mabel S. C. Smallwood. *Natural History and the American Mind*. New York: Columbia University Press, 1941.

Sprunt, Alexander, Jr., and E. Burnham Chamberlain. *South Carolina Bird Life*. Columbia: University of South Carolina Press, 1970.

Stanton, William. *The Leopard's Spots: Scientific Attitudes Toward Race in America, 1815–59*. Chicago: University of Chicago Press, 1960.

Stearns, Raymond Phineas. *Science in the British Colonies of America*. Urbana: University of Illinois Press, 1970.

Stephens, Lester D. *Ancient Animals and Other Wondrous Things: The Story of Francis Simmons Holmes, Paleontologist and Curator of the Charleston Museum*. Contributions from the Charleston Museum, No. 17, 1988.

———. "John Edwards Holbrook (1794–1871) and Lewis Reeve Gibbes (1810–1894): Exemplary Naturalists in the Old South." In *Collection Building in Ichthyology and Herpetology*, American Society of Ichthyology and Herpetology Special Publication No. 3, edited by Theodore W. Pietsch and William D. Anderson Jr., 447–67. [Lawrence, Kans.]: ASIH, 1997.

———. "The Mermaid Hoax: Indications of Scientific Thought in South Carolina in the 1840s." *Proceedings of the South Carolina Historical Association, 1983*, 45–55.

———. Review of *The Launching of Modern American Science, 1846–1876*, by Robert V. Bruce. *Science* 237 (September 18, 1987): 1515–16.

———. "Scientific Societies in the Old South." In *Science and Medicine in the Old South*, edited by Ronald L. Numbers and Todd L. Savitt, 55–78. Baton Rouge: Louisiana State University Press, 1989.

Stephens, Lester D., and Dale R. Calder. "John McCrady of South Carolina: Pioneer Student of North American Hydrozoa." *Archives of Natural History* 19 (1992): 39–54.

Taylor, W. H. "Lewis Reeve Gibbes and the Classification of the Elements." *Journal of Chemical Education* 18 (September 1941): 403–7.

Terres, John K. *The Audubon Society Encyclopedia of North American Birds*. New York: Alfred A. Knopf, 1980.

Tryon, George W., Jr. "A Sketch of the History of Conchology in the United States." *American Journal of Science*, 2nd ser., 33 (March 1862): 161–80.

Turgeon, Donna D., et al. *Common and Scientific Names of Aquatic Invertebrates from the United States and Canada: Mollusks.* Bethesda, Md.: American Fisheries Society, 1988.

Tyler, Lyon G. "Drawing the Color Line in the Episcopal Diocese of South Carolina, 1876–1890: The Role of Edward McCrady, Father and Son." *South Carolina Historical Magazine* 91 (April 1990): 107–24.

Wade, Richard C. *Slavery in the Cities: The South, 1820–1860.* New York: Oxford University Press, 1964.

Wallace, David Duncan. *South Carolina: A Short History, 1520–1948.* Chapel Hill: University of North Carolina Press, 1951.

Waring, Joseph Ioor. *A History of Medicine in South Carolina, 1825–1900.* [Charleston]: South Carolina Medical Association, 1967.

Warner, John Harley. "The Idea of Southern Medical Distinctiveness: Medical Knowledge and Practice in the Old South." In *Science and Medicine in the Old South,* edited by Ronald L. Numbers and Todd L. Savitt, 179–205. Baton Rouge: Louisiana State University Press, 1989.

———. "A Southern Medical Reform: The Meaning of the Antebellum Argument for Southern Medical Education." In *Science and Medicine in the Old South,* edited by Ronald L. Numbers and Todd L. Savitt, 206–25. Baton Rouge: Louisiana State University Press, 1989.

Warren, Leonard. *Joseph Leidy, The Man Who Knew Everything.* New Haven, Conn.: Yale University Press, 1998.

Williams, Austin B. *Shrimps, Lobsters, and Crabs of the Atlantic Coast of the Eastern United States, Maine to Florida.* Washington, D.C.: Smithsonian Institution Press, 1984.

Williams, Austin B., et al. *Common and Scientific Names of Aquatic Invertebrates from the United States and Canada: Decapod Crustaceans.* Bethesda, Md.: American Fisheries Society, 1989.

Wilson, Leonard G. *Lyell in America: Transatlantic Geology, 1841–1853.* Baltimore: Johns Hopkins University Press, 1998.

Winsor, Mary Pickard. "Louis Agassiz and the Species Question." In *Studies in History of Biology,* edited by William Coleman and Camille Limoges, 89–117. Baltimore: Johns Hopkins University Press, 1979.

———. *Reading the Shape of Nature: Comparative Zoology at the Agassiz Museum.* Chicago: University of Chicago Press, 1991.

Worthington, Richard D., and Patricia H. Worthington. "John Edwards Holbrook, Father of American Herpetology." In *North American Herpetology,* by John Edwards Holbrook, introductory essay, xii–xxvii. Reprint ed. Athens, Ohio: Society for the Study of Amphibians and Reptiles, 1976.

Wyatt-Brown, Bertram. *Southern Honor: Ethics and Behavior in the Old South.* New York: Oxford University Press, 1982.

32–36, 55–56; early years, 1–13; studies botany, 2, 13, 18; on nature, 2, 16; on slavery, 3, 166–67, 198; on southern culture, 10, 198, 214–16; family of, 14–16, 35–36, 52, 56; and interest in entomology, 16, 31, 32; publishes *Catalogue, Phanerogamous Plants and Ferns*, 18; in scientific societies, 22, 31–32, 46, 57, 138; travels to Europe, 33–35; honors received, 35, 59, 271 (n. 4), 277 (n. 31); opposes theory of evolution, 42, 176, 181, 183, 193, 202; and V. G. Audubon, 42–58 passim, 215–16; and J. W. Audubon, 45, 50–52; as College of Charleston professor, 54–55; Civil War years, 220–22, 224–26; final years, 228, 260

— and J. J. Audubon: study of birds, 17, 18–22; friendship, 17–19; work on *Quadrupeds of North America*, 26, 30, 37–59 passim, 277 (n. 29)

— on human races: unity of species, xii, 32, 94, 171–72, 175, 178–79, 180, 200–201, 204, 205, 266; varieties of, 43, 167, 176, 180–81, 198, 202; review of *Types of Mankind*, 197–207 passim

— interactions with naturalists: Holbrook, 17; Ravenel, 17; Elliott, 17, 18; L. R. Torrey, 18; L. R. Gibbes, 18, 110; Ord, 19, 20, 24, 31; Harlan, 25, 26, 27; Pickering, 26; Townsend, 27, 31, 33, 37; Nuttall, 28; Morton, 28, 167–94 passim; E. Ravenel, 31, 32, 62, 117; Richardson, 37, 39; Brewer, 37, 44; Liechtenstein, 38; Harris, 44, 57, 217; Lyell, 47; R. W. Gibbes, 52–53; Holmes, 53; Agassiz, 53, 201–3; McCrady, 147

— religious views: on pastoral role, 2, 14, 260; on Bible, 2, 185, 189; on blacks, 4, 166, 181; on Catholics, 6–7, 211–14, 270 (n. 10); on Lutheranism, 7, 9, 14, 221; on creation, 32, 142, 175, 189; *Defence of Luther*, 214

— scientific principles reflected by

comments on: Baconian method, 2, 9, 19, 24; a priori theories, 24, 176, 199; need to consult relevant sources, 43, 48, 51, 187; hoaxes, 46–47

— on species: diagnostic characters for classifying, xiii, 26, 27, 30, 39, 41, 43, 44, 167, 176–77, 179–80, 183, 191, 200, 201, 202–7; nature of, 3–4, 30, 43, 176; hybridity, 16, 172, 178–79, 182–84, 187, 188, 193, 194; discussed in *Doctrine of the Unity of the Human Race*, 172, 173, 174–76, 181, 266; opposes Agassiz's zoological zones, 199, 201–5. *See also* Species

— study of birds: color, 2, 21–22; habits and morphology, 4–5, 12, 13, 16, 17, 21–22, 24, 272 (n. 17); migration, 12, 22; vultures, 19–20; hybridization, 178, 180, 188

— study of mammals, 2, 11–12, 25, 26, 33–34, 35, 41, 42, 57; lagomorphs, 12, 25, 26, 27–28, 35, 37, 47; sciurid rodents, 12, 33, 35, 37, 42, 43, 47–48, 50; opossum, 16, 54; molting, 22, 23–25, 30, 54; murid rodents, 27, 42, 43; shrews, 28–29; moles, 37–38; bats, 42, 52, 56; fox, 57; skunks, 57

Bachman, Julia, 15, 36
Bachman, Lynch, 16, 52
Bachman, Maria Martin. *See* Martin, Maria
Bachman, Maria Rebecca, 15, 35–36
Bachman, Mary Eliza, 15, 35, 36
Bachman, Samuel Wilson, 16, 225
Bachman, William Kunhardt, 16, 220, 221, 225
Baconian view, 2, 19, 24, 117, 123
Bailey, J. W., 121
Baird, Spencer F., 57, 76, 95–96, 109, 125, 134
Barratt, John P., 87, 119
Bartram, William, 13
Bats, 42, 52, 56
Beauregard, P. G. T., 218, 223
Bellinger, John, 213–14

Belmont plantation, 85, 97, 100
Berlin, Germany: Bachman in, 34
Bibron, Gabriel, 82
Binney, Amos, 66, 67, 76, 88
Binney, W. G., 76
Birds, 2, 4–5, 12, 16, 17, 21–22, 24, 201, 272 (n. 17)
Birds of America (Audubon), 17, 18, 26
Blanding, William, 88
Blarina: *carolinensis*, 28, 29 (fig.), 30; *brevicauda*, 30;
Books and libraries in Charleston. *See* Charleston, S.C.—paucity of scientific sources in
Bosc, Louis Augustin Guillaume, 5
Boston, Mass., ix, 1, 8, 88, 114, 148, 251, 265
Boston Society of Natural History, 20, 25, 37, 92, 96, 114, 126, 300 (n. 17)
Bougainvillia carolinensis, 153, 154 (fig.)
Braun, Anton, 12, 13
Brewer, Thomas Mayo, 37, 40, 41, 44
Brooks, William Keith, 245, 258
Browne, Peter A., 190–91
Bruce, Robert V., 265–67
Brumby, R. T., 110
Buckley, S. B., 144
Bufo: *americanus*, 92; *quercicus*, 92
Bunting, painted, 24
Burden, Thomas, 134
Burke, Luke, 210, 211
Burkhardt, Jacques, 99
Buzzard, turkey, 4–5, 19–20

Cain, D. J., 182–94 passim
Catesby, Mark, 5
Cathartes aura, 19–20
Caudata, 92
Centers of origin. *See* Zoological zones
Chalmers, Thomas, 141
Charleston, S.C.: science in, ix–x, 5–6, 264–65, 266–67, 270 (n. 11); blacks in, 1, 3–10; description of, in 1815, 1–10; geological epochs of, 132–33,

139, 141, 142; during Civil War, 218–22, 224–25; in postwar era, 262–63
—paucity of scientific sources in, 9, 109–10, 264, 265; Bachman on, 8, 43, 45, 48; Ravenel on, 67, 70; Holbrook on, 89, 90, 95; L. R. Gibbes on, 110, 261; McCrady on, 155–57
Charleston Library Society, 5, 8
Charleston Museum: recognition of, ix, 144; founded, 5; donations to, 71, 133, 134, 135, 219; reestablished, 133–34; directed by Holmes, 139–40, 219; specimens in Edgefield, 219, 224; postwar fortunes of, 229–31, 232, 263
Charleston Observatory, 108–9, 157
Chemistry: in Charleston, 260, 270 (n. 11)
Cheraw, S.C., 225
Chione grus, 144
Civil War: involvement of Charleston naturalists in, 218–24; impact on Charleston naturalists, 224–28; impact on College of Charleston, 228–32
Cnidaria, 147, 150
College of Charleston, 8, 54, 131, 133–34, 219, 227, 228–32, 262. *See also* Bachman, John; Gibbes, Lewis Reeve; Holmes, Francis Simmons; McCrady, John
Comb jellies, 147, 153
Conchology, 31–32, 66–77 passim, 110
Conrad, Timothy, 66, 69, 75, 96
Confederate States Nitre and Mining Bureau, 222
Consumption. *See* Tuberculosis
Conuropsis carolinensis, 24
Conversation Club. *See* Literary and Philosophical Society
Cook, Joseph, 254–55
Cooper, Thomas, 103, 104, 171
Cooper River, 1, 66, 71
Cope, E. D., 253
Coragyps atratus, 19–20

Frogfish, 62
Frogs and toads, 92

Garden, Alexander, 5
Geddings, Eli, 235
Geddings, J. F. M., 260
Geological epochs of Charleston area, 132–33, 139, 141, 142
Geology, interest in: by Ravenel, 71; by Holbrook, 80–82; by L. R. Gibbes, 119–20; by Tuomey, 129–30, 141–42; by Holmes, 130–31, 132–33, 138
Gibbes, Anna Barnwell, 126
Gibbes, Charles, 102, 105, 126
Gibbes, John, 102, 105
Gibbes, Lewis Ladson, 101, 102
Gibbes, Lewis Reeve, 71, 73, 75, 76; early years, 101–7; studies medicine, 103, 104, 105; association with other Charleston naturalists, 103, 105, 110; visits Paris, 105–6; as College of Charleston professor, 106–7; on popularization of science, 112–14; sectionalist sympathies of, 113–14, 122, 126; and Academy of Natural Sciences of Philadelphia, 114, 126; and Boston Society of Naturalists, 114, 126; at American Association for the Advancement of Science meeting in Charleston, 115, 126; and Baconian view, 117, 123; in Elliott Society of Natural History, 117, 229; on scientific language, 124–25; association with Agassiz, 132; opposes Holmes, 139–40; Civil War years, 219, 222, 225; final years, 228, 229, 236, 252, 260–62
—scientific studies of: paleontology, 53, 111–12, 121; botany, 103–4, 106, 110, 121–22; mathematics, 104, 106, 123, 261; physics, 104, 122–23, 261; Niagara Falls, 105, 120; meteorology, 107–8; astronomy, 108–9, 262; conchology, 110; vertebrate zoology, 110; crustacea, 114–17, 150; taxa, 115–16;

salamanders, 117–19; geology, 119–20; earthquakes, 120–21, 261; chemistry, 260
Gibbes, Louisa, 106
Gibbes, Maria Henrietta Drayton, 101, 102
Gibbes, Robert W.: criticizes Koch, 52–53; as paleontologist, 71, 131, 132; helps L. R. Gibbes at South Carolina College, 103, 104; on age of earth, 142; as advocate of pluralist view, 172, 187, 194; helps Agassiz study slaves, 174, 200
Gibbesia neglecta, 116
Gill, Theodore, 95, 99
Gillespie, Neal, 68
Gilman, Charlotte, 91
Gilman, Daniel C., 247, 258, 259
Gilman, Samuel, 94
Girard, Charles, 91
Glaucomys sabrinus, 37
Gliddon, George R., 195; *Types of Mankind*, 195–211 passim
Gobineau, Joseph Arthur de, 158, 210
Godman, John, 58
Gopher, pocket, 37, 43
Gould, Augustus A., 110, 115, 124–25
Gould, Benjamin A., 112, 113
Graptolites, 153
Gray, Asa, 104, 121, 208
Green, Jacob, 87–88
Grote, A. R., 253
Grus americana, 22
Guizot, François, 158
Gurney, Ephraim Whitman, 241

Hair, types of: misuse as diagnostic character, 191
Haldeman, Samuel S., 88
Hall, James, 120
Hallowell, Edward, 88, 100
Happoldt, Christopher, 33–34
Harden, John M. B., 87
Hares. *See* Lagomorpha
Harlan, Richard, 25, 26, 27, 58, 88

Harris, Edward, 31, 45, 57, 217

Harvard University, 147, 149, 239, 242–43, 245, 247, 249–52

Haskell, Harriet Eva. *See* Bachman, Harriet Eva

Haskell, John Bachman, 221

Hayne, Paul, 147, 157

Henry, Joseph, 76, 108, 122–23, 232, 235

Hitchcock, Edward, 141

Hodgson, Telfair, 256, 258, 259

Holbrook, Daniel, 79

Holbrook, Harriott Pinckney Rutledge, 84–85, 100

Holbrook, John Edwards: association with Ravenel, 62, 78, 83; early years, 78–86; medical interests of, 79–80, 83, 93; studies abroad, 79–83; and geology and age of earth, 80–82; and French naturalists, 82–83, 92; network of collectors for, 84, 85–89, 94; and scientific societies, 84, 92–93, 94; association with Agassiz, 85, 90, 96–97; and Belmont plantation, 85, 97, 100; taxa of, 91, 92, 95, 98, 99–100; and farming, 92; honors received by, 92, 260; sells and donates specimens, 96, 134; at American Association for the Advancement of Science meeting in Charleston, 98; views on southern culture, 99, 100; supports pluralist view, 172; during Civil War and final years, 220, 225, 228, 235, 260

—studies of fishes: early work, 83; *Southern Ichthyology*, 95; paper on, 98; *Ichthyology*, 98–100

—studies of reptiles and amphibians: recognition of, 31, 73; *North American Herpetology*, 85–93

Holbrook, Mary Edwards, 78–79

Holbrook, Silas, 78–79

Holbrook, Silas Pinckney, 78–79

Holmes, Elizabeth Toomer, 128, 144

Holmes, Francis Simmons: early years, 127–31; miscellaneous scientific studies by, 130, 135, 139; interest in geology, 130–31, 132–33, 138; rivalry with R. W. Gibbes, 131, 132; friendship with Leidy, 131, 137, 140, 145; as College of Charleston professor, 133, 135, 139–40, 219, 227, 228–31; as museum curator, 133–44 passim, 219, 228–31; on sectionalism, 136, 144–45; role in Elliott Society of Natural History, 138–39, 229; on creation, 141–42; on slavery, 144; taxa, 144; Civil War years, 219–20, 222, 224, 226–27; final years, 230–32, 234–35, 260

—paleontological work: fossil collections of, 53, 128–29, 131, 141, 232; collaborates with Tuomey, 71, 129–30, 135; *Pleiocene Fossils of South Carolina*, 74, 135–37, 144; American Association for the Advancement of Science presentations, 132–33; *Post-Pleiocene Fossils of South Carolina*, 141–43, 144; taxa of, 144

Holmes, Helen Boomer, 127

Holmes, John, 127

Humboldt, Alexander von, 34, 123

Humboldt, Wilhelm von, 158

Hume, Alexander, 28, 234

Hunt, T. Storry, 253

Huxley, Thomas, 244, 254–55, 258

Hyatt, Alpheus, 251

Hybridity, 168–70, 171, 176–94 passim. *See also* Species—diagnostic characters for classifying

Hydromedusae, 150–53, 154 (fig.)

Hydrozoa, 147, 149, 150, 152

Hyena, 178, 183

Ichthyology, 94–95, 98–100

Ichthyology of South Carolina (Holbrook), 98–100

Icterus galbula, 12

James, William, 244, 245, 246, 247, 248, 250, 251, 253, 259

study of fishes, 62; visits England, 63; holds medical professorship, 63–66; medical service of, 63–66, 73, 74–75; moves to Grove, 66; on slavery, 66, 76–77; rejects theory of evolution, 69; on sectionalism, 74; later years, 228, 260
— scientific work: in conchology, 31–32, 66–77 passim, 149; and Literary and Philosophical Society, 61, 138; collects for others, 62, 69, 73–74, 110; association with other Charleston naturalists, 62, 71, 73, 78, 83, 103, 129, 147; catalogue of mollusk shells, 66–67; on species concept, 67, 69, 75; notes deficiency of resources in Charleston, 67, 70; taxa established, 67, 70, 77; on fossil echinoderms, 69–71; patronyms, 70, 74, 75, 77; guides Lyell on geology tour, 70–71; donations to museums, 71; participates in American Association for the Advancement of Science meeting, 73; in Elliott Society of Natural History, 74, 75–76; association with Agassiz, 132

Ravenel, Edmund, Jr., 64, 77
Ravenel, Emma, 64, 66, 73
Ravenel, Harriott Horry Rutledge, 85
Ravenel, Henry William, 135
Ravenel, James, 61, 63, 78
Ravenel, Louisa Catherine Ford, 63–64, 66
Ravenel, Mary Louisa, 63, 66
Ravenel, St. Julien, 85, 232, 242, 252
Ravenel, Theodosia, 64, 73
Religious views of naturalists: on nature as evidence of divine actions, 2, 12, 31, 67, 120, 158
— on creation: biblical literalists, xiii, 3, 32, 106, 166, 170, 175, 188, 189, 190; long-day theorists, 32, 81, 141–42, 185–86, 190; Pre-Adamite theorists, 141, 197; multiple-acts theorists, 170, 173

See also Bachman, John — religious views; Evolution, theory of
Report on the Geology of South Carolina (Tuomey), 111–12, 129–30, 141–42
Reynolds, J. A., 213
Reithrodontomys humulis, 42
Richard, J. H., 87, 90, 95, 99
Richardson, John, 37, 39
Rodents, 11, 45; murid, 27, 33, 42, 43, 44; sciurid, 12, 33, 35, 37, 42, 43
Ruffin, Edmund, 110, 128, 135, 216, 221, 224
Runkel, J. D., 123
Rüppell, Edward, 96

St. John's Lutheran Church (Charleston, S.C.), 2, 13–14
Salamanders, 92, 117, 118 (fig.), 119
Savannah, Ga., 223–24
Say, Thomas, 66, 69
Scapanus: townsendii, 33; *latimanus*, 38
Schwartz, John G., 14
Scientific sources in Charleston, paucity of. *See* Charleston, S.C. — paucity of scientific sources in
Sciurid rodents. *See* Rodents: sciurid
Sciurus: carolinensis, 12; *niger*, 35
Scyphomedusae, 150
Scyphozoa, 147, 150
Sea jellies. *See* Jellyfishes
Sera, J., 66, 87, 90
Shaler, Nathaniel Southgate, 242, 244, 249, 250, 251–52
Shepard, Charles Upham, 119–20, 134, 232, 270 (n. 11)
Shrews, 28, 29 (fig.), 30
Silliman, Benjamin, 119, 125
Simms, William Gilmore, 112–13
Siphonophores, 152. See also *Physalia physalis*
Skin color: misused as diagnostic character, 179–80, 191
Skulls, 70, 96, 167, 177, 181, 200, 205. *See also* Species — diagnostic characters for classifying

Skunks, 57

Slavery: supported by scientists, 9, 100, 122, 155, 159, 166, 167, 192, 198, 266. *See also* Racial views—on slavery

Slaves: daguerreotypes of, 174

Smith, Charles Hamilton, 172, 178, 181–82, 184, 205

Smith, J. Lawrence, 270 (n. 11)

Smith, James H., 32

Smithsonian Institution, 71, 73, 75, 76, 95–96, 122–23, 125, 143, 144, 232

Smyth, Thomas, 94, 170, 172, 173, 185

Snakes, 88, 91, 92

Sonrel, Auguste, 99

Sorex longirostris, 28–30

Soricidae, 28–30

South Carolina: legislative subsidies for scientific works, 89, 99, 135, 136, 155

South Carolina Executive Council, 222

Southern Ichthyology (Holbrook), 94

Sparrow, Bachman's, 21

Species: variations within, 30, 43, 67, 69, 75, 199
—definitions of: by Morton, 167, 168, 189–90, 192; by Bachman, 176; by Agassiz, 199–201
—diagnostic characters for classifying, xiii, 176–77, 183; Bachman on, 26, 27, 30, 39, 41, 43, 44; correct ones ignored by pluralists, 167, 179–80, 191, 200, 201, 202–7
See also Hybridity

Spermophilus: townsendii, 33; *annulatus*, 42

Sphaerocoryne agassizii, 152

Spilogale putorius, 57

Squamata, 92

Squirrels, 11, 12, 33, 35, 37, 42, 43

Stanton, William, xiii, 165–66

Stimpson, William, 75, 76, 143, 152, 281 (n. 27)

Storer, D. Humphreys, 88–90

Strobel, Martin, 15

Suidae, 183

Sullivan's Island, S.C., 7, 60, 61, 66, 73, 74, 77, 101, 153

Sylvilagus: palustris, 26, 47, 185; *aquaticus*, 28; *nuttalli*, 28; *bachmani*, 35; *floridanus*, 185

Synonymies, 43, 47

Talpidae, 30, 33, 37–38

Tamias minimus, 33

Taylor, Alanson, 46–47

Thomomys townsendii, 37

Toads, 92

Torrey, John, 18, 104, 121

Townsend, John K., 27, 31, 33, 37

Trenholm, George A., 128, 220

Trescott, William Henry, 252

Troost, Gerard, 87

Tuberculosis, 1, 13, 35–36, 216–17

Tuomey, Michael, 53, 71, 74, 111–12, 129–30, 134, 135, 141–42

Turritopsis nutricula, 146, 149, 291 (n. 1)

Turtles, 87, 92, 93 (fig.)

Types of Mankind (Nott and Gliddon): contents of, 195–97; reviewed by John Bachman, 197–207 passim; criticized by Cull, 211; influence of, 211

University of the South, 252–53, 256–58, 259

Usher, William, 195, 196, 197, 211

Valenciennes, Achille, 62, 82, 83, 92, 94

Velella velella, 152

Venable, Charles, 236

Vermivora bachmanii, 21, 272 (n. 17)

Viviparous Quadrupeds of North America. See Quadrupeds of North America

Vulpes vulpes, 57

Vulture, black, 4–5, 19–20

Warblers: Bachman's, 21; Swainson's, 21

Waterhouse, G. R., 35

Waterton, Charles, 20–21